MW00761067

Vibrational Spectroscopic Imaging for Biomedical Applications

About the Editor

Gokulakrishnan Srinivasan, Ph.D., is a Sr. Technical Specialist at PerkinElmer India Pvt Ltd. Before this, he was an application scientist at Bruker Optik GmbH in Ettlingen, Germany and in Mumbai, India. During his postdoctoral tenure in the Department of Bioengineering and the Beckman Institute for Advanced Science and Technology at University of Illinois at Urbana-Champaign (UIUC), USA, he was engaged in applications of infrared spectroscopic imaging to diagnosis of human cancers, studying the kinetics of self-healing polymers and other biopolymers. He received his Ph.D at the Institute for Physical Chemistry at University of Stuttgart, Germany, and his doctoral dissertation involves the Characterization of Chromatographic Column Materials by Solid-State NMR and FTIR spectroscopy. Dr. Srinivasan is a reviewer for *Applied Spectroscopy, Journal of Chromatography A,* and *Analytical and Bioanalytical Chemistry.*

Vibrational Spectroscopic Imaging for Biomedical Applications

Edited by
Gokulakrishnan Srinivasan

New York   Chicago   San Francisco
Lisbon   London   Madrid   Mexico City
Milan   New Delhi   San Juan
Seoul   Singapore   Sydney   Toronto

The McGraw·Hill Companies

Library of Congress Cataloging-in-Publication Data

Vibrational spectroscopic imaging for biomedical applications / [edited by]
Gokulakrishnan Srinivasan.
    p. ; cm.
    Includes bibliographical references and index.
    Summary: "Spectroscopic imaging revolutionizes medical imaging and diagnostics.
This book offers expert discussions on two major vibrational spectroscopic techniques—
infrared and raman spectroscopy—and research outcomes"—Provided by publisher.
    ISBN 978-0-07-159699-2 (hardback : alk. paper)   1. Infrared spectroscopy—Diagnostic
use.   2. Raman spectroscopy—Diagnostic use.   I. Srinivasan, Gokulakrishnan.
    [DNLM: 1. Laboratory Techniques and Procedures.   2. Diagnostic Imaging—methods.
3. Spectrophotometry, Infrared—methods.   4. Spectrum Analysis, Raman—methods.
QY 25 V626   2010]
    RC78.7.S65V527   2010
    616.07'54—dc22
                                                                            2010015381

McGraw-Hill books are available at special quantity discounts to use as premiums and
sales promotions, or for use in corporate training programs. To contact a representative
please e-mail us at bulksales@mcgraw-hill.com.

Vibrational Spectroscopic Imaging for Biomedical Applications

1 2 3 4 5 6 7 8 9 0 CTP/CTP 1 9 8 7 6 5 4 3 2 1 0

ISBN 978-0-07-159699-2
MHID 0-07-159699-2

The pages within this book were printed on acid-free paper.

Sponsoring Editor
Michael Penn

Acquisitions Coordinator
Michael Mulcahy

Editorial Supervisor
David E. Fogarty

Project Manager
Harleen Chopra,
Glyph International

Copy Editor
Surendra N. Shivam,
Glyph International

Proofreader
Eina Malik,
Glyph International

Indexer
Edwin Durbin

Production Supervisor
Richard C. Ruzycka

Composition
Glyph International

Art Director, Cover
Jeff Weeks

Contents

Contributors

Paul S. Bernstein   *Moran Eye Center, University of Utah, Salt Lake City, Utah* (CHAP. 7)

R. Bhargava   *Department of Bioengineering and Beckman Institute for Advanced Science and Technology, The University of Illinois at Urbana-Champaign, Illinois* (CHAP. 1)

Amy Drauch   *ChemImage Corporation, Pittsburgh, Pennsylvania* (CHAP. 6)

Igor V. Ermakov   *Department of Physics, University of Utah, Salt Lake City, Utah* (CHAP. 7)

Andrew P. Evan   *Department of Anatomy, Indiana University School of Medicine, Indianapolis, Indiana* (CHAP. 4)

Peter Gardner   *Manchester Interdisciplinary Biocentre (MIB), The University of Manchester, Manchester, United Kingdom* (CHAP. 3)

Ehsan Gazi   *Manchester Interdisciplinary Biocentre (MIB), The University of Manchester, Manchester, United Kingdom* (CHAP. 3)

Werner Gellermann   *Department of Physics, University of Utah, Salt Lake City, Utah* (CHAP. 7)

Mario Giordano   *Università Politecnica delle Marche, Ancona, Italy* (CHAP. 2)

Claudia Gohr   *Medical College of Wisconsin, Milwaukee, Wisconsin* (CHAP. 2)

Kathleen M. Gough   *Department of Chemistry, University of Manitoba, Winnipeg, Manitoba, Canada* (CHAP. 5)

Heather J. Gulley-Stahl   *Molecular Microspectroscopy Laboratory, Department of Chemistry and Biochemistry, Miami University, Oxford, Ohio* (CHAP. 4)

Carol J. Hirschmugl   *University of Wisconsin-Milwaukee, Milwaukee, Wisconsin* (CHAP. 2)

Susan G. W. Kaminskyj   *Department of Biology, University of Saskatchewan, Saskatoon, Saskatchewan, Canada* (CHAP. 5)

Christoph Krafft *Institute of Photonic Technology, Jena, Germany* (CHAP. 8)

Eric Mattson *University of Wisconsin-Milwaukee, Milwaukee, Wisconsin* (CHAP. 2)

Michael J. Nasse *University of Wisconsin-Milwaukee, Milwaukee, Wisconsin and Synchrotron Radiation Center, Stoughton, Wisconsin and University of Wisconsin-Madison, Madison, Wisconsin* (CHAP. 2)

Ute Neugebauer *Institute of Photonic Technology, Jena, Germany* (CHAP. 8)

Janice Panza *ChemImage Corporation, Pittsburgh, Pennsylvania* (CHAP. 6)

Giuseppe Pezzotti *Ceramic Physics Laboratory and Research Institute for Nanoscience, Kyoto Institute of Technology, Kyoto, Japan and The Center for Advanced Medical Engineering and Informatics, Osaka University, Osaka, Japan* (CHAP. 10)

Jürgen Popp *Institute of Physical Chemistry, University Jena, Jena, Germany* (CHAP. 8)

Eric Olaf Potma *Department of Chemistry & Beckman Laser Institute, University of California, Irvine, California* (CHAP. 11)

F. Nell Pounder *Department of Bioengineering and Beckman Institute for Advanced Science and Technology, The University of Illinois at Urbana-Champaign, Illinois* (CHAP. 1)

Gerwin J. Puppels *Center for Optical Diagnostics and Therapy, Department of Dermatology, Erasmus Medical Center Rotterdam, The Netherlands* (CHAP. 9)

Simona Ratti *University of Wisconsin-Milwaukee, Milwaukee, Wisconsin* (CHAP. 2)

Ann Rosenthal *Medical College of Wisconsin, Milwaukee, Wisconsin* (CHAP. 2)

Tom C. Bakker Schut *Center for Optical Diagnostics and Therapy, Department of Dermatology, Erasmus Medical Center Rotterdam, The Netherlands* (CHAP. 9)

Mohsen Sharifzadeh *Department of Physics, University of Utah, Salt Lake City, Utah* (CHAP. 7)

Mariya Sholkina *Center for Optical Diagnostics and Therapy, Department of Dermatology, Erasmus Medical Center Rotterdam, The Netherlands* (CHAP. 9)

André J. Sommer *Molecular Microspectroscopy Laboratory, Department of Chemistry and Biochemistry, Miami University, Oxford, Ohio* (CHAP. 4)

Shona Stewart *ChemImage Corporation, Pittsburgh, Pennsylvania* (CHAP. 6)

Preface

The renaissance of vibrational spectroscopy into an imaging technique has happened in the past 10 years, thanks to the advent of multichannel detection technology, integration of microscopy, and optimization of data acquisition time and analysis. The rich information content provided by infrared and Raman techniques make them suitable for biomedical applications. The data obtained is available through a simple univariate analysis, and in the case of complex applications like cancer diagnoses, the data acquisition, sampling, and analyses must be integrated in a coherent manner. The unique advantage of observing an entire field of view rapidly in the infrared imaging technique permitted applications that allowed for (1) monitoring dynamic processes, (2) spatially resolved spectroscopy of large or multiple samples, and (3) enhancement of spatial resolution due to retention of radiation throughput. An emerging biomedical application in infrared imaging is tissue histopathology, in which Fourier transform IR (FTIR) imaging has been proposed as a solution that can potentially help pathologists. It provides an objective and reproducible assessment of diseases in a manner that is easily understood by clinicians. The other developments witnessed in recent years are the incorporation of reflective substrates, integration of attenuated total internal reflection (ATR) elements with microscopy and large sample imaging, various sample forming, grazing angle, and multisample accessories. The utilization of ATR accessory in tissue samples provides a high-spatial resolution image, which would further assist in tissue histopathology and thus in diagnosis of diseases.

Unlike in IR spectroscopy, the Raman spectrum of water is weak, allowing good spectra to be acquired of species in aqueous solution. Owing to this unique advantage, biological samples like cells can be measured in their typical environments, for example, in a buffer solution or special culture medium. Using confocal Raman imaging, currently we are able to acquire depth profiles of spectra at a nanometer resolution. However, in some cases, long integration times are required because of the weak intrinsic Raman signal. Nanoparticle-based SERS imaging, by contrast, has proven to be a potential technique to provide a much stronger signal and hence shorter integration times to do measurements at subcellular level, such as sensing DNA hybridization, protein binding, etc. However, the reproducibility of the surface enhancement factor is still disputed. Another recent technique in Raman spectroscopy is coherent anti-Stokes Raman scattering (CARS) microscopy, which is a nonlinear imaging technique that offers chemical selectivity through vibrational sensitivity. Recent developments in ultrafast light sources and improved detection schemes have advanced CARS microscopy as a useful imaging tool for biomedical applications.

In this book, a large number of enthusiastic spectroscopists, including biochemists and clinicians, have discussed the latest developments in the aforementioned vibrational spectroscopic imaging. This book would give a broad overview of the recent progress in aspects like instrumentation, detector technology, novel modes of data collection, and data analysis (multivariate). Emphasis has been given on applications in the biomedical arena and to assess progress in the fields.

Scientific developments in FTIR and Raman spectroscopic imaging techniques, high-throughput tissue microarray (TMA) sampling, and multivariate data analysis have been instrumental in accelerating this imaging technique for the applications in histopathologic imaging for cancer diagnosis and research. Chapter 1 is about automated breast histopathology using FTIR spectroscopic imaging techniques. The authors have employed multivariate segmentation approach, based on a modified bayesian classifier to FTIR spectral images acquired from human breast tissue microarray. The results discussed here demonstrate promising results for reliable epithelium and stromal recognition. Chapter 2 describes the novel instrumentation and biomedical experiments that would provide an opportunity to measure in situ (in vivo) kinetics of pathological mineralization. Biomedical application of synchrotron IR microspectroscopy—studying calcium-containing crystals in cartilage from human samples and model systems—has been reviewed. The detailed description about the IR synchrotron beamline design and implementation of IRENI (IR Environmental Imaging) at the Synchrotron Radiation Center (Stoughton, Wisconsin) is discussed. Chapter 3 describes the preparation of tissues and cells for infrared and Raman spectroscopy and imaging. The importance of sample preparation is described in detail because the experimental design can have significant implications for the interpretation of spectra and thus for their biochemical relevance as well as the spatial distribution of biomolecules in imaging studies.

Among the different sampling IR imaging techniques, transmission mode imaging is the most common, while reflection-absorption is also widely practiced. In recent times, ATR imaging has become a common choice of measurement in some research groups. The reason being that it allows users to work with relatively thick sample sections and it does not require much sample preparation, expertise, or time. Chapter 4 discusses the historic development, theory, and biomedical applications of evanescent wave imaging (ATR imaging).

In recent years, Raman microscopy and imaging have been getting increasing attention and have been used for a variety of applications including some in the biomedical arena. Raman imaging combines Raman spectroscopy with digital imaging technology in order to visualize material chemical composition and molecular structure. Chapter 5 is about the applications of different microscopic techniques such as sFTIR and Raman in particular and surface-enhanced Raman spectroscopic imaging for elucidating the biochemistry of lifestyles of fungi, including saprotrophs, endophytes, and lichen symbionts.

Chapter 6 describes widefield Raman imaging that provides spectral information of all pixels of an entire field of view at once. The technological issues involved in the acquisition and preprocessing of data, and the methods that can be employed to analyze the large datasets that result from such experiments are discussed. The chapter also describes the state of the technology with respect to the study of cells and tissues. Chapter 7 covers resonance Raman imaging and quantification of carotenoid antioxidants in the human retina and skin. Raman scattering is used as noninvasive optical detection of carotenoids in living human tissue. Chapter 8 summarizes recent research results on fiber-optic Raman spectroscopy of tissue, Raman imaging of tissue and cells, and Raman spectroscopy of bacteria. The sections are organized from low-spatial resolution which was obtained using multimode optical fiber probes to high-spatial resolution which was obtained in tip-enhanced Raman spectroscopy using functionalized AFM tips. Chapter 9 provides detailed information on the Raman instrumental components such as laser, the microscope, the filter, the spectrograph and the detector. The differences between the different Raman imaging techniques, multivariate analysis, and its biomedical applications are discussed in detail.

Advances in Raman microscopic imaging provide insights into the micromechanical behavior of biomaterials, including the origin of improved fracture toughness in natural and synthetic inorganic biomaterials and the visualization of residual stress patterns stored on load bearing surfaces. Chapter 10 describes how to quantitatively assess in situ the microscopic stress fields developed during fracture at the crack tip of natural and synthetic biomaterials. Crack-tip toughening mechanisms are clearly visualized and assessed quantitatively. This chapter also presents results on microscopic stress analysis of ceramic biomaterials as collected by Raman microspectroscopy on the bearing surfaces of artificial hip joints.

Chapter 11 is about tissue imaging with coherent anti-Stokes Raman scattering microscopy. Theory, instrumentation, and its biomedical applications are elaborated.

I am quite convinced that this book will familiarize the readers with the state of the art in vibrational spectroscopic biomedical imaging and thus convincingly create a path toward the translation of vibrational spectroscopy to clinical applications.

I am most greatful to all the authors who have contributed chapters in this book.

Dr. Gokulakrishnan Srinivasan

Vibrational Spectroscopic Imaging for Biomedical Applications

CHAPTER 1

Toward Automated Breast Histopathology Using Mid-IR Spectroscopic Imaging

F. Nell Pounder and R. Bhargava

*Department of Bioengineering and Beckman Institute for
Advanced Science and Technology
The University of Illinois at Urbana-Champaign
Illinois, USA*

Recent technological developments in Fourier transform infrared (FT-IR) spectroscopic imaging, high-throughput tissue microarray (TMA) sampling, and multivariate data analysis have greatly accelerated efforts toward automated and reproducible cancer diagnosis. While several studies indicate the potential of tissue analysis by FT-IR imaging for clinical applications, vigorously validated protocols with rapid data acquisition and highly accurate classification are needed. Here, we report progress toward that goal by the development of a protocol for breast cancer histopathology. We first employ FT-IR imaging to acquire data from human breast TMAs. This TMA sampling permits rapid acquisition of spectral images from large sets of patients to select potentially useful spectral and spatial features, termed metrics, for subsequent classification. These metrics are applied to develop a robust classification system and results are extensively validated. This multivariate segmentation approach, based on a modified Bayesian classifier, demonstrates promising results for reliable epithelial and stromal recognition. This approach also has the advantage of providing insight into important biochemical properties of tissues, and sensitivity analysis of the method is straightforward. We extend the classification to include spatial measures of disease and propose that the highly accurate results may lead to specific applications in areas of clinical need.

1.1   Introduction

As 98 percent of localized breast cancers are treated effectively,[1] all women over 40 years old are recommended for annual mammography screening[2] and over 1.6 million breast biopsies are performed each year[3] to investigate screening abnormalities by removing a small sample of tissue for further analysis.[4] A manual examination of microscopic structure (histology) within the biopsy to determine the cancer type and grade forms the gold standard of diagnoses for most cancers.[5] Histologic examinations involve extensive human interpretation, making consistency difficult[6] and second opinions necessary.[7] Further, patients often wait days or weeks to receive a pathology report following a tumor biopsy.[8] Although 80 percent of these biopsies are eventually diagnosed as benign,[9] this extended waiting period is associated with substantial distress in all biopsy patients.[10] When a report is available, intra- and interobserver variability in diagnosis and treatment recommendations ranges from 1 to 43 percent.[11] Illustrative of these concerns is a study of 481 breast cancer patients from 1982 to 2000 at a regional cancer center that revealed that 73 percent of breast ductal carcinoma in situ (DCIS) patients were referred by a general pathologist for review by an expert pathologist. After expert pathologist's review, 43 percent of these cases received different treatment recommendations and 29 percent of these cases had a change in assessment of cancer recurrence risk.[12] A separate study found that 52 percent of patients referred to a multidisciplinary breast cancer review board at a university hospital for a second evaluation received a change in surgical treatment recommendation.[13] This delay and variability in tumor diagnosis may impact studies that guide basic science and clinical decision making. Clearly, the process is suboptimal; improvements in cancer diagnosis and prognosis prediction are of wide interest to clinicians,[14] pathologists,[15] health insurance companies,[16] and the general public.[17]

The root cause of these problems in cancer diagnoses, leading to complications in treatment and research, is the inability to universally provide rapid, accurate, and reproducible histologic determinations. Technology to address these needs for cancer histopathology can, thus, prove to be of central importance to cancer research and treatment. Imaging-based technology for this purpose is especially attractive, since visual evidence readily relates to the knowledge base of pathology and provides information in a compact form that can be universally comprehended. Simple structural imaging (e.g., optical microscopy of hematoxylin and eosin (H&E)-stained tissue) and manual recognition is already practiced in the clinic. Hence, efforts to improve this process are the logical first attempts at improving practice. More recently, molecular imaging has provided some understanding of specific epitopes' roles in cancer progression. Hence, it provides an alternative to add more information to classical images. Molecular bases for disease diagnoses are not universal, however, and there are significant numbers of patients in every cancer category, for whom the approach fails to provide any useful information. Another alternative of chemical imaging is emerging in which the contrast arises from endogenous chemical constitution of the tissue.

FT-IR spectroscopic imaging, the imaging analogue of molecular infrared spectroscopy, provides an alternative platform for histopathologic imaging.[18] Near-IR (NIR) light (14000 to 4000 cm^{-1}) is most commonly employed for biomedical imaging as it encompasses a region of low absorption and scattering within the body. NIR light can penetrate deep into samples and has been applied to develop noninvasive medical diagnostics. The primary NIR contrast mechanism is scattering as the region only consists of broad and significantly overlapped molecular vibrational overtones. Therefore, the NIR spectral region has limited utility in distinguishing biochemical features in complex tissue. The far-IR spectral region (below 400 cm^{-1}) contains absorption frequencies for atoms with a high mass. This is a common feature for metals and metal complexes with organic molecules.[19] In contrast, the frequencies in the mid-IR region (4000 to 400 cm^{-1}) correspond directly to fundamental vibrational modes of organic chemical species. Hence, the spectral response of any material is a chemical fingerprint that can uniquely identify chemical species, their local environment and their macromolecular conformation. Therefore the mid-IR spectral region is most appropriate for distinguishing biochemical features in tissues and is especially attractive for cancer pathology due to its ability to detect subtle transformations.

By measuring the intrinsic chemical composition of tissue, FT-IR imaging can provide significant biochemical information without the application of contrast agents or chemical stains. The use of nonperturbing radiation and compatibility of developed approaches with clinical practice are additional advantages that can lead to translation of this technology to pathology laboratories. In addition, a knowledge base for spectral changes corresponding to disease states exists and significant understanding of tissue spectroscopy is available.[20] Last, instrumentation is well developed and has a large user base.

A major impediment to clinical translation, however, has been both the lack of fast imaging methods and the lack of robustly validated protocols that are ready for implementation.[18]

The motivation for developing imaging methods is now clearly accepted. In the past 20 years several research groups have investigated the application of mid-IR spectroscopy for automated disease diagnosis.[20,21] The first attempt involved simply measuring the FT-IR spectrum of an extracted tissue sample.[22] Noticing that histologic composition provided a stronger variance than the benign-malignant differences,[23] practitioners moved quickly to employing microscopy approaches.[24] Since each pixel required a spectrum to be scanned, these approaches were very slow. Consequently, studies typically measured only a few spectra from a small number of samples. Thus, the validity and robustness of these studies were not clearly established due to the low statistical power of the studies. It is only during the last ~5 years that microscopy approaches became routine in rapidly providing high-quality data. Consequently, it is now increasingly recognized that this technology has the potential to provide an objective method for histopathology. The crucial question, however, has remained one of accuracy while being robust. In this manuscript, we discuss the key steps and large population feasibility in the development of a practical algorithm for clinical translation. We employ two-class models for breast histopathology to provide the essential features of, first, breast histology and, second, breast pathology.

1.1.1   FT-IR Imaging

Early efforts in providing spatially resolved FT-IR data involved a point-mapping approach.[25] Briefly, a target sample region was identified and radiation restricted to this region with an opaque aperture to achieve the desired spatial localization of the beam. A single-element detector measured the spectrum at each point and the entire sample could be measured in a mapping sense by raster scanning. Though useful for small numbers of samples at a low-spatial resolution, this technique is prohibitively slow and produces noisy data for high-resolution images of large samples.[26] A typical field of view for a sample of pathologic interest is about 0.5×0.5 mm. Point mapping of this size of sample at diffraction limited spatial resolution (~5 µm at the center wavelength) would require almost a day. Thus, even though promising results were shown by many, a practical approach to translating developments to the clinic was lacking.

The use of focal place array (FPA) detectors provided a multi-channel detection advantage that allowed simultaneous measurement of interferometric data from large fields of view and significantly increased data acquisition rates. The first FPA detectors employed step scanning interferometry,[27] which have been almost exclusively replaced by rapid scanning techniques[28] to achieve higher scanning efficiency.[29] Briefly, the process of data acquisition involves the interferometric encoding of a broadband blackbody emitter at high temperature as a source. The output is guided to a microscope equipped with all-reflecting optics. Since glass absorbs strongly, lenses are reflective and specially coated mirrors are employed to permit collection of the wide mid-IR spectral region. The focusing optics utilizes Cassegranian-type elements and typically condense the beam by a factor of ~10 to 15. Detection is accomplished by liquid nitrogen cooled mercury-cadmium-telluride (MCT) array detectors. Visible images of a specimen are collected from the same field of view using a white light source and a parfocal and collinear optical path.

1.1.2 FT-IR Spectroscopic Characterization of Cells and Tissues

Efforts to analyze tissue with IR spectroscopy began nearly 60 years ago with the first published diseased and normal breast, bladder, and blood spectra.[22] Early work in applying FT-IR spectroscopy for histopathology recognition involved examining single spectra from large tissue sections, DNA extracts and cell cultures. Although this work was groundbreaking, the field did not immediately take off due to the significant amount of tissue needed to acquire the spectra, primitive instrument sensitivity, and difficulty in reproducing data. This made meaningful spectral interpretation nearly impossible and the work was not pursued for nearly 40 years. In the late 1980s, this line of work was revived and led to notable publications on spectral abnormalities in colon cancer.[30] However, the low sample numbers, uncertain tissue heterogeneity, and lack of reproducibility of simple measures used to discriminate benign from malignant samples cast doubts on the validity of these studies. Several other studies applied FT-IR spectroscopy to discern premalignant tumor markers[31] and metastatic DNA features.[32] While these works supported the concept of monitoring cancer and its predisposition as well as understanding of the importance of the microenvironment in tumor development, they did not directly address the question of providing a diagnostic measure that would appeal to clinicians.

Following the resurrection of interest in application of IR spectroscopy for disease diagnosis, breast cancer was one of the major foci investigated using human tumor tissue samples, human tumor cell lines, and xenografted human tumor cells.[23] Most of the observed spectral differences were attributed to tissue heterogeneity due to collagen and fat content in connective tissues, emphasizing the need to deconvolve the effects of tissue histology from that of pathology. The relevance of collagen content in tumor samples due to the significant overlap of collagen and DNA/RNA phosphate spectral features was also recognized.[33] This work emphasized the importance of considering spatial heterogeneity and connective tissue contributions when probing for disease markers in tissue spectra. While not explicitly mentioned, it was clear that the first step in cancer diagnoses would be to separate the histologic units of tissue and then examine specific cell types individually for markers of malignancy.[34,35] With the microscopy resolution prerequisite, the use of FT-IR microscopy for cell level spectral acquisition was proposed.[36] It is also now generally recognized that univariate analysis of features, as reported in early studies, is unlikely to provide robust measures of disease. Hence, the focus of recent studies has been to employ microscopy approaches and multivariate spectral analyses[37] to provide clinically relevant information.[38,39] While the discussion above makes it clear that cell-level spectral data is needed and multivariate analyses should be employed, the emergence of FT-IR imaging is a critical technological development that enables both requirements to be met. An additional need is to demonstrate that the developed protocols are robustly applicable to a large sample population.

1.1.3   FT-IR Imaging for Pathology

This potential for using FT-IR imaging for pathology is illustrated in Fig. 1.1. The image on the left (Fig. 1.1a) is a low-power optical microscopy image, the standard practice in any pathology laboratory, in which the contrast arises due to H&E staining of nucleic acid regions blue and protein regions pink. The images in the center and on the right are from corresponding sections that are unstained. Figure 1.1b displays the relative absorption image at a frequency associated with glycoproteins, which are generally concentrated in secretory epithelium. Figure 1.1c highlights the tissue in a similar manner at a frequency associated with collagen, which is a significant component of

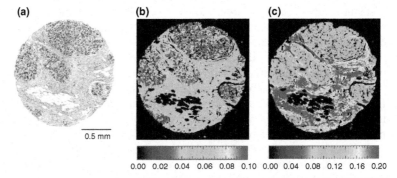

FIGURE 1.1   (a) A breast H&E-stained tissue core is compared with infrared images at (b) 1080 cm⁻¹ to highlight epithelial tissue features that correspond to hematoxylin staining and at (c) 1240 cm⁻¹ to highlight connective tissue features that correspond with eosin staining.

connective tissue. FT-IR imaging with array detectors also allows for spatial resolution near the cellular level, which provides opportunities for detailed tissue analysis. This simple example demonstrates the applicability of chemical imaging in distinguishing prominent tissue features without the use of chemical dyes or contrast agents, yet in a manner that is appreciable by practitioners who may not be experts in spectral analysis.

FT-IR imaging has been demonstrated to be a useful tool in the analyses of many tissue types.[18] For breast tissue and cancer, a number of studies have provided evidence of feasibility. One of the first successful efforts involved a cohort of 77 breast tumor samples and incorporated linear discriminant analysis with cross validation to classify tumors by grade and steroid receptor status.[20] While the classification results were fairly accurate (87 percent for tumor grade and 93 percent for steroid receptor status), the study lacked independent validation data. Further translational activities were not reported to establish classification in a clinical setting. Several subsequent trials conducted by another group involved a collection of several thousand spectra from approximately 25 breast cancer patients with fibroadenoma, DCIS, or invasive ductal carcinoma. A supervised artificial neural network (ANN) analysis was used to develop an automated classifier.[40] In a separate study using a subset of the same data, cluster analysis was performed on 96 spectra in the fingerprint spectral range to separate fibroadenoma and DCIS.[34] Although these results show good classification accuracy, a much more extensive study is needed to evaluate the diagnostic potential of this algorithm and ensure that calibration data is not overfit by the ANN. Further, it is difficult to interpret the results of a complex, nonlinear classifier. Other approaches to classify breast tissue involved the novel use of slides and staining, as practiced in clinical settings to ensure compatibility to current practice.[34,41] The results were promising in small cohorts but larger efforts are needed to ensure that the promising early results can provide a consistent and practical protocol for clinical translation.

While many of these studies applying FT-IR spectroscopy for disease diagnosis have produced interesting results, they have not widely attracted attention from clinicians due to their preliminary nature. Specifically, the small numbers of patients included in these studies and the difficulties in achieving effective large-scale validation results due to overfitting small sets of spectral data remain concerns. Various approaches to the microscopic analysis of tissue structure have been used but can be divided roughly into two major categories. In the first, spectroscopy is used to guide the visualization of tissue. For example, displaying pixels with spectral similarity with the same color codes allows a human to recognize gross structure. An example of this approach is a hierarchical clustering analysis that allows a user to choose the number of clusters to be displayed. In the second, prior clinical knowledge is employed to guide the segmentation of tissue and the end result displays the data corresponding to the chosen model without requiring any human intervention. An example is a modified Bayesian classification[42,43] method in which the universal set of pixel values is bounded by a prior clinical model. The method necessarily requires the measurement of a large number of known samples (priors) and, hence, was simply not possible in the eras of mapping and early FT-IR imaging.

1.1.4   High-Throughput Sampling

While it is generally recognized that a large number of samples are needed for calibration and validation of a prediction model, a convenient method to image such large numbers was not available. The development of tissue microarrays (TMAs) has provided a useful solution[44] and their application with rapid FT-IR imaging demonstrated wide population robustness.[45] TMAs can be obtained from numerous tissue banks and incorporate small malignant and benign tissue samples from many different patients on a single slide. This tissue source promotes the development of large FT-IR imaging studies to achieve statistically significant histologic and pathologic classification results.

1.1.5   Modified Bayesian Classification and Automated Tissue Histopathology

In this work, we follow an approach that combines the use of TMAs, FT-IR spectroscopic imaging, and supervised automated histologic segmentation.[46] As outlined in Fig. 1.2, the experimental procedure involves acquiring spectral images and examining tissue spectra to select metrics for classification. The metrics are tissue spectral features such as peak heights, ratios, areas, and centers of gravity. These features capture the essential elements of the spectra, without regard to histologic tissue type or disease state. Since the number of metrics is considerably less than the number of spectral data points, this step helps reduce the dimensionality of data and makes subsequent calculations fast. The next step is to determine the probability distribution function (pdf) for each metric and quantitatively estimate the overlap in pdfs. Pdfs are estimated from ground truth pixels that have been marked manually by referring to a corresponding section that was H&E stained and examined by a pathologist. The types of classes marked by a pathologist are restricted to the task at hand. For example, in this chapter, we report the two-class case in which epithelium is first segmented from stroma. After histological epithelium recognition is established, the epithelium is further separated into benign and malignant classes. Each cell type (class) is denoted by a false color to provide visualization. The overlap in pdfs forms the region of ambiguity in classification and provides a preliminary estimate of the

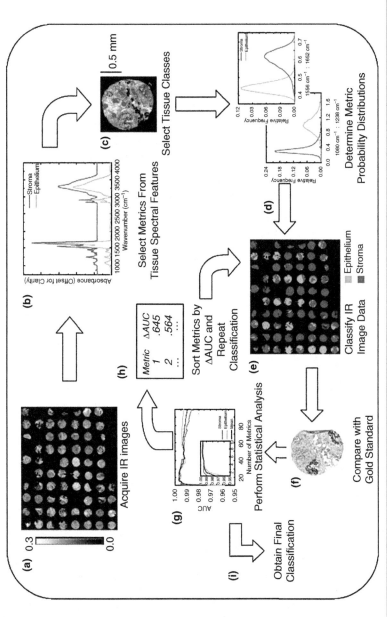

FIGURE 1.2 (a) FT-IR breast TMA image data is acquired and (b) the resulting tissue spectra are analyzed to select spectral metrics. (c) FT-IR and H&E-stained images are then compared to select stroma and epithelium tissue regions. (d) The metric value frequency distributions for stroma and epithelium are determined and (e) each pixel on the spectral image of the breast TMA is classified as stroma or epithelium. (f) The classified image is then carefully compared with an H&E-stained image to qualitatively assess histologic segmentation and (g) ROC statistical analysis is performed for quantitative classification evaluation. (h) The procedure is optimized by sorting the metrics until the classifier rapidly converges at AUC ~ 1 at which (i) an optimal set of classification metrics is selected.

9

error that would result in using that specific metric for classification. The metrics are arranged in order of increasing error and employed to classify tissue. An entire classifier is built using the first metric, the first two, the first three, and so on. The total number of classifiers is equal to the total combinations of metrics that are present. We restricted ourselves to linear combinations or singular measures of metrics to allow interpretation of results in terms of the underlying spectral data.

Statistical analysis of classification accuracy is performed by the application of receiver operating characteristic (ROC) analysis with quantitative evaluation by calculation of the area under the ROC curve (AUC). Since each classifier differs from the previous by the addition of a metric, this process has also been termed the sequential forward selection process. A plot of the AUC with the addition of specific metrics reveals those that increase or reduce classification accuracy. Classification is then optimized by sorting the metrics by the change in the AUC after the addition of a given metric and subsequently iterating the classification procedure. The classification algorithm is based on Bayes' decision rule which states that

$$p(c_1|m_i) = \frac{p(m_i|c_1)p(c_1)}{p(m_i)} \quad \text{and} \quad p(c_2|m_i) = \frac{p(m_i|c_2)p(c_2)}{p(m_i)} \quad (1.1)$$

where c is a tissue class and m is a spectral metric. Due to the limited tissue sampling for determining the distributions of $p(m_i|c_1)$ and $p(m_i|c_2)$, it is not possible to find exact values for the prior tissue class probabilities $p(c_1)$ and $p(c_2)$. Therefore, $p(c_1)$ and $p(c_2)$ are estimated during the calibration step to determine which values provide the highest accuracy.

1.2   Materials and Methods

Two paraffin-embedded breast TMAs from US Biomax Inc. with tissue samples from 40 breast cancer patients are analyzed in this study. The TMAs are fixed on barium fluoride (BaF_2) substrates to permit data collection over the entire mid-IR spectral region of interest (720 to 4000 cm^{-1}). The first array contains carcinoma and adjacent normal tissue from 40 patients (2 with a grade I tumor, 26 with a grade II tumor, 6 with a grade III tumor, and 6 with an unknown tumor grade). This array is used as a calibration dataset to develop algorithms to segment breast histology and pathology as outlined in Fig. 1.2. These algorithms are then validated on a separate cut of the same TMA containing different tissue sections from the same patients. Prior to imaging, paraffin is removed from each TMA by immersing in hexane for 48 to 72 hours at 40°C while stirring. To ensure continued paraffin removal, fresh hexane is added every 3 to 4 hours. Paraffin elimination is checked at 24 hours to monitor the disappearance of the 1462 cm^{-1} peak on several tissue cores.

A Perkin-Elmer Spotlight 300 spectrometer is used for collection of spectral images at a 6.25 µm pixel size and a 4 cm^{-1} nominal spectral resolution with 2 scans per pixel. An IR background is collected at 120 scans per pixel at a location on the array substrate with no sample present. Tissue spectral images are output as the ratio of the raw data to the background spectra. Spectral images are then compiled, analyzed, and classified using Environment for Visualizing Images (ENVI) imaging software with programs written in the interactive data language (IDL) compiler to perform the classification analysis described in Fig. 1.2.

1.2.1 Models for Spectral Recognition and Analysis of Class Data

The first model developed for breast tissue classification involves the segmentation of stroma and epithelium. This step is necessary to determine the important spectral features for breast tissue, as these are two of the most prominent tissue classes in the breast.[47] Distinguishing epithelium from stroma is particularly important, as over 99 percent of malignant breast tumors arise in epithelial tissue.[48] Therefore reliable epithelial segmentation is a prerequisite for tumor recognition. Stromal identification is also important as many recent cancer studies have highlighted the importance of the stromal microenvironment in epithelial tumor development.[49] As stroma and epithelium display significantly different biochemical properties, they should be segmented in spectral images with a high degree of classification accuracy and confidence.

To develop a classification model, spectral image regions are identified by comparing FT-IR and H&E images to select spectral image pixels that clearly correspond to stroma or epithelium. Approximately 200,000 pixel spectra are selected for calibration to eliminate errors due to variation between individual patients and inherent spectral noise. Selection of a large number of spectra for calibration is necessary to ensure classification accuracy in validation studies. Average spectra for stroma and epithelium are then computed from these selected spectral image pixels and are displayed in Fig. 1.3. Important spectral features are selected by examination of tissue class average spectra to reduce data dimensionality prior to classification. Prominent stroma and epithelium spectral features are compared with breast tissue spectral features previously identified by other groups[33,39,40] to assess the biological relevance of each metric. Stroma and epithelium spectra are then distinguished by considering spectral features associated with unique biochemical tissue properties. For example, epithelial tissue is observed to have a higher relative

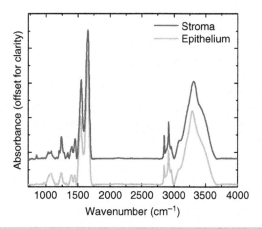

Figure 1.3   Spectral profiles from epithelial and stromal cells.

absorbance value at the 1080 cm^{-1} peak attributed to symmetric phosphate stretching vibrations in DNA.[23] Conversely, stromal tissue displays more prominent peaks at 1236 cm^{-1} and 1338 cm^{-1}, which are associated with collagen protein glycine and proline side chains.[33] Additional important spectral differences are observed at 1456 cm^{-1} due to CH$_3$ asymmetric bending and at 1556 cm^{-1} due to amide II CN stretching and NH bending.[33] Peak heights, ratios, areas, and centers of gravity associated with these and other relevant spectral features are selected for further evaluation.

Regions of interest on spectral images are manually selected by careful comparison with H&E tissue sections denoted by a trained pathologist. As emphasized in the tissue images in Fig. 1.1, a spectral image at 1080 cm^{-1} that indicates DNA chemical features is useful in identifying epithelial tissue and a spectral image at 1240 cm^{-1} that designates biochemical characteristics of proteins is useful in selecting stromal tissue. Spectral images at the amide I peak (1652 cm^{-1}), corresponding with protein C=O stretching vibrations,[33] and the amide A peak (3294 cm^{-1}), corresponding with NH bending vibrations,[50] are also useful in visualizing contrast between cell types. An H&E-stained TMA core and the corresponding spectral image with marked regions of interest are shown in Fig. 1.4a and b. Boundary spectral image pixels are not marked to avoid classification errors associated with incorrect manual identification of tissue regions.

1.2.2   Automated Metric Selection and Classification Protocol Optimization

After regions of interest are selected for stroma and epithelium, the spectral value distributions for each metric are determined. These distributions are used, first, to predict the classification error associated with each metric by evaluating the area of overlap for each

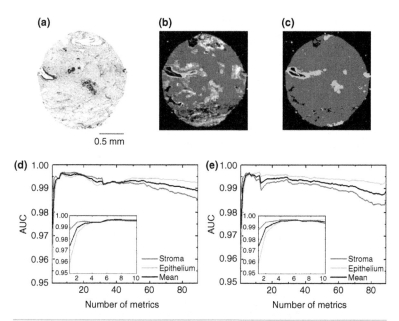

FIGURE 1.4 (a) An H&E-stained image and (b) an IR image of the amide I intensity of a typical TMA core displaying the manually marked regions of interest belonging to epithelium (green) and stroma (magenta). (c) The classified spot demonstrates a correspondence with the manually marked region. (d) The first and (e) second iteration demonstrate the quick convergence of the AUC value to a maximum of ~1 with 6 metrics.

calculated pdf. These distributions are, second, used to classify spectral image pixels as stroma or epithelium using the modified bayesian classifier described previously. Classification accuracy is assessed with ROC analysis and the spectral metrics are sorted based on the change in AUC. The classification and statistical analysis is repeated until sorting the metrics does not decrease the number of metrics required to reach a maximum AUC at ~1.

This classification technique is very accurate for the proposed two-class model, as indicated by the quick rise in the AUC value for breast stroma and epithelium tissue classification (Fig. 1.4d and e). As seen in the inset for each AUC curve, the first iteration required 7 metrics to reach a maximum AUC while the second iteration required only 6 metrics to reach this point. The rapid convergence of the classification optimization is permitted by the sorting of metrics by increasing pdf class overlap prior to beginning classification. Many valuable metrics were initially listed in the first 40 metrics, and were quickly identified by sorting the metrics by the change in AUC associated with each metric. This optimized classifier requires only six metrics, which can be rapidly applied in a clinical setting.

The classification accuracy is quantitatively measured against a gold standard of tissue regions selected by a trained pathologist. Classified images for breast TMA cores (Fig. 1.4c) are produced from the optimal six metric classifier obtained by iteration. Qualitative comparison of a classified image with the corresponding H&E image from an adjacent TMA cut in Fig. 1.4a indicates a reasonable correlation between classified images and H&E-stained images. These classified images and ROC analysis results indicate that FT-IR imaging coupled with Bayesian classification has the potential to reliably select tissue classes that correspond with H&E staining. This high-throughput spectral classification approach that incorporates millions of spectra from many patients is the first step to establish spectral image classification as a diagnostic tool in a clinical setting.

1.2.3   Spectral Metrics and Biochemical Basis

The six metrics selected for stroma and epithelium classification are displayed in Table 1.1. These spectral features can be used as an optimal classifier with an AUC of 0.995 for calibration data. Nearly all of these metrics fall in the spectral "fingerprint region." This region contains many narrow overlapping peaks, and is often useful for identifying complex molecules. Therefore, it is not surprising that

Feature	Position (cm^{-1})	Assignment[33]	Molecular Origin
Peak ratio	1080:1456	1080 cm^{-1}: symmetric PO_2^- stretching, CO stretching 1456 cm^{-1}: assymetric CH_3 bending	DNA/RNA Protein
Peak ratio	1556:1652	1556 cm^{-1}: NH bending, CN stretching 1652 cm^{-1}: CO stretching	Protein (Amide I & Amide II)
Peak ratio	1080:1238	1080 cm^{-1}: symmetric PO_2^- stretching, CO stretching 1238 cm^{-1}: assymetric PO_2^- stretching	DNA/RNA
Center of gravity	1216–1274	1236 cm^{-1}: NH bending, CN stretching, CH_2 wagging, assymetric PO_2^- stretching	DNA/RNA Protein (Amide III)
Peak ratio	1338:1080	1080 cm^{-1}: symmetric PO_2^- stretching, CO stretching 1338 cm^{-1}: CH_2 wagging	DNA/RNA Protein (Amide III)
Peak area	1426–1482	1456 cm^{-1}: assymetric CH_3 bending	Protein

TABLE 1.1   Spectral Metrics Selected by Optimization of the Stroma and Epithelium Histology Classification Model

robust metrics for tissue identification would be found in this region. In addition, the molecular origins for these metrics involve proteins and DNA, which are also partly responsible for epithelium and stroma identification by H&E staining. While these six metrics were identified as an effective classifier using the AUC optimization method described previously, they may or may not be the best possible classifier for this calibration TMA. Some useful spectral features initially listed toward the end in the initial metric order may not have been adequately considered in the metric sorting process due to the rapid convergence of the AUC value. Selection of a single optimal classifier would require more rigorous and time consuming optimization analysis, which is not necessary for this two-class model due to the quick AUC convergence using the simple classification iteration method described in this manuscript.

1.2.4 Validation and Dependence on Experimental Parameters

Validation studies are performed on a separate TMA with tissue samples from the same 40 patients to assess the robustness of the classifier. From Fig. 1.5 it is clear that the six metric classification model

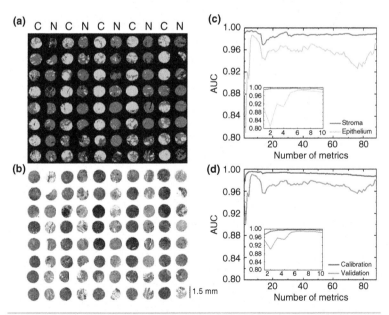

FIGURE 1.5 (a) Classified images for a validation dataset for the developed protocol demonstrate segmentation of the tissue into the two selected classes. (b) The corresponding H&E-stained image is shown for reference. (c) ROC curves for epithelium and stroma, indicating the AUC values, demonstrate high degree of confidence in the classification. (d) Mean AUC curves for calibration and validation TMAs indicate that the classifier is robust and effective on independent datasets.

developed on the calibration TMA readily translates to a separate spectral dataset. This indicates that the classifier does not overfit the spectral data and has the potential to provide reproducible results in a clinical setting.

Stroma and epithelium are easily visualized on classified images (Fig. 1.5a) and appear to correspond accurately with H&E images (Fig. 1.5b). The infiltrating tumors on malignant TMA cores are readily recognized by the extensive green epithelium visible on the classified image. The classified image provides the advantage of quick visualization of tissue heterogeneity without the necessity of adding stains or chemical dyes that irreversibly alter tissue properties. The qualitative correspondence between H&E-stained images and classified spectral images is confirmed by quantitative ROC analysis (Fig. 1.5c). This analysis is performed after manual labeling of ~50,000 pixels in the validation TMA spectral image as stroma or epithelium to serve as a gold standard for classification evaluation. The small inset plot reveals that both stroma and epithelium reach a maximum AUC of ~1 with the six metric classification model obtained by validation on a separate dataset.

A comparison of the calibration and validation studies (Fig. 1.5d) also indicates significant similarities in classification performance. In both studies, the mean AUC curve reaches a maximum of over 0.99 with only six spectral metrics. A slight decrease in the AUC for each curve past 60 metrics reveals that the data may be overfit by the addition of these metrics. This is even more noticeable in validation, which is reasonable since the classification algorithm was not designed on this dataset. A closer examination of the mean AUC curve for the first six metrics in the plot inset indicates that the contribution of two metrics to classification performance is more significant in the calibration than in the validation dataset. However, the other four metrics show similar results in both calibration and validation.

Although this quantitative evaluation provides an excellent analysis of classification accuracy for stroma and epithelium as compared with the gold standard tissue sections selected by a trained pathologist, it does not provide any indication of tissue segmentation accuracy outside these selected regions of interest. This quantitative analysis only evaluates supervised data, and does not provide a numerical indication of the potential of this algorithm on unsupervised spectral image classification. Image regions not included in this quantitative evaluation include boundary pixels neglected in selecting regions of interest and infiltrating epithelial tissue sections intermixed with malignant stroma in high-grade tumor samples, which are difficult to manually select in tissue spectral images. Nonetheless, a qualitative comparison of H&E and classified images indicates reasonable classification of spectral image pixels that are not manually mapped to a specific tissue class.

The dependence of classification accuracy on spectral resolution is also considered. The validation data originally acquired at a 4-cm^{-1}

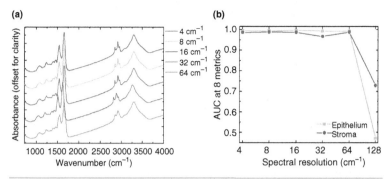

FIGURE 1.6 (a) Epithelium spectra are obtained by downsampling data acquired at 4 cm⁻¹ to lower spectral resolutions. (b) AUC analysis for stroma and epithelium segmentation for each spectral resolution demonstrates a decrease in classification accuracy only at a very course spectral resolution.

resolution is downsampled to more course spectral resolutions using a neighbor binning procedure. Average epithelium spectra (Fig. 1.6a) demonstrate the effect of downsampling on spectral features. Important spectral elements remain constant at 4, 8, and 16 cm⁻¹ resolutions, but peak locations and characteristic shapes begin to change significantly at 32 and 64 cm⁻¹. However, a significant drop in classification accuracy does not occur until the spectral resolution decreases to 128 cm⁻¹ (Fig. 1.6b).

The robust classifier performance at downsampled spectral resolutions is permitted by the significant biochemical and spectral differences between stroma and epithelium and the inherent nature of the selected spectral metrics. As reflected in the spectra in Fig. 1.3, numerous differences between these two tissue classes are visible and indicate that there are significant biochemical differences between these two types of tissue. Therefore, fine spectral resolution is not essential to distinguish stroma and epithelium. In addition, the peak height, area, and center of gravity metrics selected are not extremely sensitive to small changes in spectral features. Spectral absorbance values are generally measured accurately as long as the full width at half maximum (FWHM) is not significantly less than the spectral resolution. Therefore, many peaks are not affected by moderate decreases in spectral resolution. Also, the center of gravity metrics incorporated in the classifier depend on both peak position and shape, and are therefore less significantly affected by changes in peak location in downsampled spectra. The inherent biochemical differences between epithelium and stroma and the types of metrics selected for tissue segmentation allow the potential of faster data acquisition at lower spectral resolutions without considerable loss in classification potential.

1.2.5   Application for Cancer Pixel Segmentation

Upon successful differentiation of stroma and epithelium, the next problem of automated segmentation of cancerous and normal epithelium is addressed. All spectral image regions classified as stroma are removed prior to commencing with cancer identification. Epithelial pixels are then divided into cancer and normal classes based on identification of malignant tissue cores by a trained pathologist. These manually selected epithelial pixels are used as a gold standard to determine the frequency distribution of each spectral metric for cancerous and adjacent normal epithelium and to assess cancer identification during statistical analysis. A malignant and benign epithelium classifier is then calibrated using the procedure described in Fig. 1.2. The pixel-based ROC analysis for calibration spectral image data (Fig. 1.7a) reveals a maximum AUC of 0.79 with a nine-metric classifier. This predicted accuracy for cancer and normal segmentation is comparable with other pixel-based classification algorithms.[51]

However, this classification accuracy is significantly lower in validation spectral image data, as displayed in Fig. 1.7b. This is likely due to several factors involved in determining segmentation accuracy. First, the spectral images were acquired with only two scans per pixel to permit rapid data collection, resulting in a lower signal-to-noise ratio (SNR). In addition, malignant tissues have vastly different pathologic characteristics depending upon the organization of the tumor, the type of tumor, and the grade of the tumor. This variation in observed pathology would impact the biochemical features present in the spectra. Therefore, attempting to group all malignant spectra as one class may create difficulties in developing a reproducible segmentation algorithm to separate all malignant tissue as a single class. The metric probability distributions may not be uniform for different tumors and different patients, which would decrease predicted accuracy confidence in the validation AUC curve. To further assess classification potential, a core-based ROC analysis is performed to analyze the sensitivity and specificity of the classification (Fig. 1.7c). This resulting curve illustrates the trade-off between these two factors in determining appropriate classification parameters. The threshold for separating cancer and normal tissue could be altered based upon a cost model for false positive and false negative results. The significant decrease in AUC for cancer segmentation for validation data (Fig. 1.7b) indicates that further studies to improve cancer detection are required.

However, a comparison of calibration data H&E and classified images indicates that using a threshold that assumes an equal cost for false positives and false negative segmentation produces reasonable cancer detection. As shown in Fig. 1.7d and e, many malignant tissue cores contain a significant number of epithelial pixels classified as cancerous while the normal tissue cores do not contain a significant number of pixels classified as malignant. These images indicate that

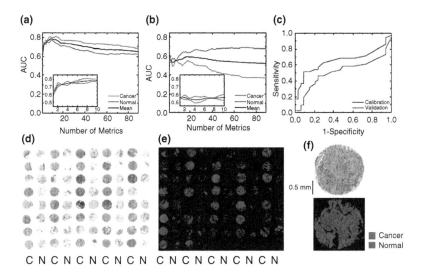

CNCNCNCNCN CNCNCNCNCN

FIGURE 1.7   (a) Training and (b) validation ROC curves to separate benign from malignant pixels. (c) A core level ROC curve demonstrates the overall sensitivity and specificity of the developed algorithm to segmenting tissue. (d) An H&E image and (e) a classified image to demonstrate the quality of classification achieved. (f) A single TMA core demonstrates heterogeneity in classification.

FT-IR imaging and classification has the potential to provide an automated indication of tumor presence, which can be subsequently reviewed by a pathologist if a significant number of malignant pixels are detected. Figure 1.7f shows a side-by-side H&E and classified image for a single cancerous tissue core. The stroma tissue removed by initial epithelium and stroma segmentation is visible in the black regions on the interior of the tissue core in the classified image. Tissue heterogeneity in tumor classification is evident, as a significant portion of the epithelial pixels are classified as benign. Notwithstanding, enough malignant pixels are identified to indicate the presence of a tumor.

The nine spectral metrics selected by classification optimization to segment malignant and benign epithelium on calibration data are listed in Table 1.2. Notably, two of these nine metrics were also included in the six metrics selected for stroma and epithelium segmentation. Likewise, most of the spectral features used to compute the other seven metrics are also used to compute the stroma and epithelium classifier metrics. This observation provides biochemical supporting evidence that tumor epithelial cells develop mesenchymal characteristics during malignant transformation. Earlier studies discussed previously noted similar spectral differences between cancerous and normal IR spectra. However, these differences were generally attributed to tumor tissue heterogeneity. While this conclusion was reasonable for these studies that did not involve the histology

Feature	Position (cm^{-1})	Assignment[33]	Molecular Origin
Peak ratio	1238:1542	1238 cm^{-1}: CN stretching, CH$_2$ wagging, assymetric PO$_2^-$ stretching, 1542 cm^{-1}: NH bending	DNA/RNA Protein
Peak area	1426–1482	1456 cm^{-1}: assymetric CH$_3$ bending	Protein
Center of gravity	1300–1358	1338 cm^{-1}: CH$_2$ wagging	Protein (Amide III)
Peak ratio	1080:1234	1080 cm^{-1}: symmetric PO$_2^-$ stretching, CO stretching 1234 cm^{-1}: NH bending, CN stretching, CH$_2$ wagging	DNA/RNA Protein (Amide III)
Center of gravity	1482–1594	1556 cm^{-1}: NH bending, CN stretching	Protein (Amide II)
Peak ratio	1204:1652	1204 cm^{-1}: NH bending, CN stretching, CH$_2$ wagging 1652 cm^{-1}: CO stretching	Protein (Amide I & Amide III)
Peak ratio	1238:1396	1238 cm^{-1}: CN stretching, CH$_2$ wagging, assymetric PO$_2^-$ stretching, 1396 cm^{-1}: COO$^-$ stretching	DNA/RNA Protein
Peak area	1324–1358	1338 cm^{-1}: CH$_2$ wagging	Protein (Amide III)
Peak ratio	1080:1456	1080 cm^{-1}: symmetric PO$_2^-$ stretching, CO stretching 1456 cm^{-1}: assymetric CH$_3$ bending	DNA/RNA Protein

TABLE 1.2   Spectral Metrics Selected by Optimization of the Cancer and Normal Epithelium Classification Model

segmentation step, the method presented here for distinguishing cancer removes stromal image pixels prior to beginning cancer classification. In addition, statistical analysis is only performed on epithelial pixels manually selected by comparison with gold standard H&E images. Therefore, the similarity in metrics selected in the optimized classifiers for segmenting epithelium and stroma and segmenting malignant and benign epithelium indicate potential epithelial to mesenchymal transition in tumor development. Conversely, less than half of the metrics selected relate to DNA content and characteristics. This indicates that focusing only on DNA biochemical features in tumor spectra may not be the optimal method for distinguishing cancerous epithelial cells in intact human tissue.

These classification results for cancerous and adjacent normal epithelium represent only an initial effort in cancer segmentation. Further studies will be conducted to analyze the impact of spectral noise on broadening malignant and benign probability distributions, which in turn decreases segmentation accuracy estimates. Alternative segmentation methods such as genetic algorithms that do not require prerequisite knowledge of metric population frequency distributions will be also considered for cancer classification, as variation between individual patients may make the application of a single set of biochemical features to all tumors unfeasible. Finally, histological discrimination of different types of stroma found in malignant and benign tissue will be evaluated to analyze changes in stromal features with malignant development.

1.2.6   Application for Patient Cancer Segmentation

Breast carcinomas are identified as a mass of epithelial cells. These epithelial tumor masses can be discriminated from normal breast epithelium by altered cellular morphology and tissue structure. We have previously demonstrated automated prostate tumor discrimination in spectral image datasets by evaluating altered cellular morphology based on cell polarity.[42] For breast tissue, we examine the potential for tumor discrimination by evaluating the changes in tissue structure in invasive tumors in which tumor cells move out of the breast ducts and lobules and infiltrate surrounding tissue to create a large mass of epithelium. Thus, malignant tissue contains more epithelium than normal tissue, as previously noted in the discussion of the false-color classified spectral images of TMA datasets (Fig. 1.5).

To quantify this change in epithelial tissue content, we have developed a spatial analysis strategy termed "multiscale neighborhood polling." In this approach, boxes ranging in size from 1×1 pixel (6.25×6.25 µm) to 12×12 pixels (75×75 µm) are evaluated by computer simulation to compute the fraction of pixels classified as epithelium for each box. This fraction is expected to be higher in cancer TMA cores than adjacent normal TMA cores. To eliminate errors associated with TMA dataset pixels that do not contain tissue, all boxes without any pixels classified as stroma or epithelium are not included in calculations. Any cancerous TMA cores that do not contain epithelium or do not have a clear pathology diagnosis are also not considered. To distinguish cancerous and normal TMA cores, the fraction of boxes containing at least 50 percent epithelium for each box size is calculated for each TMA core. Average values for cancerous and normal TMA cores are compared (Fig. 1.8*a*), and standard deviation error bars indicate that there is a clear distinction in epithelial content between cancer and normal TMA cores for all box dimensions. An optimal cutoff for selection of cancerous TMA cores is determined from the standard deviation values for each tissue class by the relationship

$$\frac{d_C}{d_N} = \frac{\sigma_N}{\sigma_C} \tag{1.2}$$

where d_C = distance of the cutoff from the mean of the cancer TMA cores
d_N = distance of the cutoff from the mean of the adjacent normal TMA cores
σ_C = standard deviation for cancer TMA cores
σ_N = standard deviation for adjacent normal TMA cores.

An optimal cutoff point is calculated for each box size from 1×1 pixel to 12×12 pixels. Optimal operating points for each box size are found by Eq. (1.2) to account for the lower variance of the average fraction of boxes with more than 50 percent epithelium in adjacent normal TMA cores. A least squares linear trendline is fit to the optimal cutoff point for each box size to compute the operating line in Fig. 1.8*a*.

The absence of overlap of standard deviations for each box size indicates that each of these metrics should provide similar separation of cancer and adjacent normal TMA cores. The standard deviations for the cancer and normal TMA classes are relatively constant, regardless of box size. Therefore, it is feasible to reduce these 12-box-size metrics to a single parameter to facilitate more rapid TMA core classification. This is accomplished by applying a least squares linear fit to each core for the fraction of boxes containing at least 50 percent epithelium versus box size dataset and computing the offset (*y*-intercept) value. A single offset value can then be selected as a cutoff, where all TMA core datasets with an offset above this value are classified as cancer. Since cancer TMA cores have a greater fraction of boxes containing at least 50 percent epithelium, these cores will also have a greater offset value. The offset cutoff for cancer determination can be altered to adjust the classification sensitivity and specificity.

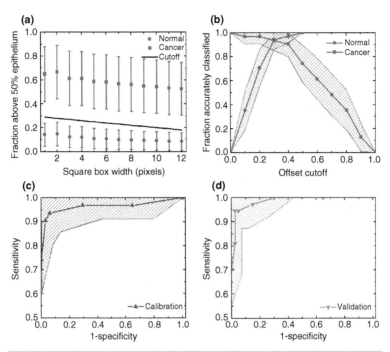

FIGURE 1.8   Square boxes are selected on each TMA core to range in size from 1 × 1 pixel (6.25 × 6.25 µm) to 12 × 12 pixels (75 × 75 µm) and the fraction of epithelium is calculated for each box. (a) A plot of the fraction of boxes containing over 50 percent epithelium vs. box size with error bars representing standard deviation indicates that a significantly larger portion of boxes contain over 50 percent epithelium on cancer TMA cores for all box sizes. An optimal cutoff is selected based on the calculated mean and standard deviation for the cancer and normal classes. A least-squares linear fit model is computed for the fraction above 50 percent epithelium vs. square box width for each TMA core, and the average offset is determined for the cancer and normal datasets. An offset value is then selected as a cutoff point for separating cancer and normal cores. (b) A plot of the fraction of accurately classified TMA cores vs. offset cutoff with shaded areas representing 95 percent confidence regions indicates an optimal operating point at an offset of 0.3. (c) Calibration and (d) validation ROC curves with 95 percent confidence regions demonstrate the effective overall sensitivity and specificity of the developed algorithm in segmenting cancer and adjacent normal TMA cores.

A plot of the fraction of TMA cores accurately classified versus selected offset cutoff (Fig. 1.8b) indicates that an offset cutoff at 0.3 achieves optimal TMA core segmentation, with true positive and true negative fractions over 0.9 for both cancer and adjacent normal TMA cores. The 95 percent confidence regions, approximated using a binomial large-sample formula,[52] indicate that the true optimal offset cutoff is in the range of 0.2 to 0.5. The relatively narrow width of the confidence bands reflects the significant difference between the offset of adjacent normal and cancer TMA core datasets. These confidence bands also reflect the narrow widths for the offset distributions for both classes, particularly for the normal class.

The trade off between sensitivity and specificity is demonstrated by ROC analysis (Fig. 1.8c and d). The ROC curve with 95 percent confidence intervals for the calibration TMA dataset in Fig. 1.8c is derived from the sensitivity and specificity curves in Fig. 1.8b. At the optimal operating point of a 0.3 offset cutoff the sensitivity is 93 percent and the specificity is 94 percent, indicating clear discrimination of both cancer and normal TMA cores. At this location on the ROC curve the 95 percent confidence interval gives a lower bound of 85 percent sensitivity and 86 percent specificity, which are minimum acceptable standards for a potential cancer diagnostic tool. The AUC value is calculated as 0.96 ± 0.02, indicating that the sample size of 31 tumor TMA cores and 34 normal TMA cores is sufficient to demonstrate confidence in cancer and normal classification potential with this algorithm.[53] Due to the minimal overlap in offset distributions for cancer and normal TMA cores and the reasonably large sample size, the statistical power is calculated as 100 percent with a z-score greater than 3.72 using standard methods to compare means for standard normal distributions.[52] Examination of the frequency distribution for adjacent normal and cancer TMA cores validates this assumption of normal distribution. This indicates that the number of sampled cancer and adjacent normal TMA cores is large enough to determine that cancer cores contain a larger fraction of boxes containing more than 50 percent epithelium pixels.

However, examination of the validation TMA dataset demonstrates some limitations in diagnostic determination associated with the number of patients in the TMA sample. A similar trend to the calibration ROC curve in Fig. 1.8c is observed in the validation TMA dataset ROC curve in Fig. 1.8d. For this slightly larger dataset with 37 cancer and 40 adjacent normal TMA cores, the optimal operating point at a 0.5 offset cutoff demonstrates a sensitivity of 95 percent and a specificity of 98 percent. At a 0.3 offset, which provided optimal cancer segmentation for the calibration TMA dataset, the sensitivity is 97 percent and the specificity is 85 percent. This would also be a reasonable operating point for the validation dataset, as this lower specificity is still acceptable for a diagnostic test. Notwithstanding, this validation study demonstrates that the optimal offset cutoff for cancer diagnosis remains uncertain.

Comparison of the calibration and validation ROC curves also demonstrates a disadvantage associated with TMA sampling. The AUC value for the validation ROC curve is 0.99 ± 0.01, which is slightly greater than that of the calibration ROC curve. This difference in the calibration and validation AUC values is attributed to the presence of a cancer TMA core in the calibration dataset from a small invasive tumor that contains only a minimal amount of epithelium in the tumor region selected for TMA core. This serves as an outlier in the calibration TMA dataset and delays the sensitivity from reaching a value of 1 on the ROC curve, resulting in a lower AUC value. However, the difference between the calibration and validation AUC values is demonstrated to be statistically insignificant due to the limited number of cores on each TMA. In order to demonstrate a statistically significant difference between the AUC values for these ROC curves, each TMA would need to contain 10 times as many cancer and normal TMA cores.[53] This indicates that although the number of TMA cores included in this study is large enough to demonstrate the feasibility of this algorithm for breast tumor discrimination, the limited sample size for the study causes the quantitative ROC results to be somewhat sensitive to outliers.

The preliminary study presented here provides evidence that breast tissue spectral images segmented into stromal and epithelial classes can be useful for tumor discrimination by the evaluation of spatial information and epithelium content. ROC analysis indicates that near-perfect discrimination of cancer and adjacent normal TMA cores is possible by computing the fraction of simulated boxes ranging in size from $6.25 \times 6.25\,\mu m$ to $75 \times 75\,\mu m$ that contain at least 50 percent epithelium pixels. However, the results of this initial study provide somewhat limited information about the application of this technique to a large population of cancer patients, as the optimal cutoff for cancer segmentation remains unclear. Notwithstanding, this study indicates that this method for cancer discrimination has potential for automated tumor recognition. Further studies are required to provide a complete evaluation and optimization of this automated method for tumor discrimination.

1.3   Conclusions

Recent technological developments in FT-IR spectroscopic imaging have enabled the possibility for applications in histopathologic imaging for cancer diagnosis and research. While many studies indicate that FT-IR has the potential for clinical applications, at this time it has not been adopted for automated histopathology in practice due to a variety of factors. The primary reasons are a lack of robustly validated protocols in which data is acquired rapidly and classification is efficient. The multivariate segmentation approach presented here demonstrates promising results for reliable epithelium and stromal recognition. The application of a breast TMA for data collection allows for rapid acquisition and analysis of large datasets to select robust classification metrics. This supervised classification approach also has the advantage of providing insight into important biochemical properties of tissues by incorporating spectral metrics based on tumor biological characteristics. This current research indicates that FT-IR imaging may make a significant contribution to developing a method for automated histopathology in a clinical setting.

References

1. M. J. Horner, L. A. G. Ries, M. Krapcho, N. Neyman, R. Aminou, N. Howlader, S. F. Altekruse, et al. (eds.), *SEER Cancer Statistics Review*, 1975–2006, NCI. Bethesda, MD, http://seer.cancer.gov/csr/1975–2006/, based on November 2008 SEER data submission, posted to the SEER web site, 2009.
2. R. A. Smith, V. Cokkinides, and O. W. Brawley, "Cancer Screening in the United States, 2009: A Review of Current American Cancer Society Guidelines and Issues in Cancer Screening," *CA Cancer Journal for Clinicians*, **59**:27–41, 2009.
3. Data provided by Thomson Reuters In-Patient and Out-Patient Views, 2008.
4. V. L. Katz, G. Lentz, R. A. Lobo, and D. Gershenson, *Comprehensive Gynecology*, 5th ed. Mosby, Philadelphia, PA, 2007.
5. D. Carter, *Interpretation of Breast Biopsies*, 4th ed., Lippincott Williams & Wilkins, Philadelphia, PA, 2004.
6. J. M. Bueno-de-Mesquita, D. S. Nuyten, J. Wesseling, H. van Tinteren, S. C. Linn, and M. J. van de Vijver, "The Impact of Inter-Observer Variation in Pathological Assessment of Node-Negative Breast Cancer on Clinical Risk Assessment and Patient Selection for Adjuvant Systemic Treatment," *Annals of Oncology*, **21**(1):40–47, 2010.
7. J. D. Kronz, W. H. Westra, and J. I. Epstein, "Mandatory Second Opinion Surgical Pathology at a Large Referral Hospital," *Cancer*, **86**:2426–2438, 1999.
8. M. Simunovic, A. Gagliardi, D. McCready, A. Coates, M. Levine, and D. DePetrillo, "A Snapshot of Waiting Times for Cancer Surgery Provided by Surgeons Affiliated with Regional Cancer Centers in Ontario," *Canadian Medical Association Journal*, **165**(4):421–425, 2001.
9. S. H. Parker, F. Burbank, R. J. Jackman, C. J. Aucreman, G. Cardenosa, T. M. Cink, J. L. Coscia, et al., "Percutaneous Large-Core Breast Biopsy: A Multi-Institutional Study," *Radiology*, **193**:359–362, 1994.
10. E. V. Lang, K. S. Berbaum, and S. K. Lutgendorf, "Large-Core Breast Biopsy: Abnormal Salivary Cortisol Profiles Associated with Uncertainty of Diagnosis," *Radiology*, **250**(3):631–637, 2009.
11. S. S. Raab, D. M. Grzybicki, J. E. Janosky, R. J. Zarbo, F. A. Meier, C. Jensen, and S. J. Geyer, "Clinical Impact and Frequency of Anatomic Pathology Errors in Cancer Diagnosis," *Cancer*, **104**(10):2205–2213, 2005.
12. E. Rakovitch, A. Mihai, J. Pignol, W. Hanna, J. Kwinter, C. Chartier, I. Ackerman, J. Kim, K. Pritchard, and L. Paszat, "Is Expert Breast Pathology Assessment Necessary for the Management of Ductal Carcinoma In Situ?" *Breast Cancer Research and Treatment*, **87**:265–272, 2004.
13. E. Newman, A. Guest, M. Helvie, M. Roubidoux, A. Chang, C. Kleer, K. Diehl, et al., "Changes in Surgical Management Resulting From Case Review at a Multidisciplinary Tumor Board," *Cancer*, **107**:2346–2351, 2006.
14. S. E. Singletary, and J. L. Connolly, "Breast Cancer Staging: Working with the Sixth Edition of the AJCC Cancer Staging Manual," *CA Cancer Journal for Clinicians*, **56**:37–47, 2006.
15. J. Meyer, C. Alvarez, C. Milikowski, N. Olson, I. Russo, J. Russo, A. Glass, B. Zehnbauer, K. Lister, and R. Parwaresch, "Breast Carcinoma Malignancy Grading by Bloom-Richardson System vs. Proliferation Index: Reproducibility of Grade and Advantages of Proliferation Index," *Modern Pathology*, **18**:1067–1078, 2005.
16. N. Bosanquet and K. Sikora, "Scenarios for Change in Cancer Treatment 2004–2010: Impacts on Insurance," *The Geneva Papers on Risk and Insurance*, **29**(4): 728–737, 2004.
17. U. Veronesi, P. Boyle, A. Goldhirsch, R. Orecchia, and G. Viale, "Breast Cancer," *Lancet*, **365**:1727–1741, 2005.
18. I. Levin and R. Bhargava, "Fourier Transform Infrared Vibrational Spectroscopic Imaging: Integrating Microscopy and Molecular Recognition," *Annual Review of Physical Chemistry*, **56**:429–474, 2005.
19. P. Griffiths, *Chemical Infrared Fourier Transform Spectroscopy*, John Wiley & Sons, New York, N.Y., 1975.

20. M. Jackson, J. Mansfield, B. Dolenko, R. Somorjai, H. Mantsch, and P. Watson, "Classification of Breast Tumors by Grade and Steroid Receptor Status Using Pattern Recognition Analysis of Infrared Spectra," *Cancer Detection and Prevention*, **23**(3):245–253, 1999.

21. R. Shaw, J. Mansfield, S. Rempel, S. Low-Ying, V. Kupriyanov, and H. Mantsch, "Analysis of Biomedical Spectra and Images: From Data to Diagnosis," *Journal of Molecular Structure-Theochem*, **500**:129–138, 2000.

22. E. Blout and R. Mellors, "Infrared Spectra of Tissues," *Science*, **110**:137–138, 1949.

23. H. Fabian, M. Jackson, L. Murphy, P. Watson, I. Fichtner, and H. Mantsch, "A Comparative Infrared Spectroscopic Study of Human Breast Tumors and Breast Tumor Cell Xenografts," *Biospectroscopy*, **1**:37–45, 1995.

24. P. Lasch and D. Naumann, "FT-IR Microspectroscopic Imaging of Human Carcinoma Thin Sections Based on Pattern Recognition Techniques," *Cellular and Molecular Biology*, **44**(1):189–202, 1998.

25. P. B. Rousch, *The Design, Sample Handling, and Applications of Infrared Microscopes*, ASTM Special Technical Publication 949, Philadelphia, ASTM, 1985.

26. R. Bhargava and I. W. Levin, *Spectrochemical Analysis Using Infrared Multichannel Detectors*, Blackwell Publishing Ltd., Oxford England, 2005.

27. E. Lewis, P. Treado, R. Reeder, G. Story, A. Dowrey, C. Marcott, and I. Levin, "Fourier Transform Spectroscopic Imaging Using an Infrared Focal-Plane Array Detector," *Analytical Chemistry*, **67**:3377–3381, 1995.

28. C. M. Snively, S. Katzenberger, G. Oskarsdottir, and J. Lauterbach, "Fourier-Transform Infrared Imaging Using a Rapid-Scan Spectrometer," *Optics Letters*, **24**(24):1841–1843, 1999.

29. R. Bhargava and I. Levin, "Fourier Transform Infrared Imaging: Theory and Practice," *Analytical Chemistry*, **73**(21):5157–5167, 2001.

30. B. Rigas, S. Morgello, I. Goldman, and P. Wong, "Human Colorectal Cancers Display Abnormal FT-IR Spectra," *Proceedings of the National Academy of Sciences*, **87**:8140–8144, 1990.

31. D. Malins, N. Polissar, K. Nishikida, E. Holmes, H. Gardner, and S. Gunselman, "The Etiology and Prediction of Breast Cancer," *Cancer*, **75**(2):503–517, 1995.

32. D. Malins, N. Polissar, and S. Gunselman, "Progression of Human Breast Cancers to the Metastatic State Is Linked to Hydroxyl Radical-Induced DNA Damage," *Proceedings of the National Academy of Sciences*, **93**(6):2557–2563, 1996.

33. M. Jackson, L. Choo, P. Watson, W. Halliday, and H. Mantsch, "Beware of Connective Tissue Proteins: Assignment and Implications of Collagen Absorptions in Infrared Spectra of Human Tissues," *Biochimica et Biophysica Acta*, **1270**:1–6, 1995.

34. H. Fabian, P. Lasch, M. Boese, and W. Haensch, "Infrared Microspectroscopic Imaging of Benign Breast Tumor Tissue Sections," *Journal of Molecular Structure*, **661**:411–417, 2003.

35. K. Anderson, P. Jaruga, C. Ramsey, N. Gilman, V. Green, S. Rostad, J. Emerman, M. Dizdaroglu, and D. Malins, "Structural Alterations in Breast Stromal and Epithelial DNA—The Influence of 8,5 '-Cyclo-2 '-Deoxyadenosine," *Cell Cycle*, **5**(11):1240–1244, 2006.

36. H. Fabian, P. Lasch, M. Boese, and W. Haensch, "Mid-IR Microspectroscopic Imaging of Breast Tumor Tissue Sections," *Biopolymers*, **67**(4–5):354–357, 2002.

37. D. Ellis and R. Goodacre, "Metabolic Fingerprinting in Disease Diagnosis: Biomedical Applications of Infrared and Raman Spectroscopy," *Analyst*, **131**(8): 875–885, 2006.

38. M. Diem, M. Romeo, S. Boydston-White, M. Miljkovic, and C. Matthaus, "A Decade of Vibrational Microspectroscopy of Human Cells and Tissue," *The Analyst*, **129**:880–885, 2004.

39. C. Petibois and G. Deleris, "Chemical Mapping of Tumor Progression by FT-IR Imaging: Towards Molecular Histopathology," *TRENDS in Biotechnology*, **24**(10):455–462, 2006.

40. H. Fabian, N. Thi, M. Eiden, P. Lasch, J. Schmitt, and D. Naumann, "Diagnosing Benign and Malignant Lesions in Breast Tissue Sections by Using IR-Microspectroscopy," *Biochimica et Biophysica Acta*, **1758**(7):874–882, 2006.

41. R. Dukor, G. Story, and C. Marcott, "Comparison of FT-IR Microspectroscopy Methods for Analysis of Breast Tissue Samples," *Institute of Physics Conference Series*, **165**:79–80, 2000.

42. R. Bhargava, D. Fernandez, S. Hewitt, and I. Levin "High Throughput Assessment of Cells and Tissues: Bayesian Classification of Spectral Metrics from Infrared Vibrational Spectroscopic Imaging Data," *Biochimica et Biophysica Acta*, **1758**(7):830–845, 2006.

43. R. Bhargava, "Towards a Practical Fourier Transform Infrared Chemical Imaging Protocol for Cancer Histopathology," *Analytical and Bioanalytical Chemistry*, **389**(4):830–845, 2007.

44. J. Kononen, L. Bubendorf, A. Kallioniemi, M. Barlund, P. Schraml, S. Leighton, J. Torhorst, M. J. Mihatsch, G. Sauter, and O. P. Kallioniemi "Tissue Microarrays for High-Throughput Molecular Profiling of Tumor Specimens," *Nature Medicine*, **4**(7):844–847, 1998.

45. D. Fernandez, R. Bhargava, S. Hewitt, and I. Levin, "Infrared Spectroscopic Imaging for Histopathologic Recognition," *Nature Biotechnology*, **23**(4):469–474, 2005.

46. F. Keith, R. Kong, A. Pryia, and R. Bhargava, "Data Processing for Tissue Histopathology Using Fourier Transform Infrared Spectra," *Proceedings of the Asilomar Conference on Systems, Signals, and Computers*, 71–75, 2006.

47. P. Rosen, *Rosen's Breast Pathology*, 2d ed., pp. 4–5, Lippincott, Williams, and Wilkins, Philadelphia, Pa. 2001.

48. D. S. May and N. E. Stroup,"The Incidence of Sarcomas of the Breast among Women in the United States, 1973–1986," *Plastic and Reconstructive Surgery*, **87**(1):193–194, 1991.

49. T. Tlsty and L. Coussens, "Tumor Stroma and Regulation of Cancer Development," *Annual Review of Pathology: Mechanisms of Disease*, **1**:119–150, 2006.

50. R. Salzer, G. Steiner, H. Mantsch, J. Mansfield, and E. Lewis, "Infrared and Ramen Imaging of Biological and Biomimetric Samples," *Frensenius Journal of Analytical Chemistry*, **366**:712–726, 2000.

51. X. Llora, A. Priya, R. Bhargava, "Observer-Invariant Histopathology using Genetics-Based Machine Learning," *Nat. Computing*, **8**:101–120, 2009.

52. G. Van Belle, L. D. Fisher, P. J. Heagerty, and T. Lumley, *Biostatistics: A Methodology for the Health Sciences*, 2d ed., John Wiley & Sons, Hoboken, N. J., 2004.

53. J. A. Hanley and B. J. McNeil, "The Meaning and Use of the Area under a Receiver Operating Characteristic (ROC) Curve," *Diagnostic Radiology*, **143**(1): 29–36, 1982.

CHAPTER 2

Synchrotron-Based FTIR Spectromicroscopy and Imaging of Single Algal Cells and Cartilage

Michael J. Nasse,[*,†] Eric Mattson,[*] Claudia Gohr,[‡] Ann Rosenthal,[‡] Simona Ratti,[*] Mario Giordano,[§] and Carol J. Hirschmugl[*]

I n this chapter we will describe novel instrumentation and initial biomedical experiments whose combination will provide the opportunity to measure in situ (in vivo) kinetics of pathological mineralization in the near future. The instrumentation includes two components: A novel IR synchrotron beamline IRENI (IR environmental imaging) at the Synchrotron Radiation Center (Stoughton, Wisconsin) and a newly designed flow chamber to maintain living cells in a hydrated and controlled environ. IRENI has been designed to extract a swath of 12 beams of radiation from the synchrotron, optically recombine them into a single bundle of collimated beams, refocusing them with a Bruker Hyperion microscope onto a sample area of 40×60 μm^2, illuminating a focal plane array (FPA)

---

[*] University of Wisconsin-Milwaukee, Milwaukee, Wisconsin
[†] Synchrotron Radiation Center, Stoughton, Wisconsin
 (University of Wisconsin-Madison, Madison, Wisconsin, USA)
[‡] Medical College of Wisconsin, Milwaukee, Wisconsin
[§] Università Politecnica delle Marche, Ancona, Italy

detector. Signal to noise, point spread functions and IR images of micrometer-sized polystyrene beads and metal grid are presented. The flow chamber incorporates several important features that make diffraction limited, spatially resolved imaging of living cells feasible as demonstrated by studies on microalgae. Biomedical experiments focusing on development of calcium-containing crystals in cartilage model systems demonstrate the role of ATP and BGP in calcium crystal growth. By combining all of these advances in instrumentation and recent experimental findings, future research will be focused on kinetics of pathological mineralization of in vivo systems monitoring the chemistry in the neighborhood of calcium crystal growth during the crystal formation, rather than examining only snapshots in time.

2.1   Introduction

Diffraction-limited imaging of samples of biological and medical interest is an area of increasing interest, including measuring biological samples and tissue cultures in vivo. Many different groups focusing on a wide range of topics[1–18] have recently reported mid-IR studies of tissues and other living cells. Also, diffraction-limited mid-IR results have been reported in several of these papers,[1–6,8,9,14–21] although to date, the combination of both in vivo and diffraction-limited results have been difficult to achieve since they require very strict experimental conditions that can be challenging to maintain and can interfere with IR measurements.[1,8,18]

Many of the examples above have utilized a broadband synchrotron source, where light is emitted from relativistically accelerated electrons that is bright and stable. Most of the synchrotron studies to date have utilized raster-scanning methods based on a *serial* collection scheme to collect the spatially dependent hyperspectral data cubes, where the spatial resolution of the images is dependent on both the effective geometric aperture size at the sample plane and the wavelength of the incident radiation. Recent efforts at synchrotron facilities, and the basis of the first part of this chapter, have focused on coupling an IR beamline[22,23] that collects a large swath of radiation with an FPA detector to obtain diffraction-limited IR maps, where pixels are collected in *parallel* at all wavelengths in the mid-IR range, with high signal-to-noise ratios.[20] The impact of this is twofold. First, images with high-spatial resolution will be collected quickly, so living biological systems that are changing, growing, or adapting to varying environs can be monitored on a relevant timescale. Second, samples with large sample areas and features of the order of the mid-IR wavelengths that require a large dataset to utilize statistical analysis methods can be studied since large quantities of high-quality data can be gathered quickly.

Another important issue with respect to working with biological samples in vivo is to maintain living samples, especially for samples that require water, such as phytoplankton. Many IR transparent materials used for commercial, conventional flow cells are hygroscopic or poisonous, and thus not suitable for this application. Furthermore, they are dependent on thick windows, which cannot be used with the high-magnification objectives required for high-quality spatially resolved images in both the IR and visible bandwidths. The Section 2.2 of this chapter presents results using a newly designed flow chamber that addresses these issues and maintains hydrated, living cells in a 15-μm layer of water.[18]

The third part of this chapter is based on a recent biomedical application of synchrotron-based IR studies of minerals embedded in cartilage tissues of osteoarthritic joints. The data allowed spectral identification of small (~1 to 10 μm) calcium containing crystals embedded in arthritic tissues and model systems, including calcium pyrophosphate dihydrate (CPPD) and basic calcium phosphate (BCP) crystals that are common components of osteoarthritic joints and contribute to the irreversible tissue destruction seen in this form of arthritis.[24]

In the summary, future experiments will be described. These experiments will combine all of the above advances, studying in vivo, biologically important samples at diffraction-limited spatial resolution for all mid-IR wavelengths. Importantly, these results will be achievable on a rapid timescale of 1 minute per 40×60 μm^2 image with 0.54×0.54 μm^2 pixels, such that dynamic processes can be followed with chemical specificity.

2.2 IR Environmental Imaging

IR Environmental Imaging (IRENI), a new IR beamline at the Synchrotron Radiation Center, Stoughton, Wisconsin, is designed to accept a swath of radiation from the synchrotron and reshape the beam to illuminate a 40×60 μm^2 sample area of an IR microscope. Traditionally, synchrotron beamlines extract one beam that illuminates the sample plane with a two-dimensional gaussian profile (FWHM of between 10×10 μm^2 to 15×15 μm^2). For the new beamline, 12 overlapped beams illuminate a Bruker Hyperion 3000 microscope that is equipped with an IR-sensitive FPA detector, where each pixel represents 0.54×0.54 μm^2 area of the sample creating oversampled images at all wavenumbers in the mid-IR. This facility will provide the opportunity to obtain chemical images with diffraction-limited resolution of the illuminated area in under a minute.

Synchrotron radiation is inherently a broadband, bright, and stable source of IR radiation.[26,27] In the mid-1990s, the first experiments using synchrotron radiation coupled to a commercial[28] and a home-built microscope[29,30] were reported. Since then, there has been a large growth in the community of IR synchrotron microspectroscopy users. The present chapter outlines the latest developments to move this technology to a new regime, to be able to monitor samples at high spatial resolution, quickly and in vivo, to allow time-resolved studies of living biological specimen.

2.2.1   Beamline Design and Implementation

In Fig. 2.1 a schematic of IRENI is shown.[22] The beamline accepts a swath of radiation from the synchrotron, splits the beam into 12 separate beams, and then recombines them into a collimated bundle of beams that illuminate a sample area at the microscope sample plane of 40 × 60 μm². In the schematic, the individual beam paths are depicted, showing that each path consists of a toroidal mirror for refocusing the beam from the source, a plane mirror to redirect the beam, a window to separate the vacuum from the electron storage ring and the rest of the beamline. The final two mirrors in each beamline are a paraboloidal mirror to collimate the beam and a final plane mirror to steer the beams into the 3 × 4 array of collimated beams. Further details of the design are found in Ref. 22.

The beamline has recently been constructed and accepted first light in August 2008.[24] In Fig. 2.2, a series of pictures showing the synchrotron illumination of the toroidal, plane, paraboloidal and

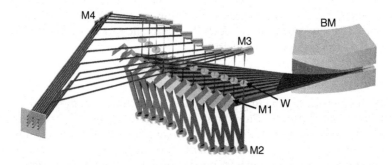

FIGURE 2.1   Schematic (not to scale) diagram of the new IRENI beamline at the SRC. To keep the system compact and limit the optical aberrations, the first optical components are 12 identical toroidal mirrors (M1) working in unity magnification. Each toroid is located 2 m from its source and together they collect the available horizontal fan of radiation. A water-cooled tube (not shown), located upstream from the mirrors, blocks the high-energy radiation emitted by the storage ring, eliminating the need for mirror cooling. The toroids deflect the beams, mostly downward, by 85° toward a set of 12 flat mirrors (M2). The flats direct the beams upward where they leave the UHV chamber through 12 ZnSe windows (W). The beam foci are above the windows. The remaining assembly, including a set of 12 paraboloids (M3) to collimate the beam and a set of 12 plane mirrors (M4) to bundle the beams along a common axis, are all outside the UHV chamber and will be in a nitrogen-purged environment. For clarity, only 4 out of the 12 M4 mirrors are included. (*Printed with permission from Ref. 22.*)

FIGURE 2.2   A series of photos showing illuminated mirrors in the beam path for IRENI. In the right photo, the array of illuminated toroids (nine of twelve are shown), the first array of mirrors in the beamline, refocus the synchrotron beam as it exits the synchrotron. Notice two bright rectangles of light just above and below the center. In addition, there is a large shadow across the center of each mirror, which is due to a water-cooled tube to absorb the higher energy soft x rays and UV radiation. The bottom, central picture shows the array of plane mirrors (11 of 12 are shown), which redirects the beams through exit windows of the ultrahigh vacuum system housing the two first arrays of mirrors. The upper-left picture shows the final 24 illuminated mirrors that collect the swath of radiation from the synchrotron and recombine it into a collimated bundle of 12 beams. (*Printed with permission from Ref. 21.*)

final plane mirrors and the beam path in air are shown. Notice on the first two mirrors, one can see shadows surrounded by illumination on either side of the shadow. The shadows are due to a water-cooling tube that is installed in the beam path to eliminate the soft x rays and UV radiation from the extracted beam to prevent overheating the first mirrors. The illuminated beam path clearly shows the recombined bundle of beams leaving the final series of plane mirrors.

The collimated beams are accepted by a Bruker Vertex 70 spectrometer, and transported to a Hyperion 3000 IR microscope. An optical arrangement that is similar to one proposed by Carr et al.[30] has been implemented. In transmission, the microscope is equipped with a 20× (modified ATR objective), 0.6 NA Schwarzchild condenser that focuses the beam to $40 \times 60 \ \mu m^2$ at the sample plane. This has been predicted theoretically and verified experimentally. Theoretically, an optical ray trace simulation of the overlapped beams, where the distances between the beams were optimized to create a homogeneous illumination at the sample plane, was performed. Thus, the gaussian tails of the individual neighboring beams are overlapped. This was confirmed with an experimental measurement, illuminating the FPA with the synchrotron source. The beam is collected by a 74×, 0.6 NA Schwarzchild objective and refocused onto a 128×128 pixel FPA detector, where the pixels are each $40 \times 40 \ \mu m^2$. The 74× magnification creates an effective geometric illumination of $0.54 \times 0.54 \ \mu m^2$/pixel at the sample plane.

In Fig. 2.3, live screen images of the illumination of the FPA are shown, where the entire effective geometric area at the sample plane is $69 \times 69 \ \mu m^2$. Since the detector is sensitive to the entire mid-IR bandwidth, the images represent integrated results over the entire bandwidth accepted by the instrument. When the optics are aligned and in focus, 12 distinct beams illuminate the area at the sample plane side by side in a 3×4 arrangement, as expected due to the optical

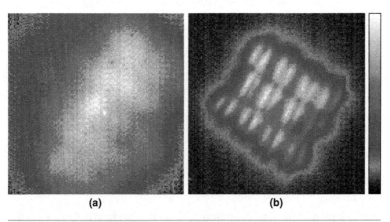

(a)   (b)

FIGURE 2.3   False color images [note color scale on (a)] of the illumination of the FPA when the condenser is in focus (a) and out of focus (b) for the transmission geometry at IRENI. The image that is in focus clearly illustrates the 3×4 array of 12 individual beams. The beams have been spaced with an approximate overlap such that when the condenser is slightly out of focus the FPA is more homogeneously illuminated, which is the condition that is used for the experiments described in the remainder of this section of the chapter.

design of the beamline. It is clear that the beams are not square to the orientation of the FPA, which is due to an optical rotation of the beams in the Bruker Vertex 70 spectrometer (Fig. 2.3). In practice, the condenser is placed slightly out of focus to homogeneously illuminate the sample plane as shown in Fig. 2.3a.

2.2.2   Initial Measurements with IRENI

In Fig. 2.4, the intensities measured through a 5-μm pinhole are shown for two different positions within the illuminated area (central position is the top set of images, and 20 μm away from the central position for the bottom set of images). Three images at different frequencies (3000, 2000, and 1500 cm^{-1} or 3.33, 5, and 6.67 μm) are presented in each case. The 5-μm aperture is similar to and smaller than the wavelength of light for the images, and thus produces the expected Airy

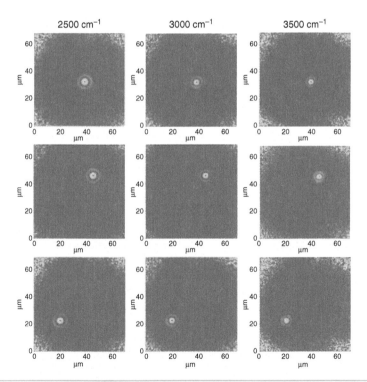

FIGURE 2.4   Color-scale (Red/blue - high/low) images of transmission at the IRENI beamline through a 2-μm aperture that is placed at the sample plane on the central optical axis, and approximately 15 and 30 μm away from the central axis of the optical path. The similarity in the images indicates that the optical system is translationally invariant—i.e., the point-spread function is similar for all positions within the field of view of the optical system.

diffraction patterns. Note that the first maximum is wavelength dependent and is located at a larger diameter for longer wavelengths. Importantly, the images clearly demonstrate the invariance of the point-spread function at different positions on the sample plane as suggested by Carr et al.[30]

First measurements using IRENI are designed to show some of the capabilities of the new instrument. A metal grid with grid bars of the order of the wavelength is used as a test sample to show frequency dependent behavior and a cluster of polystyrene beads of a similar size is used to show an image due to a polystyrene absorption band at 3025 cm⁻¹. Figure 2.5 shows the transmission of light through a metal grid with 8-μm-wide grid bars. Unprocessed images are shown at 3500, 2500, and 1500 cm⁻¹. Note, as expected due to diffraction, the images are blurrier at longer wavelengths. Above the images, three spectra are shown from three individual $0.54 \times 0.54 \ \mu m^2$ pixels within the image. This entire dataset was collected within 1 minute. Figure 2.6 shows a visible and IR image of the absorption of 6-μm polystyrene beads. It is an unprocessed image corresponding to the absorption band at 3025 cm⁻¹.

IRENI, a new facility at the SRC in Stoughton, Wisconsin, has recently been commissioned, as demonstrated by the results presented above. As described below, these advances will make it possible to take time-resolved data on biological samples in vivo in the near future.

To demonstrate the capabilities of the IRENI beam line and to facilitate a direct comparison to state-of-the-art commercially available instruments utilizing global sources, the initial data was acquired from a tissue biopsy of a benign prostate gland comprising of a layer of compressed epithelial cells (Fig. 2.7).[19] The three commercial instruments that have been used to create images for direct comparison include one that uses point mapping with an effective geometrical area at the sample plane of $10 \times 10 \ \mu m$ per pixel (Fig. 2.7a), one that uses a linear array detector that detects from an area of $6.25 \ \mu m \times 6.25 \ \mu m$ per pixel (Fig. 2.7b) and one that uses a 64×64 pixel FPA detector that detects from an area of $5.5 \ \mu m \times 5.5 \ \mu m$ per pixel (Fig. 2.7c). The IRENI beam line illuminates a 128×128 pixel FPA that detects from an effective geometrical area at the sample plane of $0.54 \ \mu m \times 0.54 \ \mu m$ per pixel (Fig. 2.7d).

2.3   Flow Cell for In Vivo IR Microspectroscopy of Biological Samples

Fourier transform IR (FTIR) microspectroscopy's capacity to nondestructively detect functional groups makes this technique a powerful tool for studying biological specimen.[1-18] To date, however, the strong absorption of liquid water in the mid-IR region as well as optical

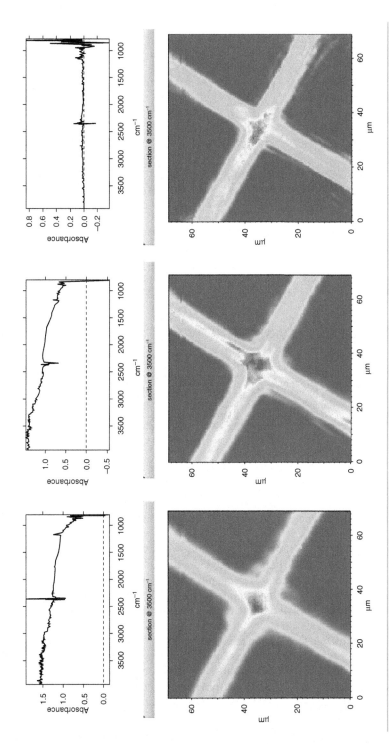

FIGURE 2.5 Color-scale images of transmission at the IRENI beamline through a metal grid with 8-μm grid bars. Images at 3500, 2500 and 1500 cm⁻¹ are shown, where red/blue indicates high/low absorption. Spectra from individual 0.54 × 0.54 μm² pixels for three different places in the field of view are also shown to demonstrate the noise level. These data were collected at 8 cm⁻¹ resolution, averaging 32 scans at 2.5 kHz data collection rate.

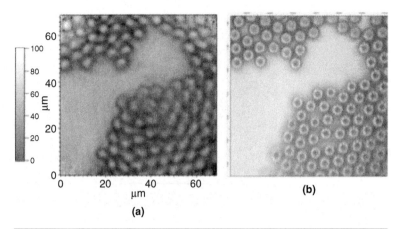

FIGURE 2.6   Unprocessed IR (*a*) and visible (*b*) grey-scale images of an agglomeration of 6- µm polystyrene beads. The IR image is generated from the absorption at 3025 cm^{-1} and 100 on the grey scale that corresponds to 0.125 absorption units. This was measured in transmission at IRENI.

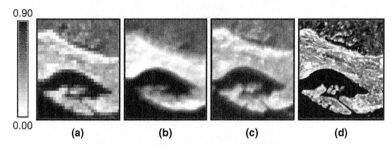

FIGURE 2.7   IR images of a benign prostate gland using (*a*) Spotlight FTIR point mapping (10 µm × 10 µm), (*b*) Spotlight FTIR imaging using a linear array detector (6.25 µm × 6.25 µm), (*c*) Varian FTIR with focal plane array (5.5 µm × 5.5 µm) and (*d*) the IRENI beam line with a focal plane array (0.54 µm × 0.54 µm). (*Printed with permission from Ref. 19.*)

aberrations due to thick-flow chamber windows has hampered high resolution in vivo IR studies. In phycology, Heraud et al.[6] have pioneered measurements on living algal cells using a custom flow-through chamber with several millimeter thick-halide windows. They acquired the data at the IR beamline 11.1 at the Synchrotron Radiation Source in Daresbury (UK). The use of synchrotron radiation as an IR light source allows researchers in principle to push the effective resolution to the diffraction limit with very good signal-to-noise ratios[21,22] but only if optical/(SNR) aberrations are negligible. This resolution is necessary to distinguish subcellular structures, e.g., the nucleus from the chloroplast.

   In this section we describe a new flow chamber that accommodates high-magnification objectives and features very low-optical aberrations. It thus permits the achievement of high-resolution data (provided by a synchrotron light source) of aqueous samples such as living specimen. Furthermore, the recent advent of multielement mid-IR detectors such as line detectors or FPAs reduces the acquisition times by several orders of magnitude. In combination with the high brightness of a synchrotron light source,[22] it opens the possibility for the in vivo acquisition of kinetic microspectroscopic maps of biological samples.

For this publication we demonstrate the performance of the flow chamber using an algal specimen (*Micrasterias* sp.), but the chamber can in principle be used with any aqueous/biological, nonaqueous or even gaseous samples.

2.3.1 Flow Chamber Design

One challenge in the design of a flow chamber is the choice of an appropriate window material: many commonly used crystals are water soluble (e.g., KBr, NaCl) or toxic (e.g., CdTe, KRS-5), absorb portions of the bandwidth of interest (e.g., Si, Al_2O_3), or exhibit dispersion (e.g., ZnS, CaF_2, BaF_2) leading to optical aberrations. The latter effect is made worse by the fact that the halide materials cannot be made into windows with a thickness of less than several millimeters due to their brittleness. This prevents the use of high numerical aperture microscope objectives with their short working distance. Furthermore, some of these substances are tinted (e.g., ZnSe) or opaque (e.g., Si, AMTIR) in the visible, which renders comparisons and co-localization of features in the visible (e.g., stained sections or parts labeled with fluorescent markers) and the IR difficult or impossible.

To circumvent the problems mentioned above, we chose diamond as a material for our windows. This material has several considerable advantages: it is water insoluble, nontoxic, transparent, and colorless in the visible as well as in the entire mid-IR spectrum of interest. Additionally, it exhibits a very low dispersion in the mid-IR region. This is illustrated by its relatively constant refractive index over the full mid-IR spectral range compared to other commonly used window materials (ZnS, BaF_2, and CaF_2) as shown in Fig. 2.8.

Multiple internal reflections inside the window materials of the order of micrometers thick can pose a serious problem because they lead to fringes on the IR spectra. This is worst when the fringe frequency is comparable to typical spectral peak widths. We overcome this by using sub-micrometer (less than one micron) thin diamond films such that the fringe frequency is comparable to the entire spectral range. The fringe(s) can then essentially be treated as an additional baseline and subtracted from the interesting spectral features. Diamond can be grown very thin by chemical vapor deposition (CVD) techniques. Thin films also help to keep the flow chamber design slim so that it can accommodate high numerical aperture objectives with short working distances.

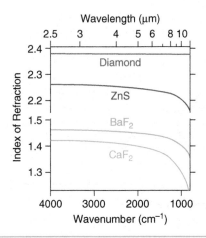

FIGURE 2.8   Indices of refraction in the mid-IR region for common window materials including diamond,[31] ZnS,[31] BaF$_2$,[32] and CaF$_2$[32] that illustrate the relative low dispersion of diamond. (*Printed with permission from Ref. 18.*)

The main components of the flow chamber (Fig. 2.9) are two diamond films (typical thickness 0.4 to 0.8 μm), hold apart by a spacer[34] with a typical thickness of 15 μm defining the flow chamber volume. They are grown on silicon wafers (diameter 32 mm, thickness 0.5 mm), which are etched away locally at seven positions around the center of the wafer. This exposes seven free-standing diamond openings or windows with a diameter of approximately 2.5 mm each through which the sample, sandwiched between the diamond windows, is imaged. We use several smaller openings over one bigger hole to improve mechanical stability of the sub-micrometer-thin diamond

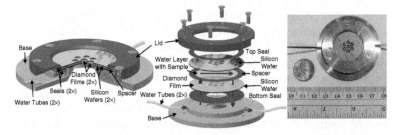

FIGURE 2.9   Design details of the flow chamber. Left panel: Three-quarter section view. Center panel: Explosion view indicating the water flow (blue arrows). Color codes for both panels: lid (blue), base (green), diamond (yellow) coated silicon wafers (gray), spacer (red), seals (purple), water tube (cyan), and screws (gray) fastening the lid to the base. Right panel: Photograph of the flow chamber (top view). The ends of the water in/outlet tubes have luer connectors for easy connecting to other equipment, e.g., a pump. (*Printed with permission from Ref. 18.*)

membranes. The bottom silicon wafer in contrast to the top wafer has two additional through holes (diameter 1.5 mm) via which the liquid enters the flow chamber volume. The bottom wafer is seated on a thin, flat silicone or Viton® seal that has two holes at the same positions. For compliance, a thin Teflon® washer is placed between the lid and the top silicon wafer. The lid is tightened to the base with the help of six screws sealing the flow chamber.

The liquid medium enters the chamber volume through a metal tube equipped with a luer lock, through an L-shaped channel in the base, the bottom seal, and the bottom silicon wafer including the diamond film (see blue arrows in Fig. 2.9). After flowing through the chamber volume, it exits through the hole on the opposite side of the chamber. We use a syringe-based push/pull pump to drive the liquid through the chamber. For the algae experiments we chose to run the pump at a flow rate of 10 µL/min, which corresponds roughly to one chamber volume per minute.

The diameter of the silicon wafers is chosen to be compatible with conventional windows from PIKE Technologies[34] which permits to use their line of round spacers with a thickness down to 15 µm. The silicon wafer thickness of 0.5 mm, on the other hand, makes it possible to use high-end microscope objectives above the flow chamber with high numerical aperture (for transmission and reflection setups). These objectives typically have a short working distance down to the sub-millimeter range. Figure 2.10, for example, shows a Micrasterias alga in the flow chamber imaged with visible light in high resolution through a 60× refractive objective with a numerical aperture of 0.70. The distance between the sample and the bottom of the flow chamber is 5.7 mm requiring the microscope condenser (assuming an upright microscope setup in transmission mode) to have a working distance exceeding this value. Most common condensers meet this condition.

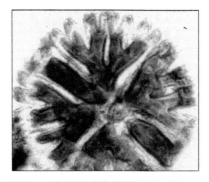

FIGURE 2.10   Example of a living Micrasterias algal cell in the flow chamber taken with a 60× refractive microscope objective (NA = 0.70) illustrating the low degree of optical aberrations introduced by the diamond windows. (*Printed with permission from Ref. 18.*)

2.3.2   Mid-IR and Vis Measurements

Figure 2.11 compares the mid-IR and visible images of a *Micrasterias* sp. cell in a conventional flow chamber with 3-mm-thick ZnS windows (Fig. 2.11*a* to 2.11*e*) and another *Micrasterias* sp. cell in the new

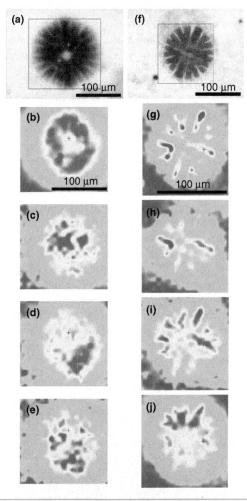

FIGURE 2.11   Comparison between living Micrasterias algal cells in a conventional flow chamber with 3-mm-thick ZnS windows (*a* to *e*) and in the new flow chamber with sub-micrometer diamond windows (*f* to *j*). Visible light images through a 32× refractive Schwarzschild objective are shown in (*a*) and (*f*). False color images of integrated peak areas (after baseline subtraction) corresponding to various functional groups (CH_n (*b*, *g*): 2961 to 2824 cm^{-1}, phospholipids (*c*, *h*): 1767 to 1723 cm^{-1}, amide II (*d*, *i*): 1571 to 1486 cm^{-1}, and carbohydrates (*e*, *j*): 1122 to 980 cm^{-1}) of the corresponding alga are presented in (*b* to *e*) and (*g* to *j*). Both the visible and IR images for the new flow chamber show more details because of reduced optical aberrations. (*Printed with permission from Ref. 18.*)

flow chamber with sub-micrometer diamond windows (Fig. 2.11f to 2.11j). ZnS windows were chosen instead of ZnSe, since the yellow tint of the ZnSe windows makes it difficult to see the green algal cells. This high-resolution data using an effective geometric aperture of 10×10 μm^2 was acquired at the mid-IR beamline 031 of the Synchrotron Radiation Center in Wisconsin that is equipped with a commercial Continuµm IR microscope and a Magna 560 FTIR spectrometer, both from Nicolet/Thermo Fisher Scientific. This microscope is used in a confocal arrangement, setting apertures before the condenser and after the objective to image the same 10×10 μm^2 area at the sample plane. Panels (a) and (f) show visible light images taken through a 32× refractive Schwarzschild objective, which had been optimized with the correction collar in both cases. The corresponding mid-IR images (Fig. 2.11b to 2.11e and 2.11g to 2.11j) depict the integrated peak areas of the CH_n, amide II, phospholipid, and carbohydrate functional groups. As expected, the visible image of the algal cell in the new flow chamber reveals more detail than the cell in the conventional chamber, due to much smaller optical chromatic and spherical aberrations. The cell wall outline, for example, that is clearly visible in image (f) is not distinguishable in image (a). The same is true for the mid-IR: images (Fig. 2.11g to 2.11j) of the new flow chamber are much better resolved than images (Fig. 2.11b to 2.11e) of the conventional chamber, which appear blurry and don't reveal any subcellular structure. The images taken through the sub-micrometer thick diamond windows, however, show a strong correlation when compared to the visible images. This is particularly apparent for the CH_n stretch images (Fig. 2.11b, g), where minimal diffraction effects allow for the best spatial resolution due to the relatively short wavelength.

Compared to Fig. 2 in Heraud et al.,[6] the flow chamber presented here yields better-resolved IR maps, which is partly due to the fact that we use a smaller aperture of 10×10 μm^2 instead of 20×20 μm^2, but mostly because of smaller optical aberrations due to the much thinner windows. Furthermore, we show a representative nonaveraged spectrum of a single pixel and maps of the CH_n stretches and the carbohydrates in contrast to Heraud et al.

Figure 2.12 gives an example of a typical in vivo mid-IR spectrum on a single *Micrasteriass* sp. algal cell (taken at the position of the red marker in Fig. 2.11f to 2.11j). The noisy areas from about 3050 to 3700 cm^{-1} and 1600 to 1700 cm^{-1} are due to the absorption of the water in the medium needed to keep the algal cells alive. This water layer also leads to fringes visible on the spectrum due to multiple reflections. The spectral regions marked in blue in Fig. 2.12 corresponding to the functional groups of interest (CH_n, phospholipids, amide II, and carbohydrates) do not overlap the water bands and can successfully be extracted. The integrated peak areas of these regions are shown as false color maps in Fig. 2.11.

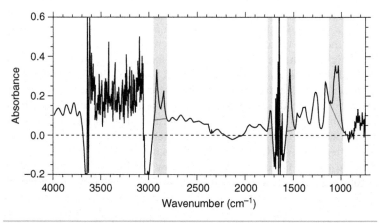

Figure 2.12   Typical in vivo spectrum in the mid-IR region of the Micrasterias alga shown in Fig. 2.11. The thin water layer leads to considerable IR absorption between 3050 to 3700 cm^{-1} and 1600 to 1700 cm^{-1}, and to fringes due to multiple reflections over the whole spectrum. The spectral regions of interest (CH$_n$, amide II, phospholipids, and carbohydrates), however, are extractable (blue regions with linear baseline in red; integrated peak areas shown in Fig. 2.11). (*Printed with permission from Ref. 18.*)

We used a pulse-amplitude-modulation (PAM) fluorescence microscope measuring the maximum photochemical quantum yield of single *Micrasterias* sp. cells as an example to determine the viability of biological cells inside the flow chamber for at least 4 hours.

2.3.3   Viability Tests: PAM Fluorescence Measurements

The chlorophyll fluorescence of *Micrasterias* cells was determined using an imaging-PAM fluorometer[35] (IMAG-CM, Walz, Effeltrich Germany). The fluorometer was coupled to an epifluorescence microscope (Axiostar plus, Zeiss, Göttingen Germany), equipped with a 20×/0.75 objective (FLUAR, Zeiss, Göttingen Germany), with an LED lamp (IMAG-L450, blue, wavelength 450 nm; Walz, Effeltrich Germany), and with a CCD camera (IMAG-K4, Walz, GmbH, Effeltrich Germany).

For this experiment the cultures were maintained in DY-V medium[35] and acclimated to continuous light (140 µmol photons m^{-2} s^{-1}) for at least four generations. The experiments were conducted on cells in the exponential growth phase.

The maximum photochemical quantum yield (F_v/F_m),[34] an indicator of photosynthetic performance, of single *Micrasterias* sp. cell was measured as a function of time. The cells were maintained in the flow chamber in the absence of medium flow, in order to monitor the photosynthetic activity of the cells in this environment. The flow chamber was assembled at time 0. The F_v/F_m value was measured every 30 minutes on cells adapted to the dark for 15 minutes. Nine measurements points at 15, 50, 80, 110, ..., 260 minutes have been recorded.

Data acquisition and analysis were conducted using the Imaging-Win v. 2.30 software (Walz, Effeltrich Germany).

Figure 2.13 presents the results of the viability study with the PAM microscope. A total of 35 "healthy looking" *Micrasterias* sp.

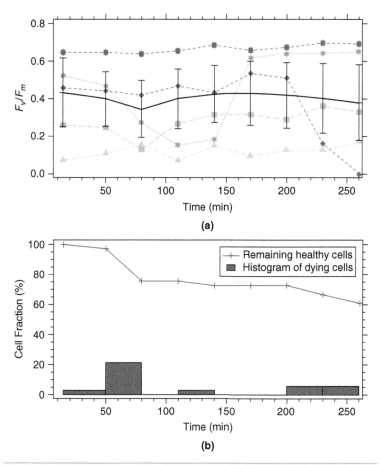

(a)

(b)

Figure 2.13 Viability study of single Micrasterias cells inside the flow chamber using a PAM microscope. The F_v/F_m value of 33 algal cells have been measured at nine different time points spaced 30 minutes apart (time = 0 corresponds to the flow chamber assembly, first measurement at 15 minute). Panel (a) shows five representative single cell PAM measurements (dashed, colored lines); the thick, solid black line is the average over 33 cells. The blue line in panel (b) illustrates the development over time of the number of viable cells (cells whose F_v/F_m stays within one standard deviation relative to their initial F_v/F_m). It demonstrates that 61 percent algal cells, after an initial decline, stay viable inside the flow chamber for at least 260 minutes. The red bars in panel (b) represent the histogram of nonviable cells. (*Printed with permission from Ref. 18.*)

cells from four separate flow chamber assemblies have been measured and analyzed (neglecting assemblies with problems). Out of 35 cells, 2 exhibited an initial F_v/F_m value of zero and were excluded from further analysis. The thick, solid black line in Fig. 2.13a corresponds to the F_v/F_m average of the remaining 33 cells with the error bars indicating the standard deviation at each time point. The overall average and standard deviation of all cells over all nine time points is $F_v/F_m = 0.40 \pm 0.17$, respectively. Since we performed the PAM fluorescence measurements on individual cells within the population the relatively high standard deviation is likely due to the variability among the cells. The standard deviation of the time development of the average (thick, solid line) is 0.03 (or 7 percent). The dashed lines show individual F_v/F_m measurements of five representative algal cells out of 33. As can be seen from the dashed lines, some cells ceased their photosynthetic activity at the end of the experiment, some had a constant activity and some even increased their activity during 260 minutes of the experiments. The thick, solid curve however confirms that on average 33 observed cells could maintain a fairly constant F_v/F_m (within 7 percent) over the entire experiment period.

The average and natural variability (standard deviation) of F_v/F_m among the 33 cells is 0.43 ± 0.18, respectively, which is derived from the measurements at the initial time point only ($t = 15$ minute.). This assumes that the influence of the flow chamber is minimal at that point in time. The cells whose F_v/F_m values over time remained within a standard deviation (42 percent) relative to the initial F_v/F_m were considered healthy. All other cells were considered unhealthy. The black line in Fig. 2.13b represents the percentage of cells that remain viable up to the corresponding point in time illustrating the development of the number of healthy cells in the chamber with time. It demonstrates that after the first hour the number of photosynthetically active cells stays virtually constant. The red bars in Fig. 2.13b show a histogram of the number of dying cells in each time slot. The elevated bar (second from the left) corresponds to cells that ceased their photosynthetic activity within roughly 1 hour after the flow chamber was filled. This might be due to cells that were stressed before they were loaded into the flow chamber. After 260 minutes 61 percent (= 20 cells) of the initial 33 cells remain photosynthetically active. These PAM measurements demonstrate the viability of algal cells in the flow chamber for an extended period of time.

In this section, we presented a new flow chamber design for in vivo mid-IR and visible measurements of biological cells. The use of sub-micrometer-thick diamond as a window material has several major advantages over conventional halide windows like ZnS. Notably, it exhibits lower optical aberrations and is transparent over an extended spectral range in the mid-IR as well as in the visible. The slim design of the flow chamber accommodates high-resolution/ numerical aperture microscope objectives, which typically have a short working distance. The optional use of a low-flow-rate pump permits to control the environs inside the chamber. As an example, we compare high-resolution mid-IR maps of single *Micrasterias* sp. algal cells acquired with a conventional and with the new flow chamber. A series of PAM measurements on 35 *Micrasterias* sp. algae demonstrate that 61 percent of the initial cells show photosynthetic activity after 4 hours and 20 minutes. This confirms that the flow chamber allows maintaining a substantial number of the cells alive for an extended period of time. Next we show initial results of infrared images of an algal cell measured with IRENI.

2.3.4 Initial Flow Cell Measurements with IRENI

Initial measurements for an algal cell maintained under controlled, hydrated conditions have been completed with IRENI (Ref. 37) using the flow cell described in the previous sections of this chapter. Importantly, the increased throughput of IR provides high signal to noise measurements and diffraction-limited spatial resolution at all wavelengths is achieved simultaneously. Diffraction-limited images at different wavelengths from one imaging dataset are shown in Fig. 2.14. Images of *Micrasterias* at 1060 cm^{-1}, corresponding to absorption by

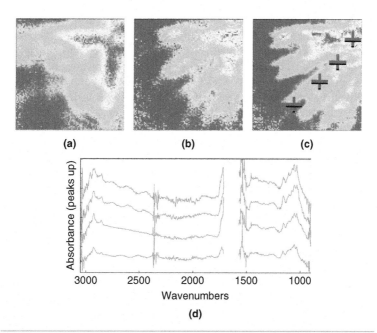

Figure 2.14 IR images of a *Micrasterias* algal cell at (*a*) 1060 cm^{-1}, (*b*) 1530 cm^{-1}, and (*c*) 2920 cm^{-1}. The spectra in (*d*) (top to bottom) correspond to the positions within the algal cell marked with the red crosses (top to bottom) within the image in (*c*). (*Printed with permission from Ref. 36.*)

functional groups in Carbohydrates; at 1530 cm^{-1}, corresponding to absorption by amide 2 in protein; and at 2920 cm^{-1}, for CH stretches found in many biological constituents are shown. This data cube was collected in 1 minute. Several spectra from individual pixels are also shown. Note that information between 1580 cm^{-1} to 1650 cm^{-1} has been removed because these spectral regions are completely dominated by 15-micron water layer absorption. The developments described in this section are crucial for future time-resolved experiments on living biological systems using IRENI.

2.4. Biomedical Application: Calcium-Containing Crystals in Arthritic Cartilage

Calcium-containing crystals, including CPPD and BCP crystals, are common components of osteoarthritic joints and contribute to the irreversible tissue destruction seen in this form of arthritis.[24] Little is known about how and why these crystals form, and consequently few effective therapies for this type of arthritis are available.

FTIR-based imaging technologies have been used to image and analyze normally mineralizing tissues such as bone.[37] Several groups of investigators assessed both matrix and mineral properties of bone using FTIR spectral analysis,[37–40] and described several key advantages over more traditional methods of biochemical analysis. In bone, FTIR-based technologies have facilitated mapping of matrix and mineral components in a small area, and provided information about the polarity of cells, the quality of mineral, and the type and integrity of fibrillar matrix components. The fact that this work can be performed in intact tissue is also a major advantage over other technologies.

We became interested in this methodology because of difficulties encountered in applying traditional methods of crystal analysis to our biologic models of calcium-crystal formation in articular cartilage. The crystals formed in these models were small and sparse and firmly embedded in a dense, well-hydrated extracellular matrix. The use of a synchrotron beam with FTIR spectral imaging improves SNR with smaller aperture sizes. We adapted synchrotron FTIR spectral microscopic analysis to crystal identification. Our success with this technology led to further studies analyzing the extracellular matrix components in and near crystals.

2.4.1   Calcium-Containing Crystals and Arthritis

Two types of calcium crystals are associated with arthritis. These include CPPD crystals and a trio of hydroxyapatite-like crystals known as basic calcium phosphate (BCP) crystals. BCP crystals are comprised of tricalcium phosphate, octacalcium phosphate and carbonate-substituted hydroxyapatite. In a normal synovial joint, the articular hyaline cartilage, which provides the smooth covering over the end of the bones, is a matrix-rich tissue composed primarily of type II collagen and large hydrophilic proteoglycans. Normally the matrix is unmineralized. During aging and with osteoarthritis, pathogenic calcium crystals deposit in the pericellular matrix around chondrocytes.

CPPD crystals cause both an acute inflammatory as well as a chronic noninflammatory polyarticular arthritis.[41] BCP crystals are associated with severe degenerative arthritis and a variety of noninflammatory articular syndromes such as Milwaukee shoulder syndrome.[42] These crystal-associated syndromes are common, often underdiagnosed, and produce irreversible joint destruction in elderly patients. No specific therapies are currently available.

2.4.2   Current Methods of Crystal Identification

Our understanding of how and why calcium crystals form in the normally unmineralized matrix of articular cartilage has been hampered by inadequate methods of crystal analysis. This is certainly problematic in the clinic where it results in missed diagnostic opportunities, but is also a major issue in the research laboratory. Standard methods of crystal identification in patient samples are tailored to the presence of relatively large numbers of crystals in synovial fluids. Typically, CPPD crystals in synovial fluids are identified morphologically under compensated polarizing light microscopy. CPPD crystals appear as weakly positively birefringent rhomboid-shaped crystals. In contrast, BCP crystals have no characteristic features under light microscopy. Alizarin red staining has been used to identify these crystals in synovial fluid, where they appear as large amorphous reddish-orange deposits. Unfortunately, alizarin red staining can be difficult to interpret, and is often misread.[43]

For research purposes, other more sophisticated and expensive techniques have been used to validate the presence of calcium-containing crystals in biologic models of crystal formation. These include FTIR spectral analysis and x-ray diffraction.[44] X-ray diffraction requires relatively pure dry samples. FTIR spectral microanalysis has proven quite useful in biologic samples with abundant crystals, but is pushed beyond its limits when crystals are very small and sparse. Using FTIR spectral microanalysis, for example, crystal identification proved impossible in models using chondrocyte monolayers, where crystals were rare and mixed with abundant complex biological material.[45]

2.4.3   Biologic Models of Calcium-Containing
Crystal Formation

No animal models of calcium-containing crystal formation currently exist. Cell and tissue culture models, however, are well described. Normal articular cartilage can be removed from an adult animal and minced into small pieces. When these cartilage pieces are incubated with ATP, they form CPPD crystals. When they are incubated with β-glycerophosphate (βGP), they form BCP crystals.[45] A similar model involves the use of chondrocytes in high-density monolayer cultures. When these cell layers are exposed to ATP, CPPD crystals are generated, and with βGP exposure they generate BCP crystals.[45] Small membrane bound extracellular organelles known as articular cartilage matrix vesicles (ACVs) can be isolated from normal articular cartilage. ACVs also generate crystals in a similar manner to chondrocytes and cartilage.[46]

2.4.4   Synchrotron-Based FTIR Microspectroscopy Spectral Analysis of Calcium-Containing Crystals

We wondered if the use of a synchrotron beam with FTIR spectral microanalysis could increase the sensitivity of this modality so that we could conclusively identify small sparse calcium-containing crystals in a variety of settings. Drop-sized samples from in vitro models or human synovial fluids were placed onto Kevley IR reflective slides and examined with plain and compensated polarized light microscopy to locate birefringent or dense materials. These areas were photographed and marked so that the same areas could be examined with synchrotron-based FTIR microspectroscopy. The samples were measured with a Thermo Fisher Continuμm Fourier Transform-IR (FT-IR) microscope coupled to the IR beamline at the Synchrotron Radiation Center (SRC) in Stoughton, Wisconsin. Measurements were taken in reflectance, acquiring reflection-absorbance results. Both individual spot-measurements and spatially resolved maps of the samples were measured with apertures ranging from 8 to 15 μm². The number of scans was selected to optimize the SNR and visible images were collected concurrently. The IR results were visually compared to reference spectra of multiple forms of calcium-containing crystals.

As shown in Figs. 2.15 and 2.16, we could conclusively identify CPPD and BCP crystals in human samples as well as in ACV and cell culture models of crystal formation.[47] The crosshairs in the visible images indicate the points at which the measurements were obtained. The top sample spectrum was collected with a 10×10 μm² aperture and 64 scans, while the bottom sample spectrum was collected with an 8×8 μm² aperture and 64 scans. For the latter crystal, the sample is clearly smaller than the 8×8 μm² aperture, but is completely within the illuminated field of view. The IR signatures produced by the synovial fluid crystals clearly match the standard spectra of BCP and CPPD crystals, respectively. The additional peaks on the synovial fluid crystal spectra suggest the presence of biologic material mixed with the crystals.

Chondrocyte monolayers were incubated with ATP for 72 hours. A crystal from these monolayers was compared with a M-CPPD

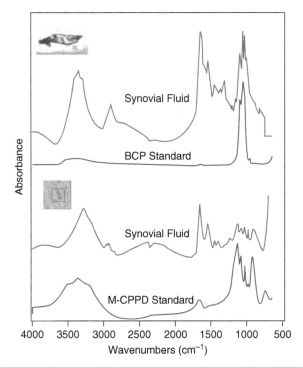

FIGURE 2.15   Representative synchrotron-based FTIR microspectroscopy spectra from human synovial fluid demonstrating CPPD and BCP crystals. (*Printed with permission from Ref. 47.*)

spectrum. The chondrocyte spectrum was collected with $12 \times 12 \ \mu m^2$ and 64 scans.

Our initial success with this methodology allowed us to make some important observations about these crystals that prompted further studies.

We noted that the spectra of the biologic crystals often did not exactly match the spectra of synthetic crystals. We asked whether this could be due to *contamination* of the crystals with biologic material such as proteoglycans. As shown in Fig. 2.17, spectra generated with combinations of synthetic CPPD crystals and cartilage proteoglycans, mimicked some of the changes observed in spectra from biologically generated CPPD crystals. The synchrotron-based FTIR microspectroscopy spectrum from a crystal generated by chondrocyte monolayers was compared with an FTIR spectrum generated by a combination of 50 percent cartilage proteoglycans, 30 percent M-CPPD, and 20 percent BCP.

We also noted that CPPD and BCP crystals were frequently found together in a single crystal deposit. This occurred in native crystals

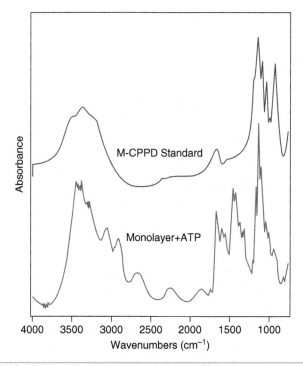

FIGURE 2.16   Representative synchrotron-based FTIR microspectroscopy spectra from porcine chondrocyte monolayers compared to M-CPPD standard. (*Printed with permission from Ref. 47.*)

from patients with arthritis (Fig. 2.18), as well as in biologically generated crystals (Fig. 2.16). As BCP and CPPD crystals can affect each other's growth and development, this finding may have major implications for our understanding of how these crystals grow and what limits their size and shape.

The data, 110 pixels over the entire area were collected with a $12 \times 12 \ \mu m^2$ aperture and 64 scans per pixel. Panel B shows a visible image of the crystal, while IR images in panels C and E are dominated by the absorbance of functional groups suggestive of CPPD (1125 to 1201 cm^{-1}) or BCP (985 to 1155 cm^{-1}). Red areas on the IR images indicate high-absorption intensity and are indicative of the presence of the crystal functional group, while blue areas indicate low absorption and relative scarcity of the functional group. The remaining panels (points 1 to 6) show standard spectra and data in a given pixel for the pixels identified in the IR images. The IR images and spectra suggest adjacent areas of CPPD and BCP in a single crystal aggregate.

The presence of brushite was also noted in both synovial fluid samples as well as in the biologic models. This interesting crystal type

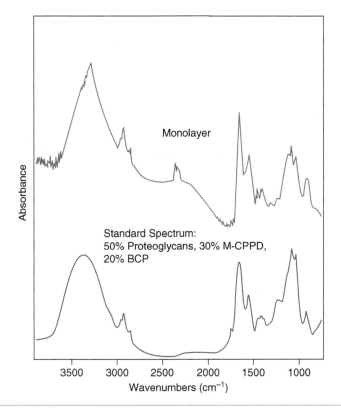

FIGURE 2.17   Representative IR spectrum from a crystal generated by chondrocyte monolayers and a mixture of BCP, CPPD, and proteoglycans. (*Printed with permission from Ref. 47.*)

is a form of calcium phosphate crystal, which may later mature into hydroxyapatite. It can be pathogenic under certain circumstances.[49]

We have also used this technology to validate crystal presence in a modified model of ACV mineralization. We were interested in designing a model of ACV mineralization that allowed us to manipulate the solid extracellular milieu in which ACVs make mineral. We adapted a model in which ACVs were mineralized in an agarose gel system. This system allowed us to manipulate the composition of the gel to include various extracellular matrix components found in normal or osteoarthritic cartilage. To validate the model, we needed to prove that ACVs embedded in agarose indeed make CPPD and BCP crystals. Synchrotron FTIR spectral analysis allowed us to identify the small, sparse crystals generated in this model, so that this important work could be completed. Using this model, we showed that the surrounding matrix had an important regulatory role in directing both the type and quantity of mineral formed by ACVs.[49]

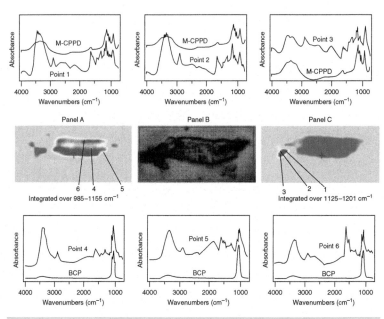

FIGURE 2.18   Synchrotron-based FTIR microspectroscopy map of a sample of synovial fluid.

We have subsequently used FTIR spectral analysis to identify crystals on the surfaces of ankle cartilage. Dr. Carol Muehleman and her associates at Rush Presbyterian St. Luke's Medical Center, Chicago, Illinois, were collecting ankle cartilages from cadaveric donors to study histologic patterns of osteoarthritis. They noted that a significant number of the cartilages examined were covered with a white powdery substance. Many of these specimens had similar deposits underneath the surface of the cartilage. Dr. Muehleman postulated that these deposits were composed of monosodium urate or CPPD crystals, and observed that in the presence of these deposits, there was a unique pattern of cartilage degeneration. She needed a way, however, to confirm the identity of these crystal deposits. We scraped small quantities of these crystals from the cartilage surface and were able to conclusively identify each sample as containing either monosodium urate or CPPD crystals using synchrotron FTIR spectral analysis. This work showed that both crystal types caused similar surface damage.[51] This added significant support to the hypothesis that crystals can contribute to cartilage degeneration by inducing mechanical wear.

We also used synchrotron FTIR spectral analysis in another clinical study testing a novel method of identifying BCP crystal0s in synovial fluids. This method was based on the observation that tetracycline, a commonly used antibiotic, binds to the hydroxyapatite mineral of bone, which is similar to the calcium phosphates in BCP crystals. Most tetracyclines are also fluorescent. Taken together, these findings led to ask if tetracycline could be used as a fluorescent marker of synovial fluid BCP crystals. Using synthetic BCP crystals, we showed that the addition of tetracycline allowed us to visualize BCP-containing particles in synovial fluid. We used synchrotron FTIR spectral analysis to prove the presence of BCP crystals in clinical synovial fluids, and then showed that these fluids also contained fluorescent particles using tetracycline staining.[52] This novel assay for BCP crystals will require further testing in the clinic, but is an exciting advance over our current identification methods. This work could not have been performed without the synchrotron FTIR analysis to prove that indeed we were identifying BCP crystals.

2.5 Future Directions: In Vivo Kinetics of Pathological Mineralization and Phytoplankton Adaptation

In the previous sections of this chapter, we have described the developments of (1) IRENI, a new synchrotron facility for rapid IR imaging at the diffraction limit covering the mid-IR frequency range from 4000 to 950 cm^{-1}, (2) a new flow chamber to maintain biological specimen in a hydrated environ that makes in vivo IR imaging feasible, and (3) a biomedical application of synchrotron IR microspectroscopy— studying calcium-containing crystals in cartilage from human samples and model systems. This combination of advances will allow collection of high-quality IR hyperspectral cubes within 1 minute, probing $40 \times 60 \ \mu m^2$ per experiment with a spatial oversampling of at least 2 to 1 for all wavelengths of interest. Future experiments will bring these developments together to study the pathological mineralization in cartilage by collecting time-resolved images of samples in vivo. Using chemometrics we can quantify collagen by using the amide I peak, denatured collagen and estimated proteoglycans, and observe crystal formation in a single area of a specific specimen. We propose to see if matrix changes precede or follow crystal formation and whether we could quantify alterations in lipid, proteoglycan, or denatured collagen. Other projects will include further studies of phytoplankton that are fully hydrated and maintained in a controlled medium, monitoring adaptation to different environmental stimuli.

Acknowledgments

This work was supported by the NSF under Award Nos. CHE-0832298 (CJH, MN, MG, SR), DMR-0619759(CJH, MN), NIH grant AR-R01-056215 (AKR), and by the Research Growth Initiative (RGI) of the University of Wisconsin-Milwaukee (CJH, MG). Part of this work is based upon research conducted at the SRC, University of Wisconsin-Madison, which is supported by the NSF under Award No. DMR-0537588.

References

1. N. Jamin, P. Dumas, J. Moncuit, W. H. Fridman, J. L. Teillaud, G. L. Carr, and G. P. Williams, "Highly Resolved Chemical Imaging of Living Cells by Using Synchrotron IR Microspectrometry," *Proceedings of the National Academy of Sciences*, **95**:4837, 1998.
2. R. Y. Huang, L. M. Miller, C. S. Carlson, and M. R. Chance, "In Situ Chemistry of Osteoporosis Revealed by Synchrotron IR Microspectroscopy," *Bone*, **33**: 514–521, 2003.
3. J. Kneipp, L. M. Miller, M. Joncic, M. Kittel, P. Laasch, M. Beekes, and D. Naumann, "In Situ Identification of Protein Structural Changes in Prion-Infected Tissue," *BioChem et Biophys Acta Molecular Basis of Disease* **139**:152, 2003.
4. P. Dumas, N. Jamin, J. L. Teillad, L. M. Miller, and B. Beccard, "Imaging Capabilities of Synchrotron IR Microspectroscopy," *Faraday Discuss*, **126**:289, 2004.
5. E. Gazi, J. Dwyer, N. P. Lockyer, J. Miyan, P. Gardner, C. Hart, M. Brown, and N. W. Clarke, "Fixation Protocols for Subcellular Imaging by Synchrotron-Based Fourier Transform IR Microspectroscopy," *Biopolymers*, **77**:18, 2005.
6. P. Heraud, B. R. Wood, M. J. Tobin, J. Beardall, and D. McNaughton. "Mapping of Nutrient-Induced Biochemical Changes in living Algal Cells Using Synchrotron Infrared Microspectroscopy," *FEMS Microbiology Letters*, **249**:219–225, 2005.
7. C. Krafft and V. Sergo, "Biomedical Applications of Raman and Infrared Spectroscopy to Diagnose Tissues," *Spectroscopy—an International Journal*, **20**:195, 2006.
8. H.-Y. N. Holman and M. C. Martin, "Synchrotron Radiation Infrared Spectromicroscopy: A Noninvasive Chemical Probe for Monitoring Biogeochemical Processes," *Advances in Agronomy*, **90**:79, 2006.
9. L. M. Miller and P. Dumas, "Chemical Imaging of Biological Tissue with Synchrotron Infrared Light," *Biochimica et Biophysica Acta—Biomembranes*, **1758**:846, 2006.
10. R. Bhargava, "Towards a Practical Fourier Transform Infrared Chemical Imaging Protocol for Cancer Histopathology," *Analytical and Bioanalytical Chemistry*, **389**:1155, 2007.
11. G. Srinivasan and R. Bhargava, "Fourier Transform-Infrared Spectroscopic Imaging: The Emerging Evolution from a Microscopy Tool to a Cancer Imaging Modality," *Spectroscopy*, **22**:30, 2007.
12. A. Boskey and N. P. Camacho, "FT-IR Imaging of Native and Tissue-Engineered Bone and Cartilage," *Biomaterials*, **28**:2465, 2007.
13. M. J. Walsh, M. J. German, M. Singh, H. M. Pollock, A. Hammiche, M. Kyrgiou, H. F. Stringfellow, E. Paraskevaidis, P. L. Martin-Hirsch, and F. L. Martin, "IR Microspectroscopy: Potential Applications in Cervical Cancer Screening," *Cancer Letters*, **246**:1, 2007.
14. M. Rak, M. R. Del Bigio, S. Mai, D. Westaway, and K. Gough, "Dense-Core and Diffuse A Beta Plaques in TgCRND8 Mice Studied with Synchrotron FTIR Microspectroscopy," *Biopolymers*, **87**:207, 2007.
15. A. Kretlow, Q. Wang, M. Beekes, D. Naumann, and L. Miller, "Changes in Protein Structure and Distribution Observed at Pre-Clinical Stages of Scrapie Pathogenesis," *Biochimica et Biophysica Acta—Molecular Basis of Disease*, **1782**:559, 2008.
16. A. K. Rosenthal, E. Mattson, C. M. Gohr, and C. J. Hirschmugl, "Characterization of Articular Calcium-Containing Crystals by Synchrotron FTIR," *Osteoarthritis and Cartilage*, **16**:1395, 2008.
17. S. Kaminskyj, K. Jilkine, A. Szeghalmi, and K. Gough, "High Spatial Resolution Analysis of Fungal Cell Biochemistry—Bridging the Analytical Gap Using Synchrotron FTIR Spectromicroscopy," *FEMS Microbiology Letters*, **284**:1, 2008.
18. M. J. Nasse, S. Ratti, M. Giordano, and C. J. Hirschmugl, "Demantable Flow Liquid Chamber for In Vivo Infrared Microspectroscopy of Biological Specimen," *Applied Spectroscopy*, **63**:1181–1186, 2009.

19. M. J. Walsh, M. J. Nasse, F. N. Pounder, V. Macias, A. Kajdacsy-Balla, C. J. Hirschmugl, and R. Bhargana, WIRMS 2009 3d International Workshop on Infrared Microscopy and Spectroscopy with Accelerator Based Sources, edited by A Pedrosi-Cross and B. E. Billingham, AIP Proceedings, 105–107, 2010.
20. E. Levenson, P. Lerch, and M. C. Martin, "Spatial Resolution Limits for Synchrotron-Based Spectromicroscopy in the Mid- and Near-Infrared," *Journal of Synchrotron Radiation*, **15**:323, 2008.
21. G. L. Carr, "Resolution Limits for Infrared Microspectroscopy Explored with Synchrotron Radiation," *Review of Scientific Instruments*, **72**:1613, 2001.
22. M. J. Nasse, R. Reininger, T. Kubala, S. Janowski and C. Hirschmugl, "Synchrotron Infrared Microspectroscopy Imaging Using a Multi-Element Detector (IRMSI-MED) for Diffraction-Limited Chemical Imaging," *Nuclear Instruments and Methods in Physics Research A*, **582**:107–110, 2007.
23. C. Hirschmugl, "IRENI," *Synchrotron Radiation News*, **21**:24, 2008.
24. Rosenthal A. "Update in Calcium Deposition Diseases," *Current Opinion in Rheumatology*, **19**:158–162, 2007.
25. W. Duncan and G. P. Williams, "Infrared Synchrotron Radiation from Electron Storage Rings," *Applied Optics*, **22**:2914, 1983.
26. G. P. Williams, C. J. Hirschmugl, E. M. Kneedler, E. A. Sullivan, D. P. Siddons, Y. J. Chabal, F. Hoffmann, and K. D. Moeller, "Infrared Synchrotron Radiation Measurements at Brookhaven," *Review of Scientific Instruments* **60**:2176–2178, 1989.
27. G. L. Carr, J. A. Reffner, and Williams, "Performance of an Infrared Spectrometer at the NSLS," *Review of Scientific Instruments*, **66**:1490–1492, 1995.
28. G. L. Carr, M. Hanfland, and G. P. Williams, "Mid Infrared Beamline at the National Synchrotron Light Source Port U2B," *Review of Scientific Instruments*, **66**:1643–1645, 1995.
29. R. J. Hemley, H. K. Mao, A. F. Goncharov, M. Hanfland, and V. V. Struzhkin, "Synchrotron Infrared Spectroscopy to 0.15 eV of H_2 and D_2 at Megabar Pressures," *Physics Review Letters*, **76**:1667–1671, 1996.
30. G. L. Carr, O. Chubar, and P. Dumas, "Multichannel Detection with a Synchrotron Light Source: Design and Potential," in *Spectrochemical Analysis using Multichannel Infrared Detectors, Analytical Chemistry Series*, In: Rohit Bhargava and Ira Levin (eds.), Blackwell Publishing Oxford, England, 2005.
31. D. F. Edwards and E. Ochoa, "Infrared Refractive Index of Diamond," *Journal of the Optical Society of America*, **71**:607–608, 1981.
32. ISP Optics Corporation. http://www.ispoptics.com/OpticalMaterialsSpecs .htm. Accessed January 27, 2009.
33. PIKE Technologies, Madison, Wis.
34. K. Maxwell and G. N. Johnson, "Chlorophyll Fluorescence—a Practical Guide," *Journal of Experimental Botany*, **51**:659–668, 2000.
35. R. A. Anderson, S. L. Morton, and J. P. Sexton, "Provasoli-Guillard National Center for Culture of Marine Phytoplankton 1997 List of Strains," *Journal of Phycology*, **33**:4–7, 1997.
36. M. J. Nasse, E. Mattson, C. J. Hirschmugl, WIRMS 2009 3rd International Workshop on Infrared Microscopy and Spectroscopy with Accelerator Based Sources, edited by A Pedrosi-Cross and B. E. Billingham, AIP Proceedings, (2010) 105–107.
37. L. Miller, C. Carlson, G. Carr, G. Williams and M. Chance, "Synchrotron Infrared Microspectroscopy As a Means of Studying the Chemical Composition of Bone: Application to Osteoarthritis," *SPIE*, **3135**:141–148, 1997.
38. M. M. W. Sato, N. Miyoshi, Y. Imamura, S. Noriki, K. Uchida, S. Kobayashi, T. Yayama, and K. Negoro, "Hydroxyapatite Maturity in the Calcified Cartilage and Underlying Subchondral Bone of Guinea Pigs with Spontaneous Osteoarthritis: Analysis by Fourier Transform Infrared Microspectroscopy," *Acta Histochem Cytochem*, **397**:101–107, 2004.
39. C. Chappard, F. Peyrin, A. Bonnassie, G. Leminer, B. Brunet-Imbault, E. Lespessailles, and C. L. Benhamou, "Subchondral Bone Microarchitectural Alterations in Osteoarthritis: A Synchrotron Micro-Computed Tomography Study," *Osteoarthritis Cartilage*, **14**:215–223, 2006.

40. E. Paschalis, E. DiCarlo, F. Betts, P. Sherman, R. Mendelsohn, and A. Boskey, "FTIR Microspectroscopic Analysis of Human Osteonal Bone," *Calcified Tissue International*, **59**:480–487, 1996.

41. A. Rosenthal and L. Ryan, "Calcium Pyrophosphate Crystal Deposition Diseases, Pseudogout, and Articular Chondrocalcinosis," In: W. Koopman and L. Moreland (eds.), *Arthritis and Allied Conditions*, Lippincott Williams & Wilkins, Philadelphia, U.S., 2004.

42. P. Halverson, "Basic Calcium Phosphate (Apatite, Octacalcium Phosphate, Tricalcium Phosphate) Crystal Deposition Diseases and Calcinosis," In: W. Koopman and L. Moreland, (eds.), *Arthritis and Allied Conditions*, Lippincott Williams & Wilkins, Philadelphia, U.S., 2004.

43. C. Gordon, A. Swan, and P. Dieppe, "Detection of Crystals in Synovial Fluid by Light Microscopy: Sensitivity and Reliability," *Annals of Rheumatic Diseases*, **48**:737–742, 1989.

44. A. Rosenthal and N. Mandel, "Identification of Crystals in Synovial Fluids and Joint Tissues," *Current Rheumatology Reports*, **3**:11–16, 2001.

45. L. Ryan, I. Kurup, B. Derfus, and V. Kushnaryov, "ATP-Induced Chondrocalcinosis," *Arthritis Rheumatology*, **35**:1520–1524, 1992.

46. B. Derfus, J. Rachow, N. Mandel, A. Boskey, M. Buday, V. Kushnaryov, and L. Ryan, "Articular Cartilage Vesicles Generate Calcium Pyrophosphate Dihydrate-Like Crystals In Vitro," *Arthritis Rheumatology*, **35**:231–240, 1992.

47. A. Rosenthal, E. Mattson, C. Gohr, and C. Hirschmugl, "Characterization of Articular Calcium-Containing Crystals by Synchrotron FTIR," *Osteoarthritis Cartilage*, **16**:1395–1402, 2008.

48. F. Higson and O. Jones, "Oxygen Radical Production by Horse and Pig Neutrophils Induced by a Range of Crystals," *Journal of Rheumatology*, **11**:735–740, 1984.

49. B. Jubeck, C. Gohr, E. Muth, M. Matthews, E. Mattson, C. Hirschmugl, and A. Rosenthal, "Type I Collagen Promotes Articular Cartilage Vesicle Mineralization," *Arthritis Rheumatology*, **58**:2809–2817, 2008.

50. C. Muehleman, J. Li, T. Aigner, L. Rappoport, E. Mattson, C. Hirschmugl, K. Masuda, and A. Rosenthal, "The Association between Crystals and Cartilage Degeneration in the Ankle," *Journal of Rheumatology*, **35**:1108–1117, 2008.

51. A. Rosenthal, M. Fahey, C. Gohr, T. Burner, I. Konon, L. Daft, E. Mattson, C. Hirschmugl, L. Ryan, and P. Simkin, "Feasibility of a Tetracycline Binding Method for Detecting Synovial Fluid Basic Calcium Phosphate Crystals," *Arthritis Rheumatology*, **58**:3270–3274, 2008.

Preparation of Tissues and Cells for Infrared and Raman Spectroscopy and Imaging

Ehsan Gazi, Peter Gardner

Manchester Interdisciplinary Biocentre (MIB)
The University of Manchester
Manchester, United Kingdom

3.1 Introduction

Vibrational techniques, Raman and Fourier transform IR (FTIR) microspectroscopy, provide structural information as well as relative quantification of lipids, proteins, carbohydrates and a variety of phosphorylated biomolecules within biological samples such as whole mammalian cells or tissue. However, the full potential of these technologies to interrogate this wide-range of biomolecules is only realized if careful consideration is given to sample preparation. This element of the experimental design can have significant implications for the interpretation of spectra and thus for their biochemical relevance as well as the spatial distribution of biomolecules in imaging studies.

Cells are naturally present in hydrated form, whereby water molecules are bound to macromolecules such as proteins, phospholipids, and carbohydrates and this contributes to their structural integrity and function. A review of the early literature concerning the application of FTIR to cell analysis for diagnostic or imaging purposes reveals that cells had generally been prepared by direct culture upon IR transparent substrates, then removal from culture medium and air-drying.[1-6] However, the air-drying process causes delocalization of biomolecules as a result of large surface tension forces associated with the passing water-air interface. Other researchers in the field had prepared cells by removing them from culture medium and then centrifugation,[7] drying under nitrogen gas[8] or cytospinning[9] with the view of minimizing the effects of these surface tension forces by increasing the rate of dehydration.

The removal of cells from pH buffered growth medium and subsequent air-drying can influence the osmotic pressure within these cells, resulting in cell shrinkage or swelling with the latter resulting in membrane rupture and leaching of intercellular components. In addition, drying of living cells can initiate autolytic processes whereby intracellular enzymes contained within lysosomes cause denaturing of proteins and dephosphorylation of mononucleotides, phospholipids and proteins. Furthermore, autolysis involves chromatin compaction, nuclear fragmentation (involving RNA and DNA nucleases) and cytoplasmic condensation and fragmentation. Thus, in FTIR-based biomechanistic studies, where researchers are interested in identifying the metabolites formed as a result of the cell's response to specific stimuli, the effects of autolysis as a consequence of inappropriate cell preparation may obscure these investigations.

In cell biology, a critical and fundamental step in any investigation is "fixation." This is used to quench autolysis, minimize leaching of biomolecular constituents, whilst at the same time using optimized dehydration protocols to bypass surface tension distortions and preserve the structural and functional chemistry of biomolecules for analysis. The common methods of cell preservation involve chemical fixation or flash-freezing for subsequent freeze-drying. Flash-freezing is appropriate for cells grown on substrates, which have good thermal contact with the freezing liquid medium and substrates that can withstand the low temperatures involved during this process. A common culture substrate for reflectance mode measurements are low-e microscope slides, for example, the MirrIR plate (Kevley Technologies). These slides are ~95 percent reflecting in the mid-IR but ~80 percent transparent to visible light. This makes them ideal for investigating biological cells and tissue, which are best observed on the microscope slide using back-illumination. They are also significantly cheaper than CaF_2 or BaF_2 plates. MirrIR slides are relatively thick (2 mm) and have a large thermal mass. Thus, the insulating effect of the MirrIR slide can slow down freezing rates, resulting in intercellular ice crystal formation during freezing. This can cause mechanical damage by rupturing cell membranes and lead to the discharge of cytoplasmic material into the extracellular matrix.

Generally, chemical fixation of cells is the most suitable sample preparation method for investigation with FTIR. In the past, however, chemical fixation has been avoided due to the potential for interference of the fixative with the IR spectrum. In this chapter, we discuss the influence of chemical fixatives on the FTIR spectrum of fixed single cells and show FTIR maps that illustrate the differences in biomolecular localizations in fixed *versus* unfixed cells. Our discussion of fixation also extends to resected tissues, where we provide a summary of the different methods employed to prepare these specimens for spectroscopic analysis, together with a review of the diagnostic information that can be obtained as a result of these preparations.

In addition to discussing fixed material, this chapter also reports on recent studies using live cells for FTIR and Raman studies, detailing the quality of spectral information obtained from these experiments, as well as the technical challenges imposed by maintaining living cells during analysis.

Another fundamental aspect of sample preparation that can influence cellular biochemistry is the surface on which they are grown. The surface can induce changes in cell adhesion and motility, in their proliferation and differentiation and in gene expression. It is desirable for in vitro cultures to mimic the in vivo environment as closely as possible and in this context, progress has recently been made in modelling cellular systems in two-dimensional cultures. Studies have also been carried out detailing the use of biomaterial surfaces (Matrigel™, fibronectin, laminin, gelatin) for this type of cell culture. The influence of these surfaces on cell morphology and the spectral information obtained is also discussed.

3.2 Tissue Preparation

3.2.1 Archived Tissue: Paraffin Embedded and Frozen Specimens

Surgically excised tissue may undergo one of two commonly used methods of preservation for long-term storage, paraffin embedding, or flash-freezing. The choice between these two methods is based on the specific purpose of the resected tissue. Currently, paraffin embedding is the preferred source for the histological examination of tissue sections by light microscopy. This method involves immersing tissue into a primary fixative, which is usually an aqueous formalin-based solution. Hydrated formalin (methylene glycol, OH-CH-OH) is a coagulative protein fixative, cross-linking the primary and secondary amine groups of proteins[10] but preserves some lipids by reacting with the double bonds of unsaturated hydrocarbon chains.[11] Following formalin fixation, the tissue is dehydrated through consecutive immersions in increasing concentrations of ethanol solution.

Displacement of water with ethanol preserves the secondary structures of proteins, however, denatures their tertiary structure. Furthermore, formalin or ethanol induces coagulation of the globular proteins present in the cytoplasm, which can result in the loss of structural integrity of organelles such as mitochondria. Another disadvantage is that ethanol precipitates lipid molecules that are not preserved through the primary fixation step. However, stabilization of intercellular proteins by formalin and ethanol localizes associated glycogen.

Following dehydration, the alcohol is replaced by an organic solvent such as xylene, which is miscible with both alcohol and molten paraffin wax. The specimen is then immersed in and permeated by molten paraffin wax. The infiltration of the wax into the intracellular spaces is promoted by the previous ethanol dehydration step that created pores in the cell's plasma membrane. The specimen is then cooled to room temperature, which solidifies the wax. This process provides a physical support to the sample enabling thin sections (usually 2 to 7 µm) to be cut without deformation of the cellular structure or architecture.

It is important to note that the process of fixation is not instantaneous and two important properties of the fixative are its penetration rate and binding time. Medawar[12] was the first to demonstrate that fixatives obey the diffusion laws, whereby the depth of penetration was proportional to the square root of time. The importance of fixative binding time was highlighted by Fox et al.[13] who investigated the binding of formaldehyde to rat kidney tissue, in which 16-µm-thin sections were used so that penetration would not be considered a factor in the kinetics of the reaction. They found that the amount of methylene glycol that covalently bound to this tissue increased with time until equilibrium was reached at 24 hours. Thus, binding time is the limiting factor for tissue stabilization. These aspects of chemical fixation (penetration rate and binding time) may be a potential source of biomolecular variance in pathological samples, since there exists a time lag in fixative exposure and binding between cells located within the core of the tissue compared with those at the extreme dimensions of the block.[13] Infact, Fox et al.[13] report that cells at the periphery of the tissue exhibit different morphological properties to cells that are a few tenths of a millimeter further within the specimen.

For molecular-based studies, snap-freezing of fresh tissue is generally preferred, since this method avoids the use of organic solvents that cause degradation or loss of some cellular components. In particular, frozen sections are used to study enzymes and soluble lipids. Furthermore, this method is used to conduct immunohistochemical analysis, since some antigens may be affected by extensive cross-linking chemical fixatives that denature their tertiary structure.

In the case of snap-freezing, the water contained within the cells acts as the supporting medium. Fresh tissue is snap-frozen in liquid-nitrogen-cooled isopentane (−170°C) to promote vitreous ice formation and to prevent ice-crystal damage, since the latter can produce holes in the tissue and destroy cellular morphology and tissue architecture. The hardened tissue can then be embedded in mounting medium such as optimal cutting temperature (OCT) compound for sectioning within a cryostat maintained at −17°C and then subsequently stained.

OCT is a viscous solution at room temperature, consisting of a resin-polyvinyl alcohol, an antifungal agent, benzalkonium chloride and polyethylene glycol to lower the freezing temperature.[14] Turbett and Sellmer[14] report that it is not advisable to store tissues in OCT for long durations, since it was found that amplification of DNA, extracted from these tissues, was significantly effected for segments of greater than 300 base pairs. However, RNA was found to be unaffected. It was suggested that snap-frozen tissues should be stored without any medium.

Although the methods outlined above represent the mainstay tissue processing techniques for paraffin embodiment/cryopreservation in the present pathology laboratories, some researchers have recently reported alternative tissue preparation protocols with the view of optimizing the assessment of specific biomolecular domains. Gillespie[15] conducted a comparative molecular profiling study in clinical tissue specimens that were fixed for long-term storage with widely used techniques (snap-frozen and formalin-fixed paraffin embedded) and a less common method of 70 percent ethanol fixation and paraffin embedding. The researchers found that although the total protein quantity was decreased in fixed and embedded tissues compared to snap-frozen tissue, 2D-PAGE analysis of proteins from ethanol-fixed, paraffin-embedded prostate, shared 98 percent identity with a matched sample from the same patient that was snap-frozen, indicating that the molecular weights and isoelectric points of the proteins were not disturbed by the tissue-processing method. The general quality and quantity of the proteins in the ethanol-fixed samples were found to be superior to formalin-fixed tissue. Furthermore, Gillespie[15] reports mRNA and DNA recovery were more pronounced in ethanol-fixed specimens compared with formalin-fixed samples. Thus, further improvements to tissue processing methodologies will play a key role toward ultimately determining the complete molecular anatomy of normal and diseased human cell types.

3.2.2   Preparation of Tissues for Diagnostic Assessment Using FTIR and Raman Microspectroscopy

Researchers working in the field of FTIR and Raman tissue diagnostics have employed a variety of methods for tissue preparation. In the first instance, FTIR spectroscopic studies have been carried out using ground samples of snap-frozen tissue for the bulk analysis of chemical composition.[16,17] Using this method, Andrus and Strickland[16] report an increasing ratio of peak areas corresponding to bands at 1121 cm^{-1} and 1020 cm^{-1} (attributed to RNA/DNA), which were associated with increased aggressiveness of malignant non-Hodgkin's lymphomas. Takahashi et al.[17] used bulk tissue analysis to study glycogen levels in tissues obtained from colorectal tumours, regions adjacent to the tumour and regions of normal colorectum. The results indicated that there was a statistically significant difference in glycogen levels (Peak area ratio 1045 cm^{-1}/1545 cm^{-1}) between cancer tissue and the other two regions.[17]

The use of ground tissue provides indiscriminate and composite measurement of both epithelial and stroma tissue compartments. However, this type of analysis must be treated with caution when molecular assignments are made for discriminatory bands. In the study by Andrus and Strickland,[16] the influence of collagen absorbance was discussed; however, other confounding variables exist in stroma tissue, namely, a variety of cell types such as endothelial cells of blood vessels, fibroblasts, ganglion, and erythrocytes in addition to possible bisecting nerves and muscle. This was clearly demonstrated by Fernandez et al.[18] who classified several prostate tissue components for diagnostic FTIR imaging.

For microspectroscopic studies, a review by Faolain et al.[19] reveals that a number of approaches have been made to prepare tissues for analysis that include fresh, frozen, air-dried, formalin-fixed, and deparaffinized formalin-fixed tissue sections. A number of papers have been published that compare the effects of these sample preparation protocols on Raman and FTIR tissue spectra[20–25] and these are discussed in the following text.

Fresh and Cryo-Preserved Tissue

More than a decade ago, Shim and Wilson[20] demonstrated that dehydration at room temperature of fresh tissue specimens (subcutaneous fat, smooth muscle, cheek pouch epithelium, esophagus) resulted in Raman spectra with a decrease in intensity of the 930 cm^{-1} (C—C stretch of proline and valine) peak relative to the peaks at 1655 cm^{-1} and 1450 cm^{-1}. Although this may be indicative of protein denaturing[21] the authors did not observe any shifts in the amide I peak. However, an increase in the lipid-protein signal was observed with increasing drying times providing evidence that the protein vibrational modes were perturbed by dehydration.

Interestingly, Shim and Wilson[20] found that Raman spectra obtained from OCT-freeze stored, snap-frozen tissue in phosphate-buffered saline (PBS) were comparable to spectra obtained from fresh tissue in PBS. Additional peaks observed at lower frequency (764 cm^{-1} and 795 cm^{-1}) in the spectra of snap-frozen adipose tissue, were attributed to the coagulation of erythrocytes. Faolain et al.[19] also conducted a comparative study of frozen and fresh tissue using parenchymal tissue

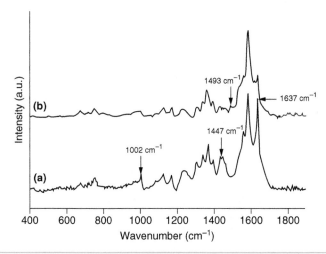

FIGURE 3.1   Raman spectra of (*a*) fresh tissue compared to (*b*) frozen tissue section. (*Reproduced from Ref. 19 with the permission from Elsevier Limited.*)

from the placenta (Fig. 3.1). Here, the authors did find significant differences in the spectra, realized through a reduction in the intensities of bands at 1002 cm^{-1} (C—C aromatic ring stretching), 1447 cm^{-1} (CH$_2$ bending mode of proteins and lipids), and 1637 cm^{-1} (amide I band of proteins) in the frozen tissue compared with fresh tissue. Additionally, frozen tissue exhibited a new peak at 1493 cm^{-1}, which was not found to be OCT contamination but was attributed to an artifact of the freezing process. Faolin et al.[19] suggested that this artifact was due to depolymerisation of the actin cytoskeleton, resulting in an increased contribution of the NH$_3^+$ deformation mode.

It is important to note that in the study by Faolain et al.,[19] the comparison of fresh and frozen tissue was carried out with prior mounting onto MirrIR plates in which the frozen tissue had been thawed before analysis. Hence, depolymerization of proteins can also result from postthawing of the frozen tissue, whereby the undesirable transition of vitreous ice into ice crystals could effect the integrity of the cytoskeletal proteins. It is well known within the structural cell biology community that this can be prevented by the application of freeze-drying. However, Shim and Wilson's[20] investigations suggest that the spectral changes in protein vibrational modes, caused by heat-induced denaturing of thawed frozen tissue, can be circumvented by thawing in PBS (maintained at room temperature). Another molecular change associated with tissue thawing and dehydration was found to include a change in the relative intensities of the amide I and methyl bending modes.[20]

At this juncture, it is important to note that the extent of protein depolymerization that has been observed in Raman spectra of freeze-dried tissue can be reduced by using an appropriate cryogen for the initial snap-freezing of the resected tissue specimen. Higher freezing rates are achieved using propane or mixtures of propane and isopentane in preference to isopentane.[26] Also, since ice crystals develop between a temperature range of 0 to $-140°C$[26] it is advantageous to maintain cryo-preserved tissues at temperatures lower than $-140°C$ during microtomy and freeze-drying. Finally, the size of the specimen also dictates the extent of ice crystal damage; in liquid-nitrogen-cooled propane, it has been reported that guinea pig liver of 0.5 mm^3 is the critical size that separates complete crystallization from partial vitrification.[26] Nevertheless, Stone et al.[27] have demonstrated on a number of different tissue types that freeze-thawed tissue, without PBS, can be used to differentiate different pathologies with greater than 90 percent sensitivity and specificity. More recently, the same method of sample preparation was used to demonstrate the biochemical basis for tumour progression of prostate and bladder cancers by determining the relative amounts of a number of tissue constituents.[28] This was carried out by obtaining basis Raman spectra of pure standards and correlating these with tissue spectra derived from each disease state using ordinary least-squares analysis. The biological explanations for these constituent fluxes through the different pathological states could be associated with known tumour behavior. Thus, freeze-thawed tissue warmed to room temperature is not only diagnostically useful but may also provide relative quantifications and qualitative biomolecular characterization of the sample.

For FTIR investigations, snap-frozen tissue has been analyzed following thawing and subsequent air-drying to avoid interference from water bands.[29–32] However, this dehydration process results in undesirable perturbations to cellular chemistry, as outlined in Sec. 3.1, therefore, some researchers have used cryosections dried under a stream of nitrogen gas to reduced oxidative and surface tension effects.[33,34] Both protocols have successfully been applied to the spectral classification of tissue pathologies[29–34] and have been shown to generate detailed biospectroscopic maps that localize tumour lesions within oral[33] and brain tissues.[34] Additionally, using an univariate analytical approach to process tissue maps, Wiens et al.[32] reported the use of dried snap-frozen sections to investigate the early appearance and development of scar tissue using FTIR signals corresponding to lipids, sugars, phosphates as well as collagen and its fibril orientation.

Chemically Fixed Tissue

Compared to fresh tissue, Raman spectra of formalin-fixed tissue[19] following 24-hour fixation, exhibit a significantly reduced intensity of the amide I peak as well as the appearance of a peak at 1490 cm^{-1}. The reduced intensity of the amide I peak is attributed to the formation of tertiary amides (and loss of secondary amide), resulting from the reaction of methylene glycol (in formalin) cross-linking the nitrogen

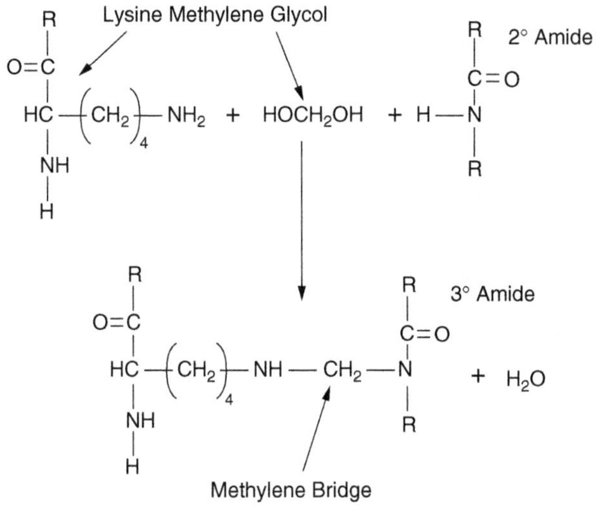

FIGURE 3.2 Scheme illustrating the formation of a methylene bridge during formalin fixation. (*Reproduced from Ref. 19 with the permission from Elsevier Limited.*)

atom of lysine with the nitrogen atom of a peptide linkage (Fig. 3.2).[19] The peak at 1490 cm^{-1} is thought to arise as a result of protein unraveling and increased activity of the NH$_3^+$ deformation mode. However, it had also been suggested that the coupling of the C—N stretching vibration with the in-phase C—H bending in amine radical cations, which may be present in the methylene bridge following formalin fixation involving lysine (Fig. 3.2), could result in a peak at 1490 cm^{-1}.[19] Thus, although previous FTIR-based studies of formalin fixation of isolated proteins have demonstrated no measurable effects on protein secondary structure (the arrangement of the polypeptide backbone),[25] the data above suggest that it does have an effect on protein intensity. However, contrary to previous findings on isolated proteins,[25] Faolain et al.[19] have found that formalin fixation of tissue produced a notable shift of 10 cm^{-1} in the amide I and II bands.

Formalin fixation was also found to preserve lipid signals in Raman spectra of lipid rich tissue such as adipose tissue and white matter of the brain.[20] However, Huang et al.[22] and Shim and Wilson[20] both report that a direct spectral contaminant from formalin in the Raman spectrum is a weak peak appearing at 1040 cm^{-1}.[20,22] This artifact was successfully removed through copious washes with PBS.[19]

Pleshko et al.[24] have shown that 70 percent ethanol fixation of 35-day-old embryonic rat femur, gave rise to FTIR spectra that exhibited amide I and II peaks that were shifted to lower frequencies (1647 cm^{-1} and 1546 cm^{-1}, respectively) compared with unfixed rat femur, which exhibited amide I and II peaks at 1651 cm^{-1} and 1550 cm^{-1}.

Plashco et al.[24] conclude that evaluation of protein structure in this tissue should be limited to snap-frozen, or formalin-fixed tissues where there were no observable shifts in the amide I and II peaks. The latter does not conform to observations made by Faolain et al.[19] in formalin-fixed tissue spectra. However, this difference may be due to the different types of tissue used in each study, mineralized tissue in Pleshko et al.[24] study and nonmineralized tissue in Faolain et al.[19] study.

3.2.3   The Effects of Xylene on Fixed Tissue and Deparaffinization of Paraffin-Embedded Tissue

As mentioned in Sec. 3.2.1, fixed tissue is immersed in xylene prior to impregnation with paraffin wax. Faolain et al.[19] have shown that Raman spectra of formalin-fixed tissue exposed to xylene produces a number of strong peaks, associated with its aromatic structure, at 620 cm^{-1} (C—C twist of aromatic rings), 1002 cm^{-1} (C—C stretching of aromatic rings), 1032 cm^{-1} (C—C skeletal stretch of aromatic rings), 601 cm^{-1} (C—C in plane bending of aromatic rings), and 1203 cm^{-1} (C—C$_6$H$_5$ stretching mode of aromatic rings).[19] As mentioned in the section "Chemically Fixed Tissue," formalin fixation reduces the intensity of the amide I band. Interestingly, upon xylene exposure, the amide I band reappears with appreciable intensity. Faolain et al. suggest that the cross-linking of proteins by methylene glycol is reversed upon xylene treatment so that the amide I band reverts back to the secondary amide. As expected, the FTIR spectrum of xylene treated formalin-fixed tissue demonstrated a loss of the lipid ester (C=O) band at 1740 cm^{-1}, due to significant removals of cellular lipids.

Presently in the fields of FTIR and Raman spectroscopy, there is lack of consensus with regard to a standard protocol for deparaffinization of paraffin-embedded sections and several approaches have been used. For example, Fernandez et al.[18] deparaffinized their prostate tissue sections by immersing in hexane at 40°C with continuous stirring for 48 hours. During this period, the vessel was emptied every 3 to 4 hours, rinsed thoroughly with acetone followed by hexane and after thorough drying, refilled with fresh hexane to promote dissolution of paraffin embedded in the tissue. The disappearance of a peak at 1462 cm^{-1} in the FTIR spectrum was used to ensure complete deparaffinization.

Sahu et al.[35] deparaffinized samples of colon tissue using xylene and alcohol. The researches washed their 10-μm paraffin-embedded sections with xylol for 10 minutes (three changes) with mild shaking. Following this, the slide was washed with 70 percent alcohol for 12 hours. To evaluate the efficacy of this procedure, FTIR spectra from the tissue were collected at each stage of the deparaffinization process—before deparaffinization, at each xylol washing step (following air-drying) and following alcohol treatment. The authors report that following two washes with xylol, a third xylol wash did not produce any significant changes to the lipid spectral regions (2800 to 3000 cm^{-1} and 1426 to 1483 cm^{-1}). Further treatment with alcohol produced changes to the region 900 to 1185 cm^{-1}, which was speculated to be the result of residue xylene removal. Alcohol treatment also showed a further reduction in lipid hydrocarbon signals in the spectral region 2800 to 3000 cm^{-1}. The authors observed that hematoxylin and eosin (H&E) sections of these deparaffinized tissues exhibited clear outlines for the cells that indicated the preservation of lipids in complex forms (membranes).

Faolain et al.[19] deparaffinized parenchymal tissue sections by immersing in two baths of xylene for 5 and 4 minutes, respectively. Followed by two baths of absolute ethanol for 3 and 2 minutes and a final bath of industrial methylated spirits 95 percent for 1 minute. This method was found through Raman microspectroscopy to be inefficient at removing all of the paraffin, since a number of strong signals from C—C and CH$_2$ vibrational modes were observed.[19] Gazi et al.[36] deparaffinized their prostate tissue sections by immersion in Citroclear (a deparaffinization agent that is less toxic than xylene) and placed on an orbital mixer for 6 minutes and then in acetone for a further 6 minutes at 4°C before being air-dried for 1 hour under ambient conditions. A commonality between the latter three procedures is the use of additional organic solvent(s) used to remove any residual deparaffinization agent. Figure 3.3 shows a typical deparaffinized FTIR spectrum obtained using the method outlined by Gazi et al.[36] Citroclear is composed of alkyl hydrocarbons and orange terpenes, its spectrum gives rise to several marker peaks that may be used to detect its presence in the tissue; these correspond to peaks at 1711, 888, and 800 cm^{-1} and are absent in the deparaffinized spectrum.

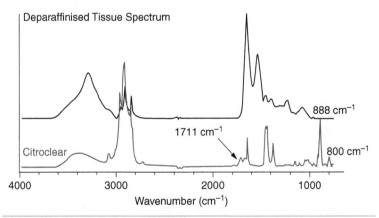

FIGURE 3.3   FTIR spectra showing the absence of Citroclear marker bands (1711, 888, and 800 cm^{-1}) in a typical spectrum of a deparaffinized prostate tissue section using the method outlined by Gazi et al.[36]

Contrary to Fernandez et al.[18] and Sahu et al.,[35] Faolain et al.[19] suggest that complete removal of paraffin from the tissue section is not possible to accurately assess with FTIR spectroscopy using ~1462-cm^{-1} signal. This was concluded following analysis of Raman and FTIR spectra obtained from deparaffinized tissue sections (using an identical deparaffinization protocol for each mode of analysis) in which spectral peaks, characteristic of paraffin, were readily resolved in Raman spectra (strong sharp bands), compared with FTIR spectra where no discernable marker peak could be seen.[19] A follow-up study by Faolain et al.[37] comprehensive evaluated the efficiency of different deparaffinization agents to remove paraffin wax from cervical tissue sections. In this study, one deparaffinization cycle involved two baths of deparaffinization agent (5 minutes and 4 minutes, respectively), followed by immersion in 2 baths of ethanol (3 minutes and 2 minutes, respectively) and a final bath in industrial methylated spirits 95 Percent. This Raman-based investigation, demonstrated that paraffin signals at 1063 cm^{-1} (C—C stretch), 1296 cm^{-1} (CH$_2$ deformation) and 1441 cm^{-1} (CH$_2$ bending) (Fig. 3.4a) were not completely removed by xylene and Histoclear even after 18 hours of treatment (Fig. 3.4b and 3.4c). However,

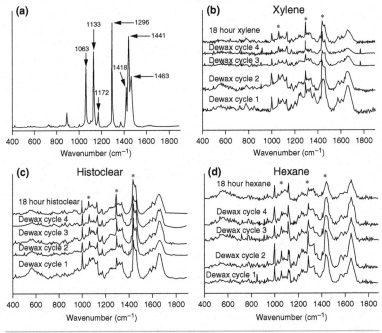

FIGURE 3.4   (a) Raman spectrum of paraffin wax, Raman spectra of paraffin-embedded tissue following treatment with repeated cycles of deparaffinization agents, (b) xylene, (c) Histoclear, and (d) hexane. The * marked peaks correspond to paraffin residue at 1063, 1296, and 1441 cm^{-1}. (*Modified with permission and reproduced from Ref. 37.*)

hexane was found to be a superior deparaffinization agent, removing nearly all of the paraffin 18 hours posttreatment (Fig. 3.4*d*).

These important findings were found to have practical implications for immunohistochemical analysis. It was found that the intensity of positive staining in tissue sections treated with hexane for 18 hours was 28 percent greater when compared to an adjacent section treated with xylene for the same period.[37]

In the light of these findings from Faolain et al.[37] the protocol used by Fernandez et al.,[18] where prostate tissue sections were treated with hexane for 48 hours, provides the most efficient method of paraffin removal used in an FTIR study to date. Infact, using this sample, preparation protocol high levels of accuracy ($\geq$90 percent) were achieved for classifying a number of different tissue components for FTIR chemomectric imaging of prostate tissue microarrays. Nevertheless, it has been shown by Gazi et al.[36] that although a less rigorous method of deparaffinization was used to process malignant prostate tissue sections, the spectral region between 1481 and 999 cm^{-1} could be used to discriminate and classify the different pathological grades of prostate cancer as well as to provide statistically significant distinction between tumors localized to the prostate gland from those showing extracapsular penetration. Moreover, the time-efficient method of less rigorous deparaffinization is suitable for other FTIR-based diagnostic parameters that do not include spectral regions that overlap with paraffin signals, such as the peak area ratio of the 1030-cm^{-1} (glycogen):1080 cm^{-1} (phosphate) bands, which differentiated malignant from benign prostate tissues in imaging studies.[9]

It may be argued that for FTIR studies, the removal of paraffin is not necessary at all as discrete frequency ranges corresponding to the lipid hydrocarbon modes are only affected. However, visualisation of the unstained tissue's anatomical features is severely hampered if the paraffin is not removed. Sahu et al.[35] report that colonic crypts in a 10-μm paraffin-embedded tissue section appeared as circular entities when viewed under light microscopy. Moreover, even between adjacent microtomed sections, tissue components can vary significantly, which in turn prevents the positioning of the IR beam upon a specific tissue location by comparison with an H&E section.

3.3   Cell Preparation

3.3.1   Chemical Fixation for FTIR and Raman Imaging

As mentioned in Sec. 3.1, to avoid potential confounding variables from autolytic processes initiated by cells during air-drying, it is important that the cells are appropriately fixed to maintain localizations of biomolecular species. To this end, Gazi et al.[38] studied the use of several chemical fixation methods for biospectroscopically mapping single prostate cancer cells with synchrotron (SR)-based FTIR micro-spectroscopy. The cells were cultured directly onto MirrIR reflection substrates for transflection mode analysis. Firstly, the cells were fixed in 4 percent formalin (in PBS) for 20 minutes at room temperature with a brief rinse in doubly deionized water (3 seconds) before being air-dried. Water rinsing was found to be an important step for removing residual PBS from the surface of the cells so that a clear distinction could be made between the nuclear and cytoplasmic compartments (Fig. 3.5a and 3.5b). This also reduces light-scattering artifact during analysis. Although tryplan blue staining of these cells demonstrated loss of plasma membrane integrity, which is likely to be due to the 1 percent methanol present in the fixative, SR-FTIR images (collected with 7×7 µm sampling-aperture-size and 3 µm step-size) revealed localizations of lipid $[v_s(C=O)]$ and phosphate $[v_{as}(PO_2)]$ domains (Fig. 3.5c). Localization of lipid to the cytoplasm was expected due to the high concentration of cytoplasmic organelles comprising lipid-rich membranes. Whereas, the most intense phosphate signal was expected to localize at the nucleolus of the nucleus due to the high concentration of phosphates constituting the backbone

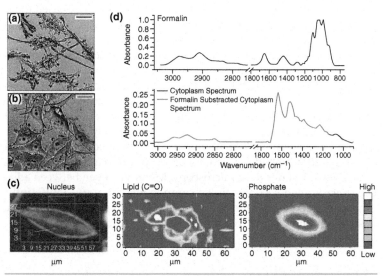

FIGURE 3.5   Photomicrographs of formalin-fixed prostate cancer cells (same magnification) (a) without subsequent rinsing in deionized water and (b) with 3-second rinse in deionized water to remove residue PBS from the surface of the cells. Scale bar in all photomicrographs = 50 µm. (c) Optical image of a single, formalin-fixed, PC-3 cell. The cells nucleus and nucleolus (N) are identified. SR-FTIR images depicting the intensity profiles of lipid ester $v_s(C=O)$ (1752 to 1722 cm^{-1} peak area and phosphate $v_{as}(PO_2)$ (1280 to 1174 cm^{-1} peak area). (d) The FTIR spectrum of formalin and overlay of the FTIR spectrum of the cytoplasm with the same spectrum processed to remove their theoretical formalin content. (*Reproduced from Ref. 38.*)

of DNA. The nucleolus also gave rise to intense protein amide I signal attributed to the densely packed histone proteins. Spectral subtraction of the neat formalin spectrum from the FTIR spectrum of the formalin-fixed cell resulted in negligible differences in the intensities of peaks across the frequency range 3000 to 1100 cm^{-1} (Fig. 3.5d). This was performed following normalization of the spectrum of formalin to the intensity value of the peak at 1000 cm^{-1} in the formalin-fixed cell spectrum, since this frequency gives rise to the most intense peak in the spectrum of formalin.

The formalin fixation protocol outlined above has successfully been used to image unstained cells in the process of mitosis and cytokinesis using both FTIR and Raman microspectroscopies.[39,40] Matthaus et al.[39] report Raman images showing the protein and phosphate scattering intensities of cells in various stages of mitosis, which were used to probe microtubules and the dense histone-packed chromatin as well as DNA condensation. Gazi et al.[40] reported SR-FTIR maps of a formalin-fixed cell in the process of cytokinesis, where features such as the contractile ring as well as organelle placement could be determined using the protein amide I and lipid ester [v_s(C=O)] signals, respectively.

Krafft et al.[41] used formalin fixation to obtain highly spatially resolved Raman microspectroscopic maps of lung fibroblast cells grown on quartz slides. These cells were analyzed at 4°C in 10 mM phosphate buffer with 1 mM sodium azide at neutral pH. Spectra from these maps could be used to identify RNA and DNA, proteins, cholesterol and phospholipids (phosphatidylcholine and phosphatidylethanolamine). Reprocessed Raman cell maps, depicting the fit coefficients for each of these biomolecules enabled an approximation of the composition of different subcellular structures: nucleus, cytoplasm, endoplasmic reticulum, vesicles, and the peripheral membrane.

Gazi et al.[38] also studied cells that had been formalin fixed and subsequently critical-point-dried (CPD). The CPD process involves several steps. First, the intercellular water molecules (from saline) in the preformalin-fixed cells must be displaced gradually with increasing concentrations of ethanol. The ethanol is then displaced by acetone, which is miscible with liquid CO_2. The acetone within the cells is then displaced by liquid CO_2, within a chamber. The chamber is heated with a simultaneous rise in pressure as liquid CO_2 enters the vapor phase. At a specific temperature and pressure, the density of the vapor equals the density of the liquid, the liquid–vapor boundary disappears, and the surface tension is zero. Thus, this method reduces any residual distortions that may occur in the prefixed cell as a result of air-drying. Since the formalin-fixed CPD dried cells were exposed to significant lipid-leaching reagents (ethanol and acetone), the cells were positive to trypan blue staining and the SR-FTIR spectrum of these cells demonstrated loss of the lipid ester v_s(C=O) peak.

A third fixation method was investigated by Gazi et al.[38] in which cells were fixed with glutaraldehyde and osmium tetroxide (OsO_4) prior to CPD. Glutaraldehyde polymerizes in solution, where dimmers and trimmers are the most abundant polymers.[42] The aldehyde groups of glutaraldehyde react with the amino groups of proteins to form imines in an irreversible reaction. The commonly used postfixative to glutaraldehyde is OsO_4, which preserves unsaturated lipids by the formation of cyclic esters and is also an irreversible reaction.[42] With the exception of a weak peak at 960 cm^{-1}, it also does not give rise to absorption bands within the spectra region of interest in the mid-IR range, where most biomolecules absorb. CPD-glutaraldehyde-OsO_4-fixed cells preserve fine structure as observed in electron microscopy studies. The cells were found to be negative to trypan blue, indicating good preservation of plasma membrane lipid molecules. SR-FTIR images of cells fixed using this method (optical image in Fig. 3.6a) are shown in Fig. 3.6b and 3.6c. The neat FTIR spectrum of glutaraldehyde is shown in Fig. 3.6d and a spectrum obtained from the SR-FTIR image of the cell is also presented.

The localizations of phosphate and lipid ester v_s(C=O) signals were consistent with SR-FTIR maps of formalin-fixed cells (see Fig. 3.5). Although no significant spectral markers from glutaraldehyde were detected in the SR-FTIR spectrum of these cells, compared to formalin-fixed cells (see Fig. 3.5d), a reduction in the intensity of peaks between 1500 and 1000 cm^{-1} was observed and the lipid ester v_s(C=O) signal appeared as a less resolved shoulder on the amide I band (Fig. 3.6d).

Unfixed cells were also investigated following preparation using a protocol outlined by Tobin et al.[7] in which cells were removed from culture medium, rinsed in PBS, and dried under centrifugation. The cells stained positive for trypan blue indicating loss of membrane integrity. SR-FTIR images of these cells revealed the expected localization of high-phosphate intensity within the nucleus; however, in other parts of the cell, phosphates were homogenously distributed in intensity. This was attributed to phosphates from PBS retained on the surface of the dried cells.[38] Interestingly, intense lipid ester v_s(C=O) signal also localized to the nucleus and decreased in intensity as a series of concentric rings between the nucleus and its periphery. This was unlike the lipid ester v_s(C=O) distributions observed for formalin (Fig. 3.5c) or CPD-glutaraldehyde-OsO_4-fixed cells (Fig. 3.6b and 3.6c) in which this signal was distributed with high-intensity surrounding the nucleus.

In a recent study by Gazi et al.[43] chemical fixation was investigated for the preparation of adipocytes for FTIR analysis. Adipocytes are specialised for the synthesis and storage of fatty acids (FAs) as triacylglycerides (TAGs) as well as for FA mobilization through lipolysis. Figure 3.7a shows the appearance of adipocytes in growth medium and illustrates the presence of numerous lipid droplets contained within their cytoplasm. Figure 3.7b shows adipocytes, prepared for FTIR analysis, following fixation in 4 percent formalin (in PBS), a brief water-rinse to remove residue salts and air-drying at ambient conditions. Although this fixation protocol

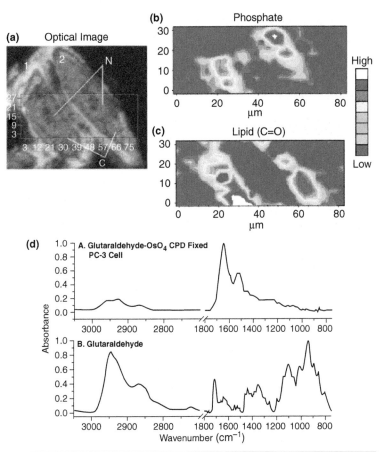

FIGURE 3.6 (a) Optical image of two glutaraldeyde-osmium tetroxide-CPD-fixed PC-3 cells (labeled 1 and 2). N designates the nucleus, C designates the cell cytoplasm; localizations and intensity profiles of (b) Phosphate (1271 to 1180 cm^{-1} peak area) and (c). Lipid ester (C=O) (1756 to 1722 cm^{-1} peak area); (d) FTIR spectrum of a glutaraldehyde-osmium tetroxide-fixed PC-3 cell; the IR spectrum of neat glutaraldehyde. (*Reproduced from Ref. 38.*)

was appropriate for the preservation of prostate cancer cells for SR-FTIR analysis, it resulted in the collapse of intracellular lipid droplet structures into an unordered lipid deposit in the adipocytes (Fig. 3.7b).

An FTIR spectrum of the intracellular lipid droplet of the formalin-fixed air-dried adipocyte [Fig. 3.7d(ii)] exhibits a lipid ester v_s(C=O) peak at 1744 cm^{-1}, which is the same frequency of absorption as the lipid ester v_s(C=O) peak in the reference TAG spectrum [Fig. 3.7d(i)]. Several other characteristic peaks of the glycerol moiety of TAG are also observed in the lipid deposit of the formalin-fixed adipocyte at frequencies >1500 cm^{-1} and these are identified in Fig. 3.7d(ii) as

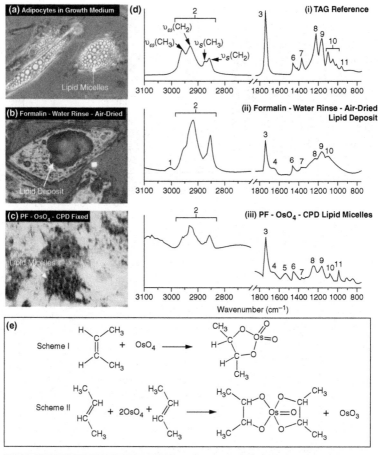

Figure 3.7 Optical photomicrographs showing (*a*) adipocytes in growth medium with prominent intracellular droplets (scale bar = 10 μm); (*b*) adipocyte following formalin fixation, water rinsing and air-drying (scale bar = 20 μm); (*c*) adipocytes fixed in paraformaldehyde (PF) and osmium tetroxide (OsO₄) and critical-point-dried (CPD) (scale bar = 30 μm); (*d*) typical FTIR spectrum of (i) triacylglyceride (TAG) reference with C_2 to C_{10} saturated hydrocarbon chains, (ii) the lipid deposit from a formalin-fixed air-dried adipocyte and (iii) the lipid droplets from a PF and osmium tetroxide (OsO₄) CPD adipocyte; (*e*) OsO₄ reaction with unsaturated hydrocarbon chains to form cyclic esters. Scheme I, reaction with a single unsaturated hydrocarbon chain. Scheme II, cross-linking reaction with adjacent unsaturated hydrocarbon chains.

peaks labeled 8 to 10. However, the peaks in this spectrum [Fig. 3.7*d*(ii)] are broader compared with the same peaks in the reference TAG spectrum [Fig. 3.7*d*(i)]. This is due to collapse of the lipid droplets that give rise to a range of bonding strengths with neighboring molecular species for those functional groups absorbing at frequencies >1500 cm⁻¹.

Figure 3.7c shows adipocytes containing well-preserved lipid droplets following paraformaldehyde (PF) fixation with OsO_4 postfixation and critical-point-drying (CPD).

There are several advantages to this sample preparation over formalin-fixation, water-rinsing and air-drying: (a) Formalin in PBS contains methanol, which permeates the plasma membrane and results in a faster fixation compared with PF that does not contain methanol. However, methanol extracts intracellular lipids, which is inappropriate for adipocyte fixation. (b) The OsO_4 postfixative preserves lipids, however, does not itself absorb in the mid-IR range where most biomolecules absorb, except for a peak at 960 cm^{-1}. (c) The three-dimensional structure of the adipocyte is retained, since the sample is dried without surface-tension effects, through CPD, and the localization of intracellular lipid droplets of the adipocyte is persevered. The disadvantage of this fixation protocol is that the mode of action by which OsO_4 preserves lipids is through complexation-reaction within the double bonds of lipid hydrocarbon chains or complexation and cross-linking between unsaturated hydrocarbon chains (Fig. 3.7e). Thus, the v_s(=C-H) signal from unsaturated hydrocarbons is present in the lipid-deposit spectrum of the spectrum of the formalin-fixed, water-rinsed, air-dried adipocyte [Fig. 3.7d(ii)], but is not observed in the lipid-droplet spectrum of the PF-OsO_4-CPD adipocyte [Fig. 3.7d(iii)]. Additionally, both methods of fixation (formalin-water rinse-air dried and PF-OsO_4-CPD) result in a decrease in peak resolution of the v_{as}(CH)$_2$ and v_{as}(CH)$_3$ modes and v_s(CH)$_2$ and v_s(CH)$_3$ modes.

The PF-OsO_4-CPD protocol outlined above was used to preserve samples of prostate cancer cells (PC-3 cell line; prostate cancer cells derived from bone metastases) that were co-cultured with adipocytes preloaded with deuterated palmitic acid (D_{31}-PA).[43] This specimen was used in an FTIR tracing experiment to determine whether PC-3 cells could uptake the fatty acids stored within adipocytes. Figure 3.8a shows an optical image of a PF-OsO_4-CPD-fixed adipocyte surrounded by PC-3 cells and stroma cells. In this figure the adipocytes are visualised as large dark bodies (designated with Adp in Fig. 3.8a), whereas PC-3 cells (1 to 4) are lighter in appearance and possess lamellipodia-pointed processes. The dark stain results from the binding of OsO_4 to the lipids. The boxed area was mapped using FTIR microspectroscopy and the v_{as}(CD)$_{2+3}$ signal intensity distribution is shown in Fig. 3.5b. As expected, there was localization of the v_{as}(CD)$_{2+3}$ signal with high intensity to the adipocyte; however, it was also found that this signal illuminated the PC-3 cells (Fig. 3.8b). Since, the only source of v_{as}(CD)$_{2+3}$ signal in the PC-3 cells is through incorporation of D_{31}-PA released by the adipocytes, this data unequivocally demonstrates the translocation of D_{31}-PA between these cell types without cell isolation or external labeling. Appropriate fixation was necessary in this experiment, since delocalization/bleeding of lipid molecules from adipocytes in the adipocyte—PC-3 cell coculture system could result

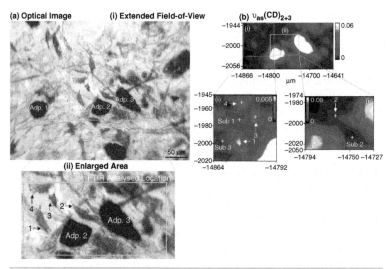

FIGURE 3.8   [a(i)] Optical photomicrograph showing PF-OsO$_4$-CPD fixed adipocytes (Adp. 1 to 3) surrounded by prostate cancer cells and stroma cells; [a(ii)] magnified region of Adp. 2 and 3 with surrounding prostate cancer cells (labeled 1 to 4). the boxed area was analyzed by imaging FTIR microspectroscopy. (b) FTIR spectral maps depicting the intensity distribution of the $v_{as}(CD)_{2+3}$ signal. The boxed areas (i) and (ii) were expanded and the color intensity threshold changed to provide better contrast of the $v_{as}(CD)_{2+3}$ signal in cells relative to the substrate.

in false-positive results concerning PC-3 uptake of adipocyte-derived D_{31}-PA.

3.3.2   Sample Preparation for Biomechanistic Studies

Unfixed cells prepared by the method outlined by Tobin et al. (drying cells by centrifugation)[7] resulted in SR-FTIR spectral maps that showed poorer localizations or contrast for FTIR signals that are expected to appear in the cytoplasmic or nuclear compartments. Nevertheless, in the study by Tobin et al.[7] this sample preparation protocol was appropriate for investigating, by FTIR, the response of cervical cancer cells to epidermal growth factor (EGF). Cells were incubated with EGF with increasing incubation times. Changes in protein conformation (noted by shifts in amide I peak position) as a result of phosphorylation by EGF (monitored by the peak area of the phosphate monoester vibration at 970 cm^{-1}), at consecutive time points were observed. The important point here is that time-course experiments such as the one carried out by Tobin et al.[7] require methods of sample preparation that are not lengthy to perform, particularly when the intervals between sampling time points are short. Gazi et al.[44] used formalin fixation to study the temporal fluctuations in phosphate, protein secondary structures and endogenous nonisotopically labeled lipid signals following stimulation with different concentrations of D_{31}-PA and deuterated arachidonic acid (D_8-AA). It was found that the shortest practical-time interval between sampling points during which the cells could be fixed was 15 minutes. As an example, Fig. 3.9a shows these biochemical fluctuations for PC-3 cells incubated with 50 µm D_{31}-PA in serum-free culture media, compared with control (PC-3 cells incubated in identical conditions but without D_{31}-PA). The endogenous lipid signal in the control PC-3 cells initially fell and is induced by metabolic/cytokine/growth factor imbalance resulting

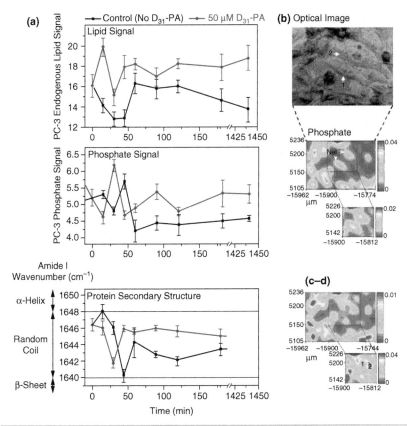

FIGURE 3.9   (a) Temporal fluctuations in various biomolecular domains probed by FTIR, for PC-3 cells exposed to 50 µM D_{31}-PA or no D_{31}-PA (control). Endogenous mean lipid hydrocarbon peak area intensities (±SE); mean phosphate diester peak area [$v_{as}(PO_2)$]] intensities (±SE) and amide I frequency shifts (±SE); (b) Optical image of PC-3 cells following incubation with D_{31}-PA for 24 hours. This area was analyzed using imaging FTIR microspectroscopy. IR biospectral maps show the intensity distributions of phosphate [nuclei are labeled (Ncl)] and $v_{as+s}(CD_{2+3})$ peak area (D_{31}-PA or its metabolites). In each image, cells 1 and 2 (see optical image) are magnified to demonstrate the intensity of IR signals in greater detail. FTIR spectra were obtained from points 1 (nucleus) and 2 (cytoplasm) in this image.

from the exchange of media to serum-free Roswell Park Memorial Institute (RPMI) media at the zero minute time-point. Conversely, cells incubated with D_{31}-PA showed an initial rise in endogenous lipids. Since the incubation media (RPMI) contains no FAs, this increase in lipid content must be due to de novo biosynthesis. This initial rise in endogenous lipid signal was followed by a fall, attributed to metabolic breakdown into adenosine triphosphate (ATP), which is a major product of lipid metabolism. This notion is supported by a phosphate spike at 30 minutes accompanied by a significant shift in the amide I frequency, indicating protein phosphorylation.

The time-efficient formalin fixation method not only suitably preserved biomolecular composition so that lipid metabolism and protein phosphorylation could be measured, but also preserved the subcellular localizations of biomolecules for imaging studies. Figure 3.9b shows an optical photomicrograph of PC-3 cells on MirrIR substrate, following exposure to 50µM D_{31}-PA for 24 hours. This area was analyzed by imaging FTIR microspectroscopy and the resulting distribution of the integrated intensity of the phosphate diester $[v_{as}(PO_2);$ (1274 to 1181 cm^{-1})] peak area is shown. As expected, for cells 1 and 2 in the optical image, it can be seen that the most intense phosphate signals localise at the nucleus. Whereas, the most intense $v_{as+s}(CD_{2+3})$ signal localized at the cytoplasm, suggesting that the subcellular localization of D_{31}-PA or its metabolites is predominately in the cytoplasm.

Another FTIR-based dose-response study had been undertaken where prior to spectroscopic examination, drug induced cells had been removed from culture media, washed in PBS and air-dried.[44] This study reports spectroscopic changes (ratio of peaks) that could be associated with exposure of the cells to increasing doses of the chemotherapeutic drug. An additional bioanalytical modality was combined with FTIR to demonstrate correlations of spectroscopic changes with cell sensitivity to the drug using the MTT [(3-(4,5-dimethylthiazol-2-yl)-2,5-diphenyltetrazolium bromide] assay. Thus, there is an evidence to suggest spectroscopic changes associated with drug exposure can be determined and this is in fact dominant over metabolite perturbations resulting from autolysis during the drying process.

3.3.3   Growth Medium and Substrate Effects on Spectroscopic Examination of Cells

Growth Medium Influences

A number of studies have investigated the use of FTIR or Raman microspectroscopies as diagnostic tools to differentiate and classify cell lines, in vitro, based on their pathological state[45–49] or resistance to drugs.[46] Interestingly, we find that some researchers have grown their different cell lines in the same culture media,[46–48] whereas others have used different media for each cell type.[49,50] The European Collection of Cell Cultures (ECACC) provides standard protocols for the optimum growth of different cell lines. In some instances, cell culture media may be different for cells of the same epithelial origin, for example, ECACC suggest PC-3 cells (prostate cancer epithelial cell line derived from bone metastases) should be grown in Ham's F-12, whereas LNCap-FGC (prostate cancer epithelial cell line derived from lymph node metastases) should be grown in RPMI 1640. The question arises: Should investigations aiming to discriminating cell types include data from cells grown in the same media or does it matter if cells are grown in different media? Taking the example of PC-3 and LNCap-FGC cell lines, both RPMI 1640 and Ham's F-12 are complex mixtures consisting of a range of inorganic salts, amino acids, vitamins, nucleotides and glucose as well as small-molecule precursors. However, differences between media can exist with respect to the relative concentrations of each component as well as compositional differences such as the presence or absence of a major biomolecular class, for instance RPMI 1640 contains no fatty acids, unlike Ham's F-12, which contains the 6-FA, linoleic acid (LA).

In a recent study by Harvey et al.[50] reflection mode FTIR photoacoustic spectroscopy (PAS) was used to obtain spectra from four different formalin-fixed prostate cell lines (BPH = benign prostatic hyperplasia; LNCap-FGC = prostate cancer epithelial cells derived from lymph node metastases; PC-3 = prostate cancer epithelial cells derived from bone metastases; PNT2-C2 = immortalized normal prostate epithelial cells by transfection with the genome of the SV40 virus). Unsupervised principle component analysis (PCA) of this spectral data set yielded separation of clusters corresponding to each of these cell lines (Fig. 3.10a). Two of these cell lines were grown in the

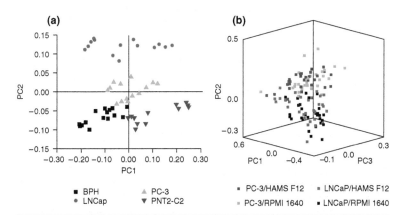

FIGURE 3.10   (a) PCA scores plot of the background-subtracted, vector normalized first derivative FTIR-PAS spectra of four different prostate cell lines (BPH, LNCap-FGC, PC-3, and PNT2-C2). (b) PCA scores plot of vector normalized, first derivative FTIR spectra of PC-3 and LNCap-FGC cell lines, grown in their "optimum" culture medium or "foreign" culture medium. (*Reproduced from Ref. 50 with permission from The Royal Society of Chemistry.*)

same media (LNCap-FGC and PNT2-C2), whereas two were grown in different media (BPH and PC-3). Importantly, the two cell lines that were grown in identical media (LNCap-FGC and PNT2-C2) showed significant separation, realized by anticorrelation on PC-2.

In a follow-up study, Harvey et al.[51] acquired conventional FTIR spectra from PC-3 cells and LNCaP cells, each grown separately in their optimum culture medium (as advised by ECACC protocols) and a "foreign medium." Unsupervised PCA of these spectra demonstrated clustering of the two cell lines that was independent of the culture medium in which they were grown, but was principally dependent on the cell type (Fig. 3.10b). Thus, it may be concluded that for at least prostate cell lines PC-3 and LNCaP, the influence of the basic media under investigation in this study (RPMI 1640 and Ham's F-12) on the cell metabolites, which may be more significant using other analytical modalities, is not the primary influence on spectroscopic measurements. However, it must be acknowledged that cells are a function of their environment (discussed in further detail below). Thus, the same cell line grown in two different media with relatively larger compositional differences (or containing potent stimuli) will effect spectroscopic classification.

If the cell is exposed to an environment that does not sustain its optimum growth and down-regulates the expression of biomolecular features (such as cell surface antigens, hormone receptors, protein expression), which characterize that cell type *in vivo*, then this may ultimately render the cell to a new class. In vivo, it is well known that stromal-cell interactions are particularly important in cancers such as of the breast, where the stromal compartment plays a critical role in directing proliferation and functional changes in the epithelium.[52] Moreover, environmental stimuli directing cell phenotype has been recently studied with imaging FTIR by Krafft et al.[53] In this study, human mesenchymal stem cells were treated with osteogenic stimulatory factors that induced their differentiation. Differentiation was detected by FTIR through changes in the amide I band shape (indicative of protein composition/structural changes) and phosphate levels (indicative of the expression of calcium phosphate salts).

Substrate Influences

In a similar manner to compositional differences in the growth media that may or may not elicit changes in cell biochemistry and thus its spectra, substrates can also induce morphological as well as functional changes in the cell. Meade et al.[54] studied the influence of a range of substrates on the normal human epithelial keratinocyte cell line (HaCaT) using a multimodal approach that included fluorescence, FTIR and Raman spectroscopies. The substrate extracellular matrix (ECM) coatings under evaluation were two glycoproteins, fibronectin and laminin and one protein, gelatin (derived from thermal denaturing of collagen). Gelatin was coated onto MirrIR slides for FTIR experiments or quartz slides for Raman experiments, by incubation for 24 hours at 4°C. Laminin and fibronectin were coated onto these substrates by incubation for 4 hours and 40 minutes, respectively, at room temperature. For the fibronectin- and laminin-coated slides, excess solution was aspirated from the substrates and washed in PBS prior to cell deposition. Whereas, for the gelatin-coated slide, excess solution was aspirated and cells were deposited for culture without prior washing in PBS. Fluorescence assays were conducted at 3 days postseeding as well as fixation using 4 percent formalin (in PBS) with water rinse for FTIR and Raman investigations.

Meade et al.[54] observed through fluorescence assays that cellular proliferation, viability as well as protein content were down-regulated when cells were grown on uncoated quartz compared with uncoated MirrIR substrates. However, increases in proliferation and viability were more pronounced when cells were grown on coated quartz than grown on coated MirrIR substrates. Additionally, it was found that quartz coated with all three ECM coatings generated significantly enhanced proliferation compared to the control (uncoated quartz). However, this was not the case for MirrIR, which resulted in a significant increase in proliferation only for the laminin-coated slide. Viability was significantly increased when cells were grown on laminin- and gelatin-coated quartz, whereas for MirrIR substrate, viability was only significantly increased when this was coated with gelatin. The authors suggest that gelatin provides a coating with similar proliferation effects on quartz and MirrIR and increases viability, which is desirable for long-term cultures.

FTIR and Raman spectroscopic analyses of coated slides demonstrate that the gelatin coating did not give rise to sufficiently high signals to significantly influence FTIR or Raman spectra of the cells cultured upon them. First derivative FTIR spectra and Raman spectra of cells on gelatin, fibronectin and laminin demonstrated spectral changes on each of these substrates that were associated with nucleic acid, lipid, and protein expression. Raman spectra provided further insight and relative quantification, since it was found that coatings that promoted proliferation gave rise to increases in spectral regions associated with DNA, RNA, and proteins, with a decrease in lipids. This has been attributed to an increase in the sustained production of signaling proteins, as a result of integrin binding to the coating, that promotes cellular proliferation. Supporting this, the authors also found through FTIR that the ratio of protein (sum of integral absorbencies corresponding to the amide I, II, and III bands) to lipid (integral absorbance 1370 to 1400 cm^{-1}) gave rise to values that could be significantly correlated with an increase in proliferation (as measured by fluorescence spectroscopy).

The experimental setup described above consisted of thin layers of ECM that were barely detectable in the FTIR or Raman spectrum; however, Lee et al.[55] studied prostate cancer cells that had been cultured onto relatively thicker layers of ECM. In their investigation, Matrigel was used as the artificial ECM. The major constituents of Matrigel are collagen type IV (a protein), heparan sulphate (a proteoglycan), laminin and entactin (glycoproteins). Figure 3.11a shows an optical photomicrograph of prostate cancer cells on Matrigel. This area was analyzed using imaging FTIR microspectroscopy. As expected, the lipid hydrocarbon signal demonstrates high intensity at the cell locations, relative to the Matrigel surroundings, due to the cumulative absorption of lipid containing biomolecules in the cells and Matrigel (Fig. 3.11b). Since the lipid background signal is nearly homogenous, it suggests that the Matrigel is of constant thickness within the analysis field-of-view. However, the protein background exhibits a heterogeneous distribution of intensity (Fig. 3.11c), which is likely to be due to concentration differences when taking into consideration the lipid intensity image. As expected, cells adhered to the low-protein concentration exhibits a higher protein intensity signal than the surrounding layer, whereas those on a high-protein concentration or thick surface revealed an unexpected lower protein intensity signal. This is illustrated in the protein cross section in Fig. 3.11d, which was plotted

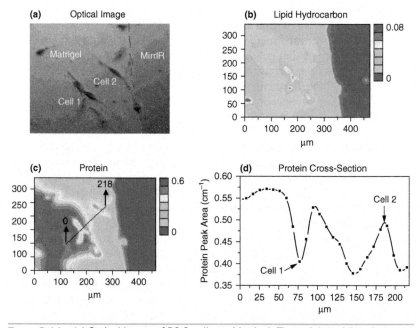

FIGURE 3.11   (a) Optical image of PC-3 cells on Matrigel. The red-dotted line denotes the Matrigel (left)–MirrIR™ substrate (right) interface. FTIR spectral maps depicting the intensity distribution of the (b) lipid hydrocarbon and (c) protein peak area signals; (d) a cross-section through the protein intensity map is displayed as a graphical plot depicting the protein peak area values from a region of high concentration of Matrigel (at 0 µm) to one of lower concentration (toward 218 µm) and bisecting cells 1 and 2. (Reproduced from Ref. 55 with permission from The Royal Society of Chemistry.)

with values taken from a region of high-protein intensity to a region of low-protein intensity and bisecting the cells.

Could the low-protein signal at the cell be due to local proteolysis of the Matrigel matrix? Time-lapse video microscopy provided valuable insight into this speculation. The final frame of the time-lapse video and the brightfield image of the same area after fixation, relocated for FTIR microspectroscopic imaging, are shown in Fig. 3.12b(i) and (ii), respectively. These optical images demonstrate that the morphology of the cells (elongated and rounded), when in culture, is suitably retained by the formalin fixation procedure. In Figure 3.12a, the video frame captured at the start of the time-lapse recording shows that the cells at internal locations on Matrigel display a rounded morphology. The cell marked with a red arrowhead was stationary throughout the course of time-lapse recording and retained its rounded appearance. It is reasonable to assume that the low-protein intensity at this cellular location [Fig. 3.12b(iii)] would be indicative of local proteolysis or mechanical degradation of the Matrigel. However, between 5 hours and the point of termination of the time-lapse study (22 hours, 22 minutes), the cell marked with the green arrowhead migrated, several times, toward and away from the cells marked with the blue and white arrowheads. Since these cells (green, blue, and white arrowhead) were motile throughout the time-lapse recording, it is unlikely that there was local digestion of the Matrigel just prior to termination of the time-lapse recording via MMPs produced by the prostate cancer cells. Moreover, if proteolytic digestion was a dominant mechanism by which the green, blue, and white arrowhead cells transversed over the Matrigel, then one would expect low-protein signals to arise from the entire path occupied by these cells. It was concluded from this, that light-scattering artefacts influenced the protein intensity maps of these cells on Matrigel, giving the illusion of protein degradation at the cell locations.[55]

A model was produced, which showed that a switch from a higher than background signal to a lower than background signal will occur at a given thickness or concentration of protein within the Matrigel layer. Importantly, the model includes light that is directly back-scattered into the microscope collection optics. These findings implicate fundamentally on research in the field of FTIR spectroscopy concerning cells on two-dimensional matrices.

3.3.4   Preparation of Living Cells for FTIR and Raman Studies

FTIR Studies

A number of studies concerning the analysis of living cells by FTIR have been performed with synchrotron radiation sources.[56–58] An early study by Holman et al.[56] reported spectral changes in HepG2

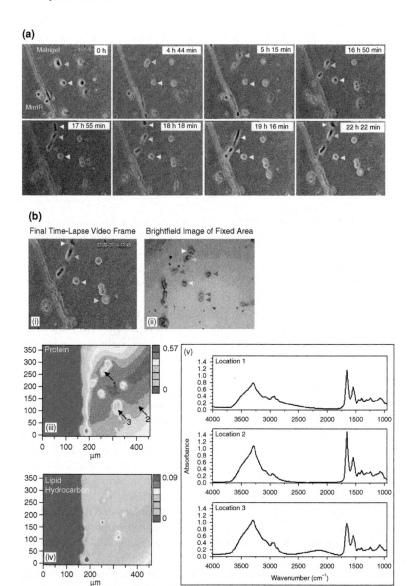

FIGURE 3.12   (a) Still frames taken from a time-lapse video of prostate cancer cells on Matrigel. The red-dotted line at zero hour designates the Matrigel–MirrIR interface. Individual cells are labeled with colored arrowheads for the ease of tracking cell migration across consecutive time frames and following figures. (b) Optical images of the (i) final frame of the time-lapse video (at 22 hours 22 minutes) and (ii) brightfield image of the same area, postfixation. Note that the image of the fixed area is rotated 22° compared with the image of the final frame such that the Matrigel–MirrIR interface is approximately vertical. The fixed area was analyzed by FTIR microspectroscopy and the intensity distribution of the (iii) protein and (iv) lipid hydrocarbon domains are shown. (v) Representative raw IR spectra taken from three locations [1 and 3 (on cell) and 2 (on Matrigel)], from the protein intensity map, are also displayed.

cells (human hepatocellular carcinoma) treated with increasing doses of an environmental toxin. In this study, posttreated cells were detached from culture substratum using trypsin, followed by two washes in PBS then kept as a suspension at 4°C and measured with SR-FTIR within 24 hours. Although the cool temperature minimizes the enzymatic effects of autolysis, without fixation there may be biochemical differences between cells at the zero time point compared to cells stored in PBS for 24 hours, particularly for glycogen stores, since the cells were in a nutrient deficient environment. Nevertheless, it has been shown that spectra from these cells showed spectroscopic changes in the ratio of peak intensities (1082 cm^{-1}/1236 cm^{-1}) that could be correlated with increasing doses of toxin exposure. In a situation where the effect of time on cell biochemistry has not been assessed, one must be careful when associating spectral changes to the direct result of a condition administered to the cell. However, if spectral discrimination is achieved following randomized sampling of cells exposed to each of the different conditions, then this may evaluate whether live cell spectra are significantly influenced by their duration in nutrient deficient media.

More recently, specialised equipment for maintaining live T-1 cells (aneuploid cells from human kidney tissues) on gold-coated slides for in situ SR-FTIR analysis has been investigated by Holman et al.[58] A mini-incubator system was used to sustain cell viability by maintaining a humidified environment, so as to retain a thin layer of growth medium around the cell during SR-FTIR measurements. The mini-incubator was temperature controlled at 37°C via circulating water from a water bath, and infrared transparent CaF$_2$ windows on the top cover were separately temperature controlled to avoid condensation. Using this incubator, the authors investigated any possible cytoxic effects that may be elicited in the cell by exposure to the SR-IR radiation. Using the Alcian blue exclusion assay, it was found that the cells showed negative staining 24 hours after exposure to 20 minutes of SR-IR radiation, which indicated that the cell membranes remained intact. The effects of 20-minute SR-IR exposure on cell metabolism was assessed using the MTT assay. This confirmed that both control cells (not exposed to SR radiation) situated nearby to exposed cells and exposed cells produced mitochondrial dehydrogenases, which is associated with glycolysis and indicates negligible effects on this metabolic pathway. Finally, colony-forming assays demonstrated that there was no long-term damage as a result of SR-IR exposure.

Although these assays could not have been carried out in situ within the mini-incubator, it is encouraging to find that the length of time (20 minutes) that these cells were placed in the incubator had no short-term or long-term effects. Furthermore, the researchers report that consecutive SR-FTIR spectra obtained at 10-minute intervals for 30 minutes exhibited an unchanging IR spectrum to within 0.005 A.U. across the entire mid-IR spectral range. This provides supporting evidence for the justification of this experimental setup for measuring single-point SR-FTIR spectra from single living cells. Since the experiment was carried out for only 30 minutes, changes in the spectrum over an extended time-period, which is required to obtain cell maps, is unknown. However, using a different experimental design for the sample compartment, Miljkovic et al. reported no spectral changes in spectra collected from live cells when data were collected every 30 minutes for 3 hours.[59]

Miljkovic et al.[59] collected FTIR images at 6.25×6.25 μm pixel resolution of living HeLa cells (cervical cancer) using the linear array detector Spotlight microspectrometer equipped with a glow bar source. In this study, different approaches were used to prepare cells for transflection and transmission mode analysis: For transmission mode, live cells in growth medium were placed into a 6-μm pathlength CaF_2 liquid cell. This preparation resulted in the compression and rupturing of larger cells; however, the smaller cells were left intact (Fig. 3.13b). For transflection measurements, cells in buffered saline solution were placed as a drop onto a MirrIR slide and a CaF_2 or BaF_2 coverslip was placed on top. This preparation also involved the use of a 5-μm Teflon spacer to prevent the coverslip touching the MirrIR slide. Raw spectra obtained from FTIR images of HeLa cells using both modes of analysis showed an unusual amide I to amids II ratio (Fig. 3.13a and 3.13c), which was more apparent in transflection mode spectra (Fig. 3.13a). This was attributed to the longer path length (10 μm) in the transflection mode measurement. The origin of this distorted amide I to amide II ratio was determined to be due to overcompensation of the water background from the cell spectrum, since the cell contains less water than the surrounding medium or buffer.

The authors suggested that correction of the amide I and II peaks could be carried out by visually fitting a scaled buffer spectrum to the raw cell spectrum, until the resulting-corrected spectrum shows a

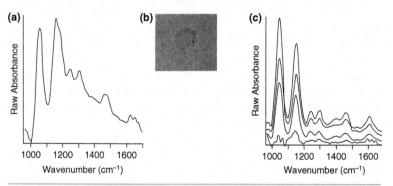

Figure 3.13 Reflection/absorption spectrum and (b) visual image of a HeLa cell in BSS buffer; (c) Absorption spectra taken from a HeLa cell in growth medium. All spectra were collected with 128 scans at 4 cm^{-1} spectral resolution. (*Adapted from Ref. 59.*)

normal amide A envelope. Although subjective, Miljkovic et al.[59] demonstrated that it was possible to obtain a protein intensity image in which the HeLa cell displayed expected high-protein intensity, centered at its nucleus.

In contrast to using cells in suspension as in Miljkovic's et al. Study,[59] cells used by Moss et al.[57] were cultured directly onto CaF_2 plates. This plate was placed into a liquid cell consisting of a 15-μm Teflon spacer, providing a pathlength of 11 to 12 μm and maintained at 35°C. A constant flow of cell culture medium was passed through the cell at a rate of 230 μL/h. As in Miljkovic's et al. Study,[59] a background spectrum of growth medium was collected in a cell-free region of the sample and ratioed to the cell spectrum. There was high reproducibility between SR-FTIR spectra obtained from 10 individual fibroblast cells when a spectrum of each cell was acquired every 24 minutes for 2 hours. Although intrasampling differences were observed between cells, these were very much smaller than the standard deviation of repeated measurements for each cell. Moss et al.[57] provides further support for the low-spectral variance observed for live cell FTIR spectra, when collected within the first few hours of transfer to the sample analysis chamber.

In agreement with the study by Miljkovic et al.[59] a distorted amide I to amide II intensity ratio was observed by Moss et al.[57] However, since the spectrum of background water is different to that of water bound to macromolecules, it was suggested that it is not possible to accurately eliminate this background absorbance. The authors also suggest that if the goal of the experiment is to obtain a spectrum from the same position of the exact same cell, before and after administration of a stimulus, then the difference between the spectra can be resolved even in the presence of background water. Additionally, it was found in this study that nonconfluent cells could migrate out of the measuring SR beam. Moss et al.[57] suggest that this could be minimized by placing the cell into a well.

Raman Studies

The spatial resolution of Raman spectroscopy is inherently higher than that of FTIR due to the shorter wavelength of the excitation radiation (the diffraction limit is generally given as $\sim\lambda/2$). An image obtained by Raman microspectroscopy requires raster scanning a focused laser beam across the cell. Using this mode of data collection, an increase in spatial resolution, which is a function of step size and beam diameter, also increases the time for chemical mapping. Although in FTIR studies it has been shown that for up to 3 hours, spectral changes are not observed at the *whole-cell level* (see section "FTIR Studies"), previously reported Raman maps of living cells have required ≥3 hours collection times.[60] At the *subcellular level* one might expect that biochemical changes could occur within this period for a given sampling point. However, it has been shown that localization and spectral distinction between cytoplasmic and nuclear compartments in living cells was not effected in Raman maps of two different cell types (human osteogenic sarcoma cell and human embryonic lung epithelial fibroblast) that required long collection times (up to 20 hours).[60] In this study by Krafft et al.,[60] difference spectra of the cytoplasm and nucleus identified the important discriminatory variables that distinguished these compartments: nucleic acids and lipids. Infact, analysis of living cells grown on quartz and analyzed in media, provided Raman spectra containing features of subcellular components that were more pronounced than those obtained from frozen-hydrated cells. This was attributed to conformational changes and aggregation of biomolecular constituents caused by the freeze-drying process and which are not present when the cells are analyzed hydrated.

The acquisition time for a Raman map of a cell can be improved by increasing the sensitivity of the technique. Kneipp et al.[61–62] have demonstrated that enhanced Raman signals (10 to 14 orders of magnitude) for the native constituents of a cell can be achieved by incorporating colloidal gold particles into the cell. The gold nanoparticles give rise to surface-enhanced Raman scattering (SERS), where Raman molecules close to the vicinity of the nanoparticles experience electronic interaction with enhanced optical fields due to resonances of the applied optical fields with the surface plasmon oscillations of the metallic nanostructures. This process results in an increase in the scattering cross section of the Raman molecules, which enabled Raman maps to be collected at 1-μm lateral resolution (1 second for one mapping point), where each spectrum in the map consisted of the spectral region 400 to 1800 cm^{-1}.[61]

Delivery of the nanoparticles into the cell interior can be carried out in two ways, sonication or fluid-phase uptake.[61] The fluid-phase uptake method involves supplementing the culture medium with colloidal gold suspensions (60 nm in size), 24 hours prior to experiments. The cells internalize the nanoparticles through endocytosis and without further induction (Fig. 3.14a).[62] This can result in the formation of colloidal aggregates inside the cell that may be 100 nm to a few micrometers in size.[61] The cells are washed in buffer to remove nonincorporated nanoparticles and replaced in fresh buffer for SERS analysis. The second method of delivering nanoparticles into the cell is by sonication, where rupture of the cell membrane enables an influx of nanoparticles before self-annealing within a few seconds. However, in low-intensity ultrasound mediated gene transfection, it has been found that sonication can induce stress responses in the cell[63] and so should be carried out 24 hours prior to experiment to allow enough time for the cell to repair any damage.

The authors report that incorporation of the nanoparticles into the cell using the fluid-phase uptake method did not yield any visible changes in growth characteristics such as signs of apoptosis or cell

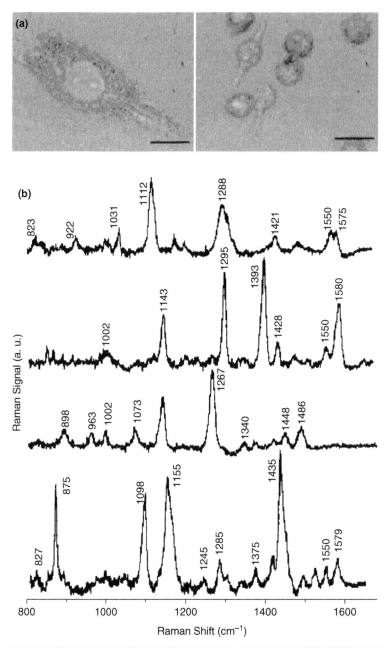

FIGURE 3.14   (a) Cells of a fibroblast cell line, NIH/3T3 (nonphagocytic) (left), and a macrophage cell line, J774 (phagocytic) (right), after uptake of gold nanoparticles; particle accumulations are visible as black dots inside the cells (scale bars = 20 μm). (b) Examples of SERS spectra acquired from NIH/3T3 cells after 3-hour incubation with gold nanostructures, excitation wavelength 830 nm, 1 second collection time. (*Reproduced from Ref. 62 with permission from American Chemical Society.*)

detachment when compared to a control monoculture.[61] Raman spectra obtained from different locations in the nanoparticle-doped cell gave rise to very different spectral profiles, illustrating the biochemical heterogeneity of the cell (Fig. 3.14*b*).

As well as the application of Raman microspectroscopy to live cell imaging, the technique has also been applied for the phenotypic typing of live cells. Krishna and colleagues[46] collected Raman spectra of two different cell lines and their respective drug resistant analogues: Breast cancer cell line MCF7 and its subclone resistant to verapamil (MCF7/VP) and promyelocytic leukemia HL60S cell line and its multi-drug-resistant phenotypes (HL60/DOX: resistant to doxorubicin; HL60/DNR: resistant to daunorubicin). PCA analyses of these Raman spectra were able to generate score plots that showed clustering and separation for each cell line and its drug-resistant clone. The authors also carried out these experiments using FTIR and found that classification, discrimination as well as reproducibility was greater using this method. However, with the view of translating this type of analysis to clinical application, it would be desirable for the chosen method to incorporate minimal sample preparation for high-throughput screening. For the Raman study, a cell pellet consisting of 1.10^6 cells (washed in 0.9 percent NaCl) was used directly for spectroscopic analysis, whereas for the FTIR experiments a time-limiting step was required that consisted of drying a cell suspension under mild vacuum onto a zinc selenide sample wheel.

The sample preparation method used by Krishna et al.[46] for the Raman study requires fast data acquisition times, since live cell pellets surrounded by a thin layer of aqueous buffer may undergo biochemical changes over time. In their study, 25 spectra were collected for each pellet, where one spectrum took 4.5 minutes to collect. Thus, between the first and final spectrum there was a time lag of 1 hour and 53 minutes. If biochemical changes did occur during this period, then it may have contributed to the lower discriminatory power achieved using Raman spectra in this study. Comparatively, the FTIR spectra were obtained from dried cells, providing perhaps a background interference that is constant over all cells and so differences due to MDR or drug sensitive phenotypes could be more readily resolved. This provides further evidence that possible artifacts from the drying process do not hamper spectroscopic differentiation between cells of differing phenotypes (as mentioned in Sec. 3.3.2).

3.4   Summary

It is clearly evident that sample preparation is a key aspect of the experimental design, where thorough dissection of the issues involved in the preparation of cells or tissue for spectroscopic analysis is essential to yield reproducible and biochemically relevant results. The continuing developments in tissue preservation for optimum detection of specific biomolecules using emerging bioanalytical approaches will shape the tissue repositories of the future. These developments will also impact biomedical vibrational spectroscopy, since this technology can play an important role in determining the biochemical basis underpinning disease progression. Nevertheless, it is apparent that existing tissue banks have proven adequate for FTIR and Raman studies of tissue pathologies, providing high-classification power. This is despite the fact that spectral artifacts exist as a result of tissue processing or section postprocessing. These spectral artifacts can be due to protein depolymerisation or a change in the lipid to protein ratio for dried cryosections or the case of deparaffinized specimens, due to residual paraffin, coagulation of proteins and loss of lipids.

Some of these artifacts can be minimized. Protein depolymerization of freeze-dried/thawed cryosections can be reduced by careful attention to the cryogen used for initial tissue snap-freezing as well as cryomicrotomy and freeze-drying environmental temperatures. Other artifacts such as residual paraffin can now be confidently removed in the light of work carried out by Faolain et al. [37] It appears that deparaffinization using hexane for >24 hours is an appropriate method for this purpose and has wider implications in immunohistochemical pathology. However, this protocol can be time limiting and so less rigorous protocols may be sufficient where spectroscopic markers for pathological assessment do not overlap with paraffin signals.

The early work of Fox et al.[12] investigating the binding time of formalin to tissue, may be of significance to those vibrational spectroscopists presently using formalin-fixed cells in imaging or biomechanistic studies, since the effects of formalin-binding time on cell spectra have not been assessed. Chemical fixation with formalin has been shown to produce spectral images of cells, where Raman or FTIR signals of various biomolecules localize to subcellular compartments that are expected to give rise to these signals. Tailored chemical fixation protocols for vibrational spectroscopy may, however, be necessary in some instances. For example, in Sec. 3.3.1, the lipid component of adipocytes is volatile in air and requires fixation with OsO_4, which itself does not absorb in the mid-IR region. Coupled to paraformaldehyde, which would influence the FTIR spectrum to a lower degree than to the relatively larger molecular weight polymers of glutaraldehyde, one can obtain a well-preserved sample for spectroscopic analysis. The lengthy procedure of OsO_4 paraformaldehyde with critical-point-drying is appropriate where experiments are capturing a cellular event at time frames that are far apart. However, for shorter time frames (intervals of 15 minutes), faster fixation methods are required and formalin has so far been proven to be adequate. Interestingly, it has also been demonstrated that air-dried cells following exposure to a pharmacological drug or stimulus can also produce spectral changes that may be associated with response to the condition administered. This is assuming that the underlying stress responses of the cell may be a constant between conditions, a hypothesis that requires further testing.

The effects of basic growth media (RPMI and Ham's F-12) on two prostate cancer epithelial cell lines does not affect its phenotypic classification using FTIR. Whether this is the case for other cell types grown in these media requires further study. However, commonly used substrates for spectroscopy do influence the cells at the whole cell level as well as at the molecular level. It was concluded that gelatin-coated quartz and MirrIR slides may provide the best approach for long-term cell viability. On thicker/highly concentrated protein-based biological supports, it was found that optical artifacts can manifest and these were supported by time-lapse observations.

There have been encouraging results reported within the context of live cell experiments using FTIR. The collective demonstration of biochemical stability for different cell types (T-1, HeLa and fibroblasts) by the various research groups working within this field,[57–59] together with Moss et al.[57] suggestion that correction for water absorbance may not be necessary, suggests that future FTIR studies may be able to measure early biochemical responses of single living cells to stimuli. Raman-based live cell studies have shown excellent prospects for cell phenotyping as well as probing the distributions of native biomolecules of a cell with high sensitivity and spatial resolution and without the requirement for exogenous labeling.

Acknowledgments

Support was received from the Association for International Cancer Research (AICR Grant number 04-518) and The Prostate Cancer Foundation during the writing of this article and some of the experiments described within it. We gratefully thank Dr. Stephen Murray (Paterson Institute for Cancer Research, UK) for use of the time-lapse video microscope.

References

1. L. Chiriboga, P. Xie, H. Yee, V. Vigorita, D. Zarou, D. Zakim, and M. Diem, "Infrared Spectroscopy of Human Tissue. I. Differentiation and Maturation of Epithelial Cells in the Human Cervix," *Biospectroscopy*, **4**:47–53, 1998.
2. H. Y. N. Holman, M. C. Martin, E. A. Blakely, K. Bjornstad, and W. R. McKinney, "IR Spectroscopic Characteristics of a Cell Cycle and Cell Death Probed by Synchrotron Radiation Based Fourier Transform IR Spectromicroscopy," *Biopolymers (Biospectroscopy)*, **57**:329–335, 2000.
3. P. Lasch, M. Boese, A. Pacifico, and M. Diem, FT-IR Spectroscopic Investigations of Single Cells on the Subcellular Level," *Vibrational Spectroscopy*, **28**:147–157, 2002.
4. P. Lasch, A. Pacifico, and M. Diem, "Spatially Resolved IR Microspectroscopy of Single Cells." *Biopolymers (Biospectroscopy)*, **67**:335–338, 2002.
5. D. Yang, D. J. Castro, I. E. El-Sayed, M. A. El-Sayed, R. E. Saxton, and N. Y. Zhang, "A Fourier-Transform Infrared Spectroscopic Comparison of Cultured Human Fibroblast and Fibrosarcoma Cells: A New Method for Detection of Malignancies," *Journal of Clinical Laser Medicine & Surgery*, **13**:55–59, 1995.

6. A. Salman, J. Ramesh, V. Erukhimovitch, M. Talyshinsky, S. Mordechai, and M. Huleihel, "FTIR Microspectroscopy of Malignant Fibroblasts Transformed by Mouse Sarcoma Virus," *Journal of Biochemical Biophysical Methods*, **55**:141–153, 2003.

7. M. J. Tobin, M. A. Chesters, J. M. Chalmers, F. J. M. Rutten, S. E. Fisher, I. M. Symonds, A. Hitchcock, R. Allibone, and S. Dias-Gunasekara, "Infrared Microscopy of Epithelial Cancer Cells in Whole Tissues and in Tissue Culture, Using Synchrotron Radiation," *Faraday Discussions*, **126**:27–38, 2004.

8. H. P. Wang, H. C. Wang, and Y. J. Huang, "Microscopic FTIR Studies of Lung Cancer Cells in Pleural Fluid," *The Science of Total Environment*, **204**:283–287, 1997.

9. N. Jamin, P. Dumas, J. Moncutt, W.-H. Fridman, J.-L. Teillaud, G. L. Carr, and G. P. Williams, "Highly Resolved Chemical Imaging of Living Cells by Using Synchrotron Infrared Microspectrometry," *Proceedings of the National Academia of Science USA*, **95**:4837–4840, 1998.

10. J. A. Kiernan, "Formaldehyde, Formalin, Paraformaldehyde and Glutaraldehyde: What They Are and What They Do," *Microscopy Today*, **00-1**:8–12, 2000.

11. D. Jones, "Introduction," in: *Fixation in Histochemistry*, P. J. Stoward (ed.), Chapman and Hall, London, 1973, pp. 2–7.

12. P. B. Medawar, "The Rate of Penetration of Fixatives", *Journal of the Royal Microscopy Society*, **61**:46, 1941.

13. C. H. Fox, F. B. Johnson, J. Whiting, and P. P. Roller, "Formaldehyde Fixation," *The Journal of Histochemistry and Cytochemistry*, **33**:845–853, 1985.

14. G. R. Turbett and L. N. Sellner, "The Use of Optimal Cutting Temperature Compound Can Inhibit Amplification by Polymerase Chain Reaction," *Diagnostic Molecular Pathology*, **6**:298–303, 1997.

15. J. W. Gillespie, "Evaluation of Non-Formalin Tissue Fixation for Molecular Profiling Studies," *American Journal of Pathology*, **160**:449–457, 2002.

16. P. G. L. Andrus and R. D. Strickland, "Cancer Grading by Fourier Transform Infrared Spectroscopy," *Biospectroscopy*, **4**:37–46, 1998.

17. S. Takahashi, A. Satomi, K. Yano, H. Kawase, T. Tanimizu, Y. Tuji, S. Murakami, and R. Hirayama, "Estimation of Glycogen Levels in Human Colorectal Cancer Tissue: Relationship with Cell Cycle and Tumour Outgrowth," *Journal of Gastroenterology*, **34**:474–480, 1999.

18. D. C. Fernandez, R. Bhargava, S. M. Hewitt, and I. W. Levin, "Infrared Spectroscopic Imaging for Histopathological Recognition," *Nature Biotechnology*, **23**:469–474, 2005.

19. E. O. Faolain, M. B. Hunter, J. M. Byrne, P. Kelehan, M. McNamara, H. J. Byrne, and F. M. Lyng "A Study Examining the Effects of Tissue Processing on Human Tissue Sections Using Vibrational Spectroscopy," *Vibrational Spectroscopy*, **38**(1–2):121–127, 2005.

20. M. G. Shim and B. C. Wilson, "The Effects of Ex Vivo Handling Procedures on the Near-Infrared Raman Spectra If Normal Mammalian Tissues," *Photochemistry and Photobiology*, **63**:662–671, 1996.

21. A. T. Tu "Peptide Backbone Conformation and Microenvironment of Protein Side Chains," in: *Spectroscopy of Biological Systems*, Vol. 13., R.J.H. Clark and R.E. Hester (eds.), John Wiley & Sons, New York, 1986, pp. 47–112.

22. Z. W. Huang, A. McWilliams, S. Lam, J. English, D. I. McLean, H. Lui, and H. Zeng, "Effect of Formalin Fixation on the Near-Infrared Raman Spectroscopy of Normal and Cancerous Human Bronchial Tissues," *International Journal of Oncology*, **23**:649–655, 2003.

23. S. Aparicio, S. B. Doty, N. P. Camacho, E. P. Paschalis, L. Spevak, R. Mendelsohn, and A. L. Boskey, "Optimal Methods for Processing Mineralized Tissues for Fourier Transform Infrared Microspectroscopy," *Calcified Tissue International*, **70**:422–429, 2002.

24. N. L. Pleshko, A. L. Boskey, and R. Mendelsohn, "An FT-IR Microscopic Investigation of the Effects of Tissue Preservation on Bone," *Calcified Tissue International*, **51**:72–77, 1992.

25. T. J. Mason and T. J. O'Leary, "Effects of Formaldehyde Fixation on Protein Secondary Structure: A Calorimetric and Infrared Spectroscopic Investigation," *Journal of Histochemistry and Cytochemistry*, **39**:225–229, 1991.

26. J. L. Stephenson, "Ice Crystal Growth During the Rapid Freezing of Tissues," *The Journal of Biophysical and Biochemical Cytology*, **2**:45–52, 1956.
27. N. Stone, C. Kendall, J. Smith, P. Crow, and H. Barr, "Raman Spectroscopy for Identification of Epithelial Cancers," *Faraday Discussions*, **126**:141–157, 2003.
28. N. Stone, M. C. H. Prieto, P. Crow, J. Uff, and A. W. Ritchie, "The Use of Raman Spectroscopy to Provide an Estimation of the Gross Biochemistry Associated with Urological Pathologies," *Analytical and Bioanalytical Chemistry*, **387**: 1657–1668, 2007.
29. M. Jackson, J. R. Mansfield, B. Dolenko, R. L. Somorjai, H. H. Mantsch, and P. H. Watson, "Classification of Breast Tumours by Grade and Steroid Receptor Status Using Pattern Recognition Analysis of Infrared Spectra," *Cancer Detection and Prevention*, **23**:245–253, 1999.
30. M. Meurens, J. Wallon, J. Tong, H. Noel, and J. Haot, "Breast Cancer Detection by Fourier Transform Infrared Spectrometry," *Vibrational Spectroscopy*, **10**:341–346, 1996.
31. G. Muller, W. Wasche, U. Bindig, and K. Liebold, "IR-Spectroscopy for Tissue Differentiation in the Medical Field," *Laser Physics*, **9**:348–356, 1999.
32. R. Wiens, M. Rak, N. Cox, S. Abraham, B. H. J. Juurlink, W. M. Kulyk, and K. M. Gough, "Synchrotron FTIR Microspectroscopic Analysis of the Effects of Anti-Inflammatory Therapeutics on Wound Healing in Laminectomized Rats," *Analytical and Bioanalytical Chemistry*, **387**:1679–1689, 2007.
33. C. P. Schultz, "The Potential Role of Fourier Transform Infrared Spectroscopy and Imaging in Cancer Diagnosis Incorporating Complex Mathematical Methods," *Technology in Cancer Research and Treatment*, **1**:95–104, 2002.
34. C. Beleites, G. Steiner, M. G. Sowa, R. Baumgartner, S. Sobottka, G. Schackert, and R. Salzer, "Classification of Human Gliomas by Infrared Imaging Spectroscopy and Chemometric Image Processing," *Vibrational Spectroscopy*, **38**:143–149, 2005.
35. R. K. Sahu, S. Argov, A. Salman, U. Zelig, M. Huleihel, N. Grossman, J. Gopas, J. Kapelushnik, and S. Mordechai, "Can Fourier Transform Infrared Spectroscopy at Higher Wavenumbers (Mid IR) Shed Light on Biomarkers?" *Journal of Biomedical Optics*, **10**:05017–05027, 2005.
36. E. Gazi, M. Baker, J. Dwyer, N. P. Lockyer, P. Gardner, J. H. Shanks, R. S. Reeve, C. A. Hart, M. D. Brown, and N. W. Clarke. "A Correlation of FTIR Spectra Derived from Prostate Cancer Tissue with Gleason Grade and Tumour Stage," *European Urology*, **50**:750–761, 2006.
37. E. O. Faolain, M. B. Hunter, J. M. Byrne, P. Kelehan, H. A. Lambkin, H. J. Byrne, and F. M. Lyng, "Raman Spectroscopic Evaluation of Efficacy of Current Paraffin Wax Section Dewaxing Agents," *Journal of Histochemistry and Cytochemistry*, **53**:121–129, 2005.
38. E. Gazi, J, Dwyer, N. P. Lockyer, P. Gardner, J. Miyan, C. A. Hart, M. D. Brown, and N. W. Clarke, "Fixation Protocols for Sub-Cellular Imaging Using Synchrotron Based FTIR-Microspectroscopy," *Biopolymers*, **77**:18–30, 2005.
39. C, Matthaus, S. Boydston-White, M. Miljkovic, M. Romeo, and M. Diem, "Raman and Infrared Microspectral Imaging of Mitotic Cells," *Applied Spectroscopy*, **60**:1–8, 2006.
40. E. Gazi, J. Dwyer, N. P. Lockyer, P. Gardner, J. Miyan, C. A. Hart, M. D. Brown, and N. W. Clarke, "A Study of Cytokinetic and Motile Prostate Cancer Cells Using Synchrotron-Based FTIR Microspectroscopic Imaging," *Vibrational Spectroscopy*, **38**:193–201, 2005.
41. C. Krafft, T. Knetschke, R. H. W. Funk, and R. Salzer, "Identification of Organelles and Vesicles in Single Cells by Raman Microspectroscopic Mapping," *Vibrational Spectroscopy*, **38**:85–93, 2005.
42. J. A. Kieran, "Fixation," Chap. 2, in: *Histological and Histochemical Methods: Theory & Practice*, Pergamon Press, Oxford, UK, 1990, pp. 10–35.
43. E. Gazi, P. Gardner, N. P. Lockyer, C. A. Hart, N. W. Clarke, and M. D. Brown "Probing Lipid Translocation between Adipocytes and Prostate Cancer Cells with Imaging FTIR Microspectroscopy," *Journal of Lipid Research*, **48**:1846–1856, 2007.

44. E. Gazi, T. J. Harvey, P. Gardner, N. P. Lockyer, C. A. Hart, N. W. Clarke, and M. D. Brown, "A FTIR Microspectroscopic Study of the Uptake and Metabolism of Isotopically Labelled Fatty Acids by Metastatic Prostate Cancer," *Vibrational Spectroscopy*, 2008, in preparation.

45. J. Sule-Suso, D. Skingsley, G. D. Sockalingum, A. Kohler, G. Kegelaer, M. Manfait, and A. J. El Haj "FT-IR Microspectroscopy as a Tool to Assess Lung Cancer Cells Response to Chemotherapy," *Vibrational Spectroscopy*, 38:179–184, 2005.

46. M. C. Krishna, G. Kegelaer, I. Adt, S. Rubin, V. B. Kartha, M. Manfait, and G. D. Sockalingum, "Characterisation of Uterine Sarcoma Cell Lines Exhibiting MDR Phenotype by Vibrational Spectroscopy," *Biochimica et Biophysica Acta*, 1726:160–167, 2005.

47. P. Crow, B. Barrass, C. Kendell, M. Hart-Prieto, M. Wright, R. Persad, and M. Stone, "The Use of Raman Spectroscopy to Differentiate Between Different Prostatic Adenocarcinoma Cell Lines," *British Journal of Cancer*, 92:2166–2170, 2005.

48. C. M. Krishna, G. D. Sockalingum, G. Kegelaer, S. Rubin, V. B. Kartha, and M. Manfait, "Micro-Raman Spectroscopy of Mixed Cancer Cell Populations," *Vibrational Spectroscopy*, 38:95–100, 2005.

49. E, Gazi, J. Dwyer, P. Gardner, A. Ghanbari-Siahkali, A. Wade, J. Miyan, N. P. Lockyer, et al., "Applications of FTIR-Microspectroscopy to Benign Prostate and Prostate Cancer," *Journal of Pathology*, 201:99–108, 2003.

50. T. J. Harvey E. Gazi, N. W. Clarke, M. D. Brown, E. C. Faria, R. D. Snook, and P. Gardner, "Discrimination of Prostate Cancer Cells by FTIR Photo-Acoustic Spectroscopy," *Analyst*, 132:292–295, 2007.

51. T. J. Harvey, E. Gazi, R. D. Snook, N. W. Clarke, M. Brown, and P. Gardner, "The classification of Prostate Cancer Cell Lines Using FTIR Microspectroscopy and Multivariate Chemomectric Analysis," *Analyst*, DOI: 10.1039/b903249e, 2009.

52. S. Z. Haslam and T. L. Woodward, "Host Microenvironment in Breast Cancer Development: Epithelial-Cell–Stromal-Cell Interactions and Steroid Hormone Action in Normal and Cancerous Mammary Gland," *Breast Cancer Research*, 5:208–215, 2003.

53. C. Krafft, R. Salzer, S. Seitz, C. Ern, and M. Schieker, "Differentiation of Individual Human Mesenchymal Stem Cells Probed FTIR Microscopic Imaging," *Analyst*, 132:647–653, 2007.

54. A. D. Meade, F. M. Lyng, P. Knief, and H. J. Byrne, "Growth Substrate Induced Functional Changes Elucidated by FTIR and Raman Spectroscopy in In-Vitro Cultured Human Keratinocytes," *Analytical and Bioanalytical Chemistry*, 387:1717–1728, 2007.

55. J. Lee, E. Gazi, J. Dwyer, N. P. Lockyer, M. D. Brown, N. W. Clarke, and P. Gardner, "Optical Artifacts in Transflection Mode FTIR Microspectroscopic Images of Single Cells on a Biological Support: Does Rayleigh Scattering Play a Role?" *Analyst*, 132:750–755, 2007.

56. H. Y. N. Holman, R. Goth-Goldstein, M. C. Martin, M. L. Russell, and W. R. McKinney, "Low-Dose Responses to 2,3,7,8-Tetrachlorodibenzo-*p*-Dioxin in Single Living Human Cells Measured by Synchrotron Infrared Spectromicroscopy," *Environmental. Science and Technology*, 34:2513–2517, 2000.

57. D. Moss, M. Keese, and R. Pepperkok. IR Microspectroscopy of Live Cells," *Vibrational Spectroscopy*, 38:185–191, 2005.

58. H. Y. N. Holman, M. C. Martin, W. R. McKinney, "Synchrotron-Based FTIR Spectromicroscopy: Cytotoxicity and Heating Considerations," *Journal of Biomedical Physics*, 29:275–286, 2003.

59. M, Miljkovic, M. Romeo, C. Matthaus, and M. Diem, "Infrared Microspectroscopy of Individual Human Cervical Cancer (HeLa) Cells Suspended in Growth Medium," *Biopolymers*, 74:172–175, 2004.

60. C. Krafft, T. Knetschke, A. Siegner, R. H. W. Funk, and R. Salzer, "Mapping of Single Cells by Near Infrared Raman Microspectroscopy," *Vibrational Spectroscopy*, 32:75–83, 2003.

61. K, Kneipp, A. S. Haka, H. Kneipp, K. Badizadegan, N. Yoshizawa, C. Boone, K. E. Shafer-Peltier, J. T. Motz, R. R. Dasari. and M. S. Feld, "Surface-Enhanced

Raman Spectroscopy in Single Living Cells Using Gold Nanoparticles," *Applied Spectroscopy*, **56**:150–154, 2002.

62. K. Kneipp and H. Kneipp, "Surface-Enhanced Raman Scattering in Local Optical Fields of Silver and Gold Nanoaggregates—From Single-Molecule Raman Spectroscopy to Ultrasensitive Probing in Live Cells," *Accounts of Chemical Research*, **39**:443–450, 2006.

63. L. B. Feril, T. Kondo, Y. Tabuchi, R. Ogawa, Q. L. Zhao, T. Nozaki, T. Yoshida, N. Kudo, and K. Tachibana, "Biomolecular Effects of Low-Intensity Ultrasound: Apoptosis, Sonotransfection, and Gene Expression," *Japanese Journal of Applied Physics*, **46**:4435–4440, 2007.

CHAPTER 4

Evanescent Wave Imaging

Heather J. Gulley-Stahl, André J. Sommer

Molecular Microspectroscopy Laboratory
Department of Chemistry and Biochemistry
Miami University
Oxford, Ohio, USA

Andrew P. Evan

Department of Anatomy
Indiana University School of Medicine
Indianapolis, Indiana, USA

4.1 Introduction

Optical microscopy has been employed for well over 344 years to study tissue specimens at the cellular level.[1] The optical microscope aids the pathologist in this task by permitting the analysis of spatial domains as small as 300 nm (0.3 μm). In diagnosing a disease, the pathologist looks for structural changes in the cells or tissue. Alternatively, one can look for chemical agents that enhance the contrast for a given structure or signal the presence of a disease. This latter method of detection has been employed for well over 298 years by using dyes or stains that are specific for a disease state or chemical variant associated with the disease.[2] Several problems associated with histopathology stem from the fact that there may not be a stain specific for the disease. In addition, the staining procedure usually involves multiple steps during which the material of interest may be lost or destroyed. In an effort to circumvent these problems, infrared methods of detection have been employed to gain similar information. Here the chemical which signals a disease is detected directly and all that is required is the preparation of a thin tissue section. However, infrared wavelengths are an order of magnitude longer than visible wavelengths so the spatial domains accessible using this method are typically an order of magnitude larger (~3 μm). To address this short-coming, infrared microspectroscopists have employed immersion methods commonly

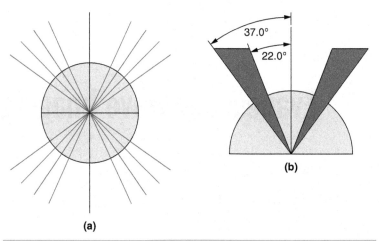

(a)

(b)

FIGURE 4.1   Immersion (*a*) and ATR (*b*) configurations for infrared microspectroscopy.

used by optical microscopists to improve the spatial resolution of the method. Instead of liquid immersion oils, however, the infrared variant employs ZnSe or Ge hemispheres as the immersion medium (see Fig. 4.1). In these attempts, the sample is interleaved between two hemispheres for transmission measurements or placed at the plano surface of a single hemisphere for attenuated total internal reflection (ATR) measurements. Although immersion transmission infrared methods have been reported, sample thickness requirements and difficulty with coupling light through the sample do not make the method optimal for thin tissue sections. A method which is inherently an immersion method and solves the requirement for specially prepared samples is ATR imaging, which is the topic of this chapter. Finally, although this chapter focuses on infrared microspectroscopy, parallel developments in visible and fluorescence microscopy using solid immersion lenses took place at or about the same time. When possible, references will be given to highlight these developments.

4.2   Theoretical Considerations

The spatial characteristics of a focused beam of light can be estimated from diffraction theory. Equation (4.1) gives the diffraction limited diameter (*x*, *y*) for light focused to a point with a lens or objective

$$d = \frac{1.22\lambda}{n_1 \sin\theta} \tag{4.1}$$

where λ = wavelength of light
θ = half angle acceptance of the optic
n_1 = refractive index of the medium in which the sample is immersed

The depth over which this focus is maintained is given by z

$$z = \frac{4.0\lambda}{n_1^2 \sin^2 \theta} \qquad (4.2)$$

Using values typical of an infrared microscope (e.g., $\sin \theta = 0.6$) in air, d becomes $\sim 2\lambda$ and $z \sim 11\lambda$. Neglecting "z" for the moment, if one were to immerse the sample in germanium ($n_1 = 4.0$), d could be reduced to $\sim 0.5\lambda$, which is a significant improvement. Thus, for a transmission measurement with the sample immersed between two germanium hemispheres, one would expect a 4X improvement in spatial resolution. However, when taking z into consideration, the short wavelength limit (2.5 µm) dictates that the sample thickness be less than 1.7 µm in order that the focused beam width is not degraded when the radiation transmits through the sample. This thickness is difficult to achieve via normal methods used to prepare thin tissue sections. More problematic is that coupling of light through the hemispheres becomes more difficult as the index increases. Carr and Lavalle et al. demonstrated the benefits of transmission immersion infrared microspectroscopy using two ZnSe hemispheres.[3,4] However, in the work of Lavalle, the method required Nujol oil to efficiently couple light through the hemisphere/sample/ hemisphere interface.

A solution for many of these problems is to employ ATR reflection. As depicted in Fig. 4.1, light from the objective is brought into the hemisphere beyond the critical angle [$\theta_c = \sin^{-1} (n_{sample}/n_{hemisphere})$]. In doing so the light is internally reflected at the hemisphere/sample interface. Although commonly referred to as "total internal reflection," a better term is frustrated internal reflection since some of the light penetrates into the sample where it can undergo absorption. The depth to which light penetrates into the sample is given by

$$d_p = \frac{\lambda}{2\pi n_{hemisphere}(\sin^2 \theta - (n_{sample}/n_{hemisphere})^2)^{1/2}} \qquad (4.3)$$

where n_{sample} = refractive index of the sample
  $n_{hemisphere}$ = refractive index of the internal reflection element
    (IRE)
    θ = incident angle of light coupled into the hemisphere

It should be noted that penetration depth is an arbitrary value, which corresponds to the point where the electric field intensity drops to $1/e$ of its value at the surface (interface).[5]

Several major benefits arise from the use of the ATR configuration. First, sample thickness is not an issue, since the penetration depth over the range of wavelengths employed with the above parameters is no greater than 2.2 µm. This limited path length through the sample means that highly absorbing materials can be studied, as well as samples that are relatively thick. The only limiting requirement is that the internal reflection element be in intimate contact with the sample. Another benefit arises from the fact that the volumetric resolution of the measurement is extremely good. The diffraction-limited volume of sample illuminated can be estimated as a cone, where the base of the cone is the diffraction-limited spot size (x,y) of the focused beam and the height of the cone is the penetration depth d_p. Based on these considerations, the limiting volume is no greater than 10 femto-liters over the mid infrared region. One final benefit of immersion is that the flux of light that can be collected is given by

$$F \propto n^2_{\text{hemisphere}} \sin^2 \theta \qquad (4.4)$$

Relative to a measurement conducted in air, 16× more light can be collected when using a Ge hemisphere. As a result, the optical conductance of the method is significantly improved.[6,7]

4.3   Historical Development

The history of ATR imaging is somewhat fragmented as it draws on developments in different fields over the last 50 years. The concept of immersion has long been known and employed by optical designers to collect more of the available light and focus that light onto a small area detector. In conventional detection systems the size of the detector could be reduced by a factor equal to the refractive index if the detector was placed in optical contact with the plano surface of a hemisphere. With the detector at the center of curvature, the lens does not introduce any spherical aberration or coma.[8] In 1976, hemispherical lenses were employed by Chen et al. to observe "surface-electromagnetic wave enhanced Raman scattering" in an ATR configuration.[9] Mansfield and Kino were the first to employ a solid immersion lens (SIL) to improve the imaging capabilities of a white-light microscope.[10] Using a SIL with a refractive index of 2, they were able to resolve features with a spatial frequency of 100 nm at a wavelength of 436 nm. These authors also proposed the use of a silicon SIL to exploit the methods advantage in the infrared region. Shortly thereafter, Mansfield et al. demonstrated these capabilities in a visible imaging microscope outfitted with a CCD detector.[11] From this point on there were numerous publications that employed solid immersion lenses, both hemispheres and hyper-hemispheres, to improve the spatial resolution in optical microscopy, fluorescence microscopy, Raman microspectroscopy, and optical data storage systems.[12–16]

With respect to attenuated total internal reflection, the early pioneers included Harrick and Fahrenfort who developed the method to study infrared spectra of organic materials.[17,18] Fahrenfort employed hemicylinders of alkali halides to demonstrate the ATR method. The main benefits of the new found technique included the ability to analyze materials with little or no sample preparation and the ability to analyze highly absorbing materials. With ATR, the only sample requirement is that it must be placed into optical (intimate) contact with the IRE. In addition, the limited depth to which the evanescent beam penetrated the sample meant that spectra of strongly absorbing materials could be obtained without total absorption of the infrared radiation at a particular wavelength. Microscopic ATR methods did not become available until 1991 when Harrick developed the Split-pea infrared microscope[19] and Spectra-Tech independently developed a specialized ATR objective for their IRPLAN microscope.[20] The Split-pea employed a germanium or silicon hemisphere with a beveled tip to improve contact with the sample and the appearance of the IRE led to the name of the device. The Spectra-Tech ATR objective employed a zinc selenide IRE and later a diamond IRE so that the user could observe the sample in white light prior to conducting an ATR analysis. Perkin Elmer later developed a dropdown accessory for their microscope, which was based on a germanium hemisphere possessing a beveled tip. The user simply aligns the sample in white light viewing mode and then lowers the IRE onto the sample for subsequent infrared analysis. An added benefit of these devices stems from the fact that the pressure applied to a given sample is the force divided by the area. Since the contact area is on the order of 100 to 200 μm for each device, the pressure and therefore the contact of the IRE with the sample increased tremendously as compared to a macro sampling accessory. At that time, the major focus of the devices was on the ability to collect infrared spectra from intractable samples and not necessarily the improvement in spatial resolution.

The first reports to study the improved spatial resolution of an infrared ATR measurement using a germanium IRE was that by Nakano and Kawata.[21,22] The authors built a specialized evanescent wave microscope that incorporated a confocal aperture for both the source and primary image of the sample to spatially isolate the sample of interest (Fig. 4.2). The hemisphere with attached sample was translated beneath the microscope using a piezoelectrically controlled stage. As shown in Fig. 4.2, when the hemisphere is on axis, rays enter the hemisphere normal to its surface and, as such, are focused at the center of the plano surface. Moving the hemisphere off-axis to either side, the rays enter at a slight angle, are refracted, and come to a focus at off-axis positions, thereby allowing different sample points to be interrogated. The authors demonstrated an improvement in spatial resolution equal to the refractive index of germanium (4×) and the ability to scan over an area of approximately 100 μm. The limited scan length was the result of spherical aberrations introduced by scanning the hemisphere off-axis. In 1995, Esaki et al. employed a chevron-shaped internal reflection element (Fig. 4.2) on a conventional microscope.[23] Esaki et al. demonstrated the ability to obtain ATR maps as large as 400×400 μm. However, since a hemisphere

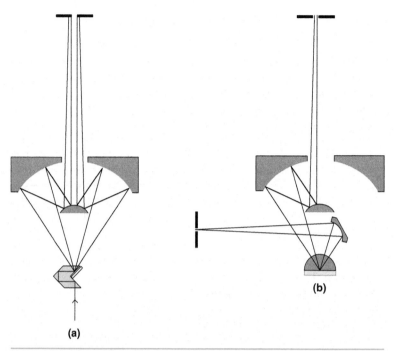

FIGURE 4.2   Adapted from Nakano and Kawata and Esaki et al. Configuration employed by Nakano and Kawata (*b*) and that by Esaki et al. (*a*). [L. L. Lewis and A. J. Sommer, *Applied Spectroscopy*, Vol. 54, No. 2, page 325, figure 1 (Society for Applied Spectroscopy, Frederick, Md., 2000).]

was not employed, no improvement in spatial resolution was realized. Tajima, of the Shimadzu Corporation, demonstrated ATR mapping with an automated microscope.[24] In this case, the sample was raised into the hemisphere for sample analysis then lowered to access subsequent sampling points. While this method was acceptable for hard stable surfaces, another method was needed to study soft surfaces due to the fact that some material from one sample point could transfer to the hemisphere and contaminate subsequent spectra. In 1999 and 2000, Lewis and Sommer reported on the approach taken by Nakano and Kawata, but on a commercial Perkin Elmer *i*-series microscope.[25,26] In this microscope, the hemisphere with attached sample was scanned off-axis, but the sample was illuminated globally and only the primary image plane possessed a confocal aperture. Lewis and Sommer demonstrated that one-dimensional ATR maps could be obtained with improved spatial resolution over a transmission measurement, but due to diffraction effects associated with the confocal aperture, the theoretical improvement in spatial resolution was not realized. Further, ATR analysis was capable of measuring a sample 4 times smaller than that associated with transmission with equal signal to noise.[25] Lewis and Sommer also demonstrated that the spherical aberrations could be compensated for by collecting a background at the exact same off-axis position as the sample and, more importantly, that the penetration depth changed for each off-axis position as a result of the changing incident angle within the crystal. Should quantitative information be required, the change in penetration depth would need to be accounted for. The work of Nakano and Kawata demonstrated the usefulness of ATR microspectroscopy, but the drawback to their system was its complexity. The work of Lewis and Sommer showed that the method could be employed on a conventional system, but with a sacrifice in the theoretical spatial resolution albeit better than a transmission measurement. Earlier, Sommer and Katon demonstrated that the main cause for the degradation in spatial resolution in an infrared microscope was the use of the confocal aperture employed to isolate the sample of interest.[27]

At about the same time these rudimentary microscopic ATR mapping experiments were conducted, array detectors became available, principally in the near-infrared region, but not too long after mid-infrared detectors also became available. Lewis et al. reported on a near–infrared microscope outfitted with an InSb array detector.[28] Soon thereafter (1996), Biorad introduced the first commercial system with an InSb array detector interfaced to a microscope, the Stingray 1. In 1997, Kidder et al. reported on a similar system, but with a mid-infrared mercury cadmium telluride (MCT) detector interfaced to the microscope.[29] In both, the report of Lewis et al. and Kidder et al. diffraction-limited performance was not achieved. However, Lewis et al. anticipated diffraction-limited performance and Kidder et al. came within a factor of 2. In theory, the degradation in spatial resolution should be greater in a conventional microscope rather than an array-based system. In a conventional infrared microscope, the majority of diffraction occurs from the confocal aperture (high-contrast edge) located at the primary image plane of the sample. Radiation diffracted by the aperture then propagates onward to a relatively large area detector (i.e., 100×100 µm), where it is detected and degrades the theoretical spatial resolution. By removing the aperture, the most significant source of diffraction is eliminated. In the array-based system, true diffraction-limited performance can be observed so long as the diffraction-limited beam diameter at the sample, when imaged onto the detector, is greater than the pixel size on the array (vide infra).

In 1997, Biorad introduced an infrared microscope outfitted with an MCT array detector, the Stingray. In the spring of 1999, Sommer attempted to repeat the ATR experiments of Lewis on the Stingray system in the 3M laboratory of Rebecca Dittmar. However, due to operational problems with the Stingray, the experiments were unsuccessful. Later in the spring of 2000, Sommer attempted the same experiments in the Procter and Gamble laboratory of Curtis Marcott. These experiments were successful and demonstrated ATR imaging

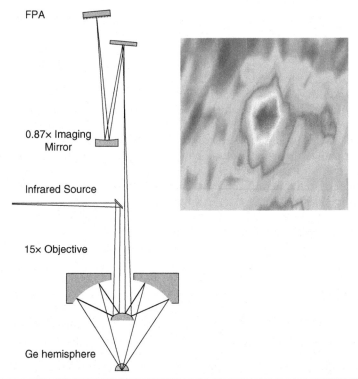

FPA

0.87× Imaging
Mirror

Infrared Source

15× Objective

Ge hemisphere

FIGURE 4.3   ATR Imaging using an on-axis configuration with an array detector and the amide I image of a human erythrocyte. [A. J. Sommer, L. G. Tisinger, C. Marcott, and G. M. Story, *Applied Spectroscopy,* Vol. 55, No. 3, page 253, figure 1 (Society for Applied Spectroscopy, Frederick, Md., 2001).] (Permission granted.)

using an MCT-array-based infrared microscope coupled to a step scan interferometer. As depicted in Fig. 4.3, the germanium hemisphere was held on-axis. In this mode, the sample was globally illuminated and the pixel size of the detector served to spatially isolate a given point on the sample. A comparison of several different sampling modes was conducted, which demonstrated that the ATR mode using the array-yielded near-diffraction-limited performance. The theoretical and measured spatial resolutions differed only by a factor of 1.3. Taking into account the magnification of the system from sample to detector, an area approximately 75×75 μm could be imaged in a matter of minutes. Sommer and co-workers demonstrated the capabilities of the system by measuring the surface image of a single human red blood cell. They further showed that the signal sensed by one pixel arose from a sample volume of 11 femto-liters. This volume relates to a mass detection limit of 13 femto-grams, assuming a density of 1.2 g/cm^3. The results of this work were presented at the 2000 Pittsburgh Conference and later published in Applied Spectroscopy.[30,31] In October of 2000, Biorad was issued a patent for ATR imaging which was principally based on microscopic on-axis measurements done with an array detector.[32]

Although ATR imaging had been demonstrated, it was not considered routine mainly due to the cost and complexity of the associated step-scan interferometer and array detector. The necessity to use a step-scan interferometer was a result of the relatively slow read out capabilities of the MCT arrays.[33] At FACSS in 2001, Perkin Elmer introduced the Spotlight 300 infrared imaging microscope which employed a linear array detector and a conventional rapid scan interferometer. Perkin Elmer engineers asked the question: "At what point does the size of the array dictate the use of a step-scan interferometer?" They settled on a 16 element linear array. The so-called "push broom" mapping was implemented through the careful synchronization of the detector, "rapid" scan interferometer and the mapping stage. With this system, off-axis ATR imaging could be conducted as proposed by Lewis and Sommer. The next significant development came in 2006 when Patterson and Havrilla realized that the spherical aberrations, which limited the total sample area, were directly related to the radius of the hemisphere.[34] This realization was also made independently by Perkin Elmer. Whereas Nakano and Kawata employed a 4-mm radius hemisphere, Lewis and Sommer employed a 1.5-mm radius hemisphere, Patterson and Havrilla employed a 12.5-mm radius (25-mm diameter) germanium hemisphere. In conjunction with the off-axis scanning on the Spotlight 300, the pair was able to obtain ATR images over an area of 2500×2500 μm. The larger radius hemisphere also provided a more constant penetration depth across the image, while maintaining the spatial resolution. Patterson et al. later employed the same hemisphere on a two-dimensional array system with a mapping stage in the off-axis imaging mode.[35] The basis for the experiment was that the 4096 element array could generate images faster than a 16 element array. Their efforts produced marginal results due to the fact that the image acquisition and stage synchronization was not optimal among other factors. In 2006, Perkin Elmer developed and introduced an ATR accessory based on the off-axis imaging concept of Nakano and Kawata and Lewis and Sommer. The device shown in Fig. 4.4 permits routine ATR imaging to be conducted on sample areas as large as 400×400 μm.

4.4 Experimental Implementation

Most infrared microscopes employ reflecting objectives of the Schwarzchild design to focus light onto the sample, or in this case the hemisphere. This requirement stems from the wavelength range associated with the mid-infrared region (2.5 to 17 μm) and the fact that reflecting

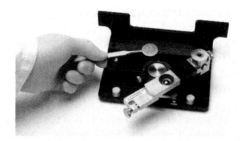

FIGURE 4.4   Perkin Elmer ATR imaging accessory.

objectives are achromatic. Since ATR imaging is a reflectance method, typically one-half of the objective is employed to direct radiation into the hemisphere, while the other half is employed to collect radiation internally reflected to the detector (a configuration which is commonly referred to as an aperture splitting beam splitter). Since only half of the microscope aperture is employed (sin θ = 0.3), the improvement in spatial resolution is normally "$n/2$" instead of "n." One could employ the entire aperture of the objective by using a conventional beam splitter; however, in this case a compromise in signal to noise may be experienced. In the aperture splitting beam splitter, if 100 photons were incident on a non-absorbing sample, nearly 100 photons would be observed at the detector. In the case of a conventional beam splitter, only 25 photons would be observed. That is, 50 percent of the radiation is lost on the first reflection at the beam splitter and another 50 percent being lost on the second reflection.

Another important consideration for the use of a hemispherical IRE on a conventional infrared microscope is that the critical angle be met. As is shown in Fig. 4.1, radiation entering the hemisphere spans a range of angles that are dictated by the design of the reflecting objective. The most extreme ray can be found from the numerical aperture of the objective and the lesser ray can be found by experiment or by contacting the manufacturer.[36] However, because the optical design of the objectives have been optimized for N.A. = 0.6 the most extreme ray entering the hemisphere is ~37° and the lesser ray is ~17°. In order for internal reflection to occur at the IRE/sample interface, radiation entering the hemisphere must be incident beyond the critical angle given by Eq. (4.5).

$$\sin \theta^{-1} = \frac{n_{\text{sample}}}{n_{\text{IRE}}} \tag{4.5}$$

Table 4.1 lists several common IRE materials along with the required critical angle assuming a sample refractive index of 1.5

The data in Table 4.1 demonstrate that only germanium permits all the radiation to be internally reflected. If ZnSe or diamond were

Material	Refractive Index	Critical Angle θ_c	Spatial Resolution*
Air	1.0	Infinity	4.1λ
Zinc selenide (ZnSe)	2.4	39	1.7λ
Diamond	2.4	39	1.7λ
Si	3.4	26	1.2λ
Ge	4.0	22	λ

*Using an aperture splitting beam splitter.

TABLE 4.1   Critical Angle Required for Various IRE Materials

employed, the light would transmit through the IRE/sample interface and no ATR spectrum would be observed. For Si, a portion of the light would be transmitted and a portion would be internally reflected. Germanium has the highest refractive index and provides the best improvement in spatial resolution of all materials. In practice, germanium is the preferred material but the material is not transparent to visible light, which prevents direct viewing of the sample. Spectra-Tech, SENSIR (Smith's Detection), and Varian have opted to design specialized objectives based on either diamond or a combination of zinc selenide and diamond. These objectives allow the user to view the sample and diamond is almost indestructible as an IRE material.

In general, two approaches have been taken in ATR imaging, on-axis imaging and off-axis imaging. With on-axis imaging the hemisphere/sample composite is centered at the microscope's focus and the hemisphere/sample is illuminated globally. Radiation that is internally reflected is then imaged onto a two-dimensional array detector. The detector size defines the sample area that can be imaged and the pixel size defines the spatial element on the sample, commonly referred to as the pixel resolution. For example, Sommer employed a 64×64 MCT array possessing a pixel size of 64×64 μm.[31] Based on these values and the magnification from the detector to the sample, the area that could be imaged was 76×76 μm with a pixel resolution of 1.2×1.2 μm. To increase both the area imaged and the pixel resolution, one can employ a larger array with smaller pixels.

For off-axis imaging, the hemisphere/sample composite is initially centered at the microscope's focus and then imaging is conducted by moving the composite off-axis as discussed earlier (Fig. 4.2). Lewis and Sommer demonstrated that for a germanium hemisphere a 1- μm stage displacement will displace the beam in the hemisphere by 0.3 μm.[26] This off-axis mode is employed with either a single point detector or a linear array detector. The pixel resolution at the sample is defined by the size of the remote aperture, in the case of a single element detector, or the pixel size in the case of a linear array. For example, in the Perkin Elmer Spotlight 300 the pixel size on the array is 30×30 μm, which results in a pixel resolution at the sample of 1.6×1.6 μm. The total area that can be imaged is dependent upon the size of the hemisphere and is limited by spherical aberrations associated with large off-axis positions. For a 3-mm diameter hemisphere, the sample area is on the order of 100×100 μm and that for a 25-mm diameter hemisphere is on the order of 2500×2500 μm. Whether on-axis or off-axis imaging is employed, the incidence angle changes slightly for each sample position. As a result, the penetration depth (optical path length) is different for different sample locations.[26] However, this change in penetration depth can normalized much like a macro-ATR spectrum is normalized for penetration depth as a function of wavelength.

In the previous discussion of ATR imaging, the pixel resolution was quoted for each method. Pixel resolution gives no indication of the spatial resolution inherent with the method. In the introduction to this chapter, Eq. (4.1) gives the diffraction limited diameter of a beam of light focused to a point with a lens or objective. The radial intensity distribution of the focused beam from the optic axis has the form of a Bessel function given by:[8,37]

$$I(P) = \frac{1}{(1-\varepsilon^2)^2} \left[\left(\frac{2J_1(kaw)}{kaw} \right) - \varepsilon^2 \left(\frac{2J_1(kaw)}{kaw} \right) \right]^2 I_0 \qquad (4.6)$$

The distribution is also known as the Airy pattern or point spread function (PSF) for an optical system. A plot of the distribution for the annular aperture present in a reflecting objective is given in Fig. 4.5. Values employed to obtain the distribution include $\sin \theta = 0.3$, $n = 4.0$, $\lambda = 6.0$ and an obscuration value of 0.31. The distribution shows that the distance between the first minima from the origin is approximately 6 μm, which is d given by Eq. (4.1). Contained within this diameter is 84 percent of the original energy from the source, with the remaining 16 percent distributed over larger diameters.

By integrating the PSF one obtains the step function also shown in the plot. To evaluate the spatial resolution of a microscope, one usually translates the edge of a polymer film through the focus of the microscope and monitors the intensity of an absorption as a function of position. After normalization of the intensity values, the distance between those abscissa values with associated intensity values 0.08 and 0.92 (8 and 92 percent) is taken as the spatial resolution. For convenience, the 5 and 95 percent or 10 and 90 percent distance has been reported. Equation (4.1) is employed as the measure of spatial resolution for infrared microspectroscopy because for a sample of that size (6 μm) one would expect 16 percent contamination

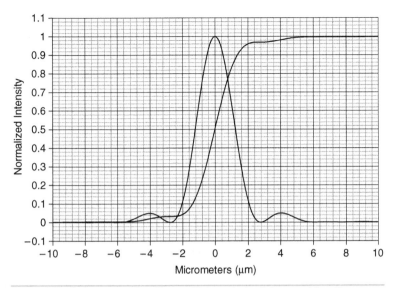

FIGURE 4.5   Point spread function and integrated point spread function.

from neighboring samples provided each sample had a similar extinction coefficient. When using the Rayleigh criterion (i.e., "$d/2$"), as some authors have reported, one would expect significant contributions from near neighbors which could prevent the material from being identified. In the case of an ATR measurement, where the sample is immobile relative to the hemisphere, a cross-sectioned laminate with a sharp interface is usually employed. The absorption of a given peak for one or both of the laminate materials is then monitored as a function of position, from which the spatial resolution can be determined.

4.5   Benefits of ATR Microspectroscopic Imaging for Biological Sections

Although there are many reports on the use of infrared analysis for the detection of disease states in tissue biopsies, probably the most challenging sample type is where the disease state involves a mineral inclusion or crystalline deposit within the tissue itself. A very good example of this type of situation are those mineral inclusions commonly found in kidney disease. As such, this type of biopsy will be employed to highlight the difficulties of an infrared analysis based on transflection (TF) and how ATR microspectroscopy overcomes those limitations.

Two important factors to be kept in mind when using infrared microspectroscopy for disease detection are that spectra with very high signal-to-noise ratios (SNRs) are required and that those spectra should be free from optical artifacts. Optical artifacts complicate spectral interpretation and could prevent an accurate analysis and\or diagnosis. Last, in any microspectroscopic analysis the sample itself becomes a critical component in the optical system. Efforts to incorporate infrared microanalysis into a protocol for disease detection have settled on the use of low-E slides and TF analysis.[38] These efforts have been the result of combining two disparate disciplines, namely, infrared microspectroscopy and histology. While the preferred sample support for infrared analysis is usually a hygroscopic alkali halide material, like sodium chloride and potassium bromide, the histologist prefers glass substrates. However, glass is not infrared transparent and the incorporation of alkali halide materials into a histological preparation would be difficult, since they are highly water soluble. Barium fluoride supports were initially employed, but these are expensive and not in the microscope slide format that the histologists prefer. A solution to these problems is low-E glass slides which are transparent to visible light and reflecting for infrared light. These slides are easily incorporated into any histological preparation and allow the pathologist to view the sample using white-light microscopy and the infrared microspectroscopist to study the sample in a TF analysis. The only limitation is that the thickness of the tissue sample should be no more than 6 µm. This thickness ensures that the features in an infrared spectrum are not totally absorbing and that the spectra are photometrically accurate from which quantitative data might be extracted.

In a typical TF analysis, light enters the sample from the objective at an average incident angle of ~27°. The light transmits through the sample to the substrate, where it is reflected back through the sample and collected by the objective. Based on Eq. (4.1), spectra from spatial domains as small as 4λ (~24 µm for radiation possessing a wavelength of 6 micrometers) with SNRs of 1000/1 can be easily recorded. Further, the average optical path length through the sample is 13.5 µm, based on the sample thickness and incident angle given above. The analysis is straight forward provided that the tissue sample is a continuous film possessing low-contrast interfaces. Low-contrast interfaces are characterized as those having optically similar materials present on either side of the interface. Probably the most important parameter in this regard is the refractive index of both materials. However, the majority of tissue preparations do not meet these criteria. Discontinuities within the sample (e.g., blood vessels, vesicles, mineral inclusions) present interfaces with relatively high-contrast edges. From an optical standpoint, a high-contrast edge can promote scattering, diffraction, reflection, and dispersion. These effects are further amplified due to the size and shape of the sample and the high convergence of the impinging infrared radiation. Even worse, is the case of a mineral inclusion which presents a high-contrast edge and a highly scattering point defect. An additional artifact associated with this later sample type is known as the reststrahlen effect, in which the sample becomes a perfect reflector near an absorption band. Last it should be remembered that the refractive index of a sample changes dramatically in and around an absorption. These effects manifest themselves in the spectra in a variety of ways and make the interpretation of the spectra and the identification of disease states very difficult. From a quantitative perspective, the adherence of the Beer Lambert law dictates that the sole mechanism for the attenuation of light must be absorption and that the optical path length through the sample be well known.

Starting out with the simple case of a blood vessel or vesicle within the tissue, the interface is comprised of air and tissue and the difference in refractive index between these two materials is ~0.40 units. When a spectrum is obtained on such an interface, portions of the light undergo specular reflection and dispersion which manifest themselves in the spectrum as derivative shaped peaks Sommer and Katon illustrated these dispersive band shapes in infrared microspectroscopy.[27] Later, Stewart and Sommer demonstrated that these optical nonlinearities increase with a greater difference in refractive indices between two materials and that the magnitude of the effect increases with a decreasing spatial domain of one material embedded

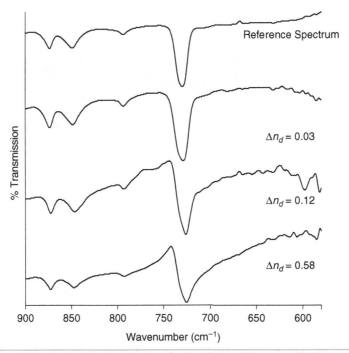

FIGURE 4.6   Band distortion due to refractive index differences. [J. M. Chalmers and P. R. Griffiths (eds.), *Handbook of Vibrational Spectroscopy*, Vol. 2, page 1381, figure 14 (John Wiley & Sons, Inc., West Sussex, UK 2002).] (Waiting on permission).

in the other.[39,40] Figure 4.6 illustrates the effects of dispersion in the absence of scattering. The spectra were collected on a polyethylene terephthalate sample ($n = 1.58$) whose cross section was 17×17 μm. The data demonstrate that as the refractive index difference increases between the sample and its surroundings, the infrared band shapes become asymmetric. This asymmetry is usually observed when specular reflection dominates the measurement. Sommer discussed how these effects and the structure of the sample could have adverse effects on a quantitative analysis.[40] Bhargava studied these anomalies in phase separated polymer systems and showed how they impacted the study of the interface.[41]

High-contrast edges can also produce scattering, and/or diffraction, which can manifest in a spectrum as a sloping baseline. The left side of Figure 4.7 illustrates a TF spectrum collected at an air/tissue interface in addition to an ATR spectrum collected at the same location. The TF spectrum exhibits a positive slope on going from short wavelengths (high energy) to long wavelengths (low energy).

Romeo and Diem studied these effects specifically for tissue sections in TF analyses and developed a computational method to correct them.[42] A solution to the problem was recognized prior to the advent of infrared microspectroscopy, where the sample was embedded in a matrix possessing a similar refractive index. Nujol mulls and

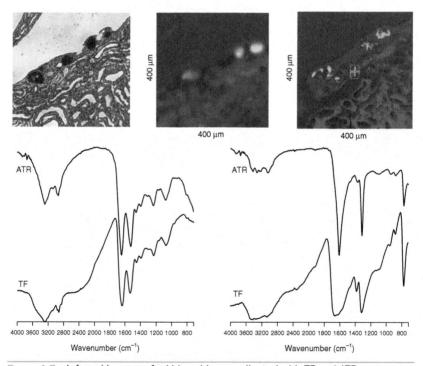

FIGURE 4.7   Infrared images of a kidney biopsy collected with TF and ATR.

KBr pellets were employed in the early days of infrared to eliminate these artifacts. A similar approach can be taken for tissue samples, where the tissue is immersed in Nujol. This approach can readily be implemented by applying a few drops of Nujol to the sample and placing a 1-mm-thick barium fluoride cover slip on top. Although the addition of Nujol may not be an optimal solution, most tissue sections are mounted in paraffin and then subsequently deparaffinized. A consideration might be to leave the paraffin intact for those samples to be studied via infrared microspectroscopy, or to use the ATR method.

In the more complicated case of a mineral inclusion in the tissue, scattering, diffraction, and the reststrahlen effect come into play. Grahlert has addressed some issues related to scattering in TF measurements of silicon carbide fibers.[43] Figure 4.8 illustrates spectra of a calcium oxalate inclusion in a kidney biopsy. The top spectrum was collected using the TF method. Features observed in the spectrum are predominantly those of the protein matrix; however, positive absorptions can be observed near 1700, 1322, and 780 cm^{-1}. These features are reststrahlen bands from the calcium oxalate inclusion. The ATR spectrum of the same inclusion site (middle) and a reference ATR spectrum of calcium oxalate (bottom) are also illustrated in Fig. 4.8. These spectra are free of the Reststrahlen effect. Not knowing what the inclusion material was, one would have difficulty in identifying its composition.

Many of the above mentioned artifacts are path length dependent. By reducing the optical path length, one can minimize their effects. In a TF measurement, the optical path length through the sample is approximately 13.5 μm for a 6-μm-thick sample. Using the

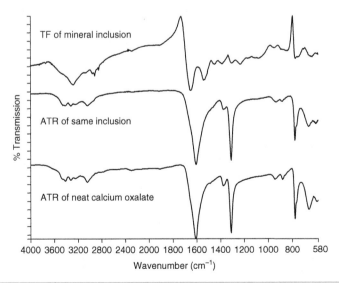

FIGURE 4.8   Spectra obtained on a calcium oxalate mineral inclusion in a kidney biopsy.

ATR approach, one can limit the optical path length to less than 1 μm. Tisinger has demonstrated that ATR can eliminate artifacts associated with edges, specifically for high-contrast edges.[44] The ATR spectrum collected at the air/protein interface as illustrated in Fig. 4.7 exhibits no such artifacts. Further, the increased spatial resolution of the ATR method and the limited penetration depth are also useful for the study of mineral inclusions in the tissue.[45] The ATR spectrum of the mineral inclusion illustrated in Fig. 4.8 exhibits features that are solely calcium oxalate with no protein absorptions. In addition, no reststrahlen features are present. Last, Fig. 4.7 illustrates a TF spectrum collected on an oxalate inclusion. These spectra not only exhibit scattering and dispersive effects, but poor photometric accuracy as well. This latter problem is a result of the optical path length through the sample. The ATR spectrum of the same inclusion exhibits no such artifacts and is photometrically accurate.

Since it is anticipated that this chapter will be read by histologists and pathologists, and they appear to be more visually oriented, the phrase "a picture is worth a 1000 words" is pertinent. Figure 4.7 illustrates infrared images of the same tissue section collected in TF mode and ATR mode. Infrared images were collected and then a principal component regression analysis was conducted to determine how many unique chemical components were present in the field of view. The different components were then color coded and plotted as a function of position and intensity. The difference between the images is quite striking in that the ATR image has better spatial definition and clarity. The fuzziness in the TF image is a direct result of the many artifacts discussed above. Spectra extracted from both images demonstrate, that from a photometric standpoint, the ATR spectra are clearly superior and could lend themselves to a less problematic quantitative analysis.

One remaining benefit of the ATR approach is the ability to detect samples whose sizes are well below the diffraction limit. Although this can be done with transmission infrared microspectroscopy, ATR should prove much better. Patterson has shown that the detection limit for a micro-ATR measurement is 20 ppm for a moderate infrared absorber dissolved in solution.[46] Drawing a similar parallel to a small particle in a surrounding matrix, one can calculate a particle size related to this detection limit. For example, the spectrum shown in Fig. 4.9 was collected in transmission mode on a 3-μm diameter polystyrene particle using a 100×100 μm aperture. Based on volume arguments the detection limit for the sphere is ~500 ppm. For the ATR measurement the sampled area is diffraction limited, so in theory a 6×6 μm area would be sampled to a depth of ~0.7 μm. Using a similar argument, one might expect to be able to obtain an ATR spectrum of similar quality on a sample as small as 0.3 μm in diameter at 1000 ppm. Although the 20 ppm detection limit may not provide sufficient information to identify

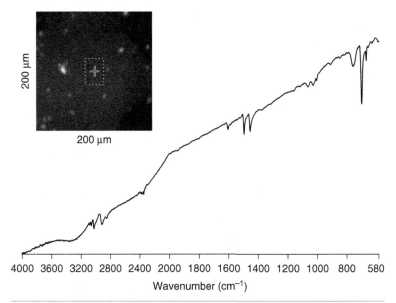

FIGURE 4.9 Transmission infrared spectrum of a 3-μm polystyrene sphere and ATR image of 1.5-μm polystyrene spheres.

the sample, at 200 ppm the associated particle size would be 0.1 μm in diameter.

4.6 Macro ATR Imaging

The methodology discussed thus far involves the use of a microscope to focus light into a hemisphere, which enables one to investigate rather small sample areas with high-spatial resolution. The concept of macro (centimeter-sized areas) ATR was first demonstrated by Harrick using the device shown in Fig. 4.10 and today serves as the basis for many inkless fingerprinting technologies.[5] The concept is the same in the infrared except the viewing screen is replaced with an

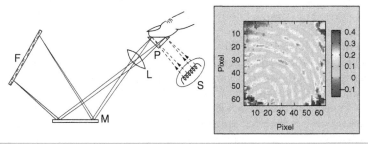

FIGURE 4.10 Macro-ATR imaging device of Harrick and amide I image of a finger print collected on a similar infrared device. [N. J. Harrick, *Internal Reflection Spectroscopy* page 4, figure 2, (John Wiley & Sons, Inc., Ossining, N.Y., Third Printing 1987).] (Permission granted.)

MCT focal plane array and the source emanates from a step-scan interferometer. Tisinger and Sommer built an identical system for the infrared region using an MCT array, a zinc selenide prism, and a step-scan interferometer.[44,47] The amide I image of a fingerprint collected with the device is illustrated in Fig. 4.10, which demonstrates the instrument's capabilities. For this system, the smallest spatial domain sampled is dependent on the magnification from the sample to detector and the size of the individual pixels on the array. The total area sampled (field of view, FOV) is dependent on the size of the array. For example, Tisinger and Sommer employed a 1:1 magnification from sample to detector and a 64 × 64 element array with a pixel size of 61 × 61 μm. In this particular case, the pixel resolution of the instrument was 61 μm with a FOV of approximately 4 × 4 mm. Tisinger later improved on the pixel resolution by replacing the zinc selenide prism with a hemisphere.[44] Tisinger demonstrated that a magnification factor of 2.4 could be achieved with a resultant pixel resolution of 25.4 μm. However, the FOV was reduced by a similar factor to ~1.6 × 1.6 mm.

At that time, the standard MCT detector was a 64 × 64 array with a 61 x 61 μm pixel size, but larger arrays with smaller pixel sizes soon became available thereby improving both the pixel resolution and the total area sampled. Marcott later reported on the use of Harrick Fast-IR accessory with a zinc selenide prism, but with a 256 × 256 array possessing a pixel size of 40 × 40 μm.[48] He was able to increase the FOV to 7.5 × 7.5 mm with a pixel resolution of 30 μm. Following the concept of Tisinger, the prism could be replaced with a zinc selenide or a germanium hemisphere which would increase the pixel resolution to 12.5 and 7.5 μm, respectively. In effect, microscopic measurements over a large area could be conducted with these instruments without the use of a microscope. Chan and Kazarian have investigated this potential principally for the study of pharmaceuticals using a SPECAC Golden Gate accessory with a diamond ATR element.[49] The choice of diamond is a compromise between FOV, pixel resolution, intimate contact considerations and IRE longevity. Diamond has an identical refractive index to that of zinc selenide, but is more robust. However, due to cost, the size of the IRE and thus the FOV are limited. The smaller IRE area and the greater penetration depth are less problematic when it comes to achieving intimate contact with the sample. An average penetration depth for diamond and germanium at 6 μm wavelength is 1.4 μm and 0.4 μm, respectively.

Whether or not these devices could be employed for the diagnosis of disease states in biopsied samples remains to be seen. Tissue biopsies are rather small and the spatial resolution required for an analysis should be very high. Other applications of these devices include the investigation of skin surfaces, materials applied to skin, and transdermal drug uptake.

4.7 ATR Microspectroscopic Raman Imaging

Although ATR Raman spectroscopy was first reported in 1976, the first report and several that followed employed very high-excitation powers at the sample, long collection times and large samples that were relatively strong scatterers.[50-53] Since these initial reports, the field of Raman spectroscopy has seen many technological advances. The application of the ATR method to microscopic investigations has taken place only recently. The primary impetus for all previous studies was again based on the surface sensitivity of the method. In addition, the ATR method has an added benefit for axial discrimination over traditional confocal Raman microspectroscopy. Figure 4.11 presents diagrams for both confocal Raman (left) and ATR Raman (right) microspectroscopy. Under typical conditions, the former method possesses an axial z resolution that is on the order of 3.1 μm (assuming 632 nm excitation and 0.9 N.A.). Tisinger calculated that by using a ZnSe hemisphere IRE and a 45° incident angle, the same wavelength would yield a penetration depth d_p of only 0.1 μm.[44,54] However, a significant difference exists between the two methods for the analysis of a thin film on a much thicker substrate. In the confocal Raman case, Millister points out that the excitation is still considered far field and a remote aperture is relied upon to spatially isolate scattered light in the confocal volume from scattered light emanating in the far field.[16] More recently, Everall demonstrated the problems associated with confocal Raman microspectroscopy for depth profiling through stratified structures.[55] However, in the ATR case, the Raman scattering is excited evanescently and, as such, there is no far field scattering induced. Tisinger calculated that the evanescent volume was on the order of 10 atto-liters.[54] In addition to this benefit, the hemisphere improves the spatial resolution by n and the

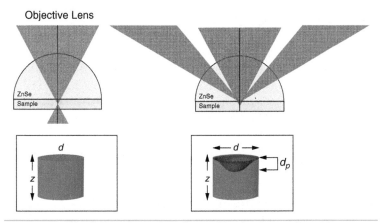

FIGURE 4.11 Far field and near field Raman illumination modes with associated illumination volumes.

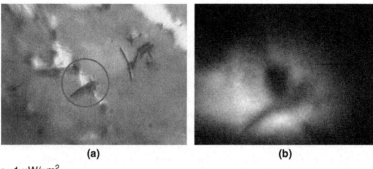

(a)   (b)

• ~1 µW/µm²
• 1 second integration time
• ~5 atto-liters/pixel

FIGURE 4.12   Visible (*a*) and ATR raman image (*b*) of polydiacetylene film.

collection efficiency by n^2. More recently, Sommer demonstrated ATR imaging for strong Raman scatterers in addition to ATR Raman spectroscopy using low incident powers on moderate Raman scatterers.[56] Figure 4.12 illustrates a visible image and an ATR image of a polydiacetylene film deposited on the IRE. The image was collected in 1 second with an incident power of 1 µW/µ²m. The signal sensed by each pixel represents a 5-atto-liter volume of the sample. A defect in the polymer film is imaged as the dark region in the center of the ATR image. Finally, Fig. 4.13 illustrates spectra of a 200-nm-thick polystyrene film

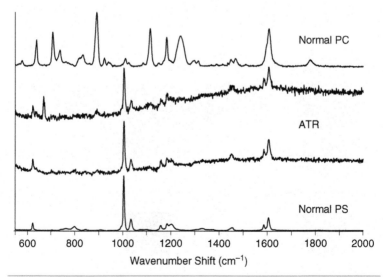

FIGURE 4.13   ATR Raman spectra of a 200-nm-thick polystyrene film on a 2-mm-thick polycarbonate substrate.

deposited on a 1-mm-thick polycarbonate substrate. The spectra illustrate that ATR Raman can be conducted on moderate scattering materials with excellent axial and volumetric resolution.

4.8 Conclusions

Evanescent imaging using visible light was first demonstrated by Harrick in 1963.[5] Harrick's justification for developing the method cited several maladies (e.g., mongolism and Turner's syndrome) and viral infections that modify the nature of the human skin pattern. He theorized that by studying the abnormal infant's handprint, it may be possible to detect these maladies at an early stage. Although this method was based on physical means, the extension to vibrational (molecular) spectroscopy enables the detection of chemical differences related to disease. In the short span of 15 years, infrared evanescent imaging has gone from concept to routine analysis in the analytical laboratory providing the researcher with a powerful tool for the study of disease etiology and detection. The major benefit of the method, relative to conventional TF infrared microspectroscopy, is that it provides enhanced spatial and volumetric resolution. In addition, it overcomes many of the spectral artifacts associated with a TF analysis. Due to these advantages, it is anticipated that pathologists will come to accept the method more readily than the current means. The next big step for ATR imaging will be the application of the method for quantitative studies. Finally, although ATR Raman imaging is by no means routine, technological innovations over the next few years and the fact that this method provides even better spatial and volumetric resolution than infrared methods, could make routine ATR Raman imaging a technology to pursue.

References

1. R. Hooke, *Micrographia*, 1665.
2. J. F. Ford, *The van Leeuwenhoek Specimens*, The Royal Society, London, 1981.
3. L. Carr, Thermo Nicolet Symposium Series, Boulder, Colo., 2001.
4. L. E. Lavalle, A. J. Sommer, G. M. Story, A. E. Dowrey, and Marcott C, "A Comparison of Immersion Infrared Microspectroscopy to Attenuated Total Internal Reflection Microspectroscopy," *Microscopy and Microanalysis*, **10**(Suppl. 2):1298–1299, 2004.
5. N. J. Harrick, *Internal Reflection Spectroscopy*, John Wiley & Sons, Inc., New York, 1987.
6. M. Pluta, "Advanced Light Microscopy," Vol. 1: *Principles and Basic Properties*, Elsevier, Amsterdam, 1988.
7. A. F. Sohn, L. G. Tisinger, and A. J. Sommer, "Combined ATR Infrared Microspectroscopy and SIL Raman Microspectroscopy," *Microscopy and Microanalysis*, **10**(Suppl. 2):1316–1317, 2004.
8. W. J. Smith, *Modern Optical Engineering: The Design of Optical Systems*, McGraw-Hill, New York, 1966.

9. Y. J. Chen, W. P. Chen, and E. Burstein, "Surface-Electromagnetic-Wave-Enhanced Raman Scattering by Overlayers on Metals," *Physical Review Letters*, **36**:1207–1210, 1976.

10. S. M. Mansfield, G. S. Kino, "Solid Immersion Microscope," *Applied Physics Letters*, **57**:2615–1616, 1990.

11. S. M. Mansfield, W. R. Studenmund, G. S. Kino, and K. Osato, "High-Numerical-Aperture Lens System for Optical Storage," *Optics Letters*, **18**:305–307, 1993.

12. B. D. Terris, H. J. Mamin, D. Rugar, W. R. Studenmund, and G. S. Kino, "Near-Field Optical Data Storage Using Solid Immersion Lens," *Applied Physics Letters*, **65**:388–390 1994.

13. Q. Wu, G. D. Feke, R. D. Grober, and L. P. Ghislain, "Realization of Numerical Aperture 2.0 Using a Gallium Phosphide Solid Immersion Lens," *Applied Physics Letters*, **75**:4064–4066, 1999.

14. K. Koyama, M. Yoshita, M. Baba, T. Suemoto, and H. Akiyama, "High Collection Efficiency in Fluorescence Microscopy with a Solid Immersion Lens," *Applied Physics Letters*, **75**:1667–1669, 1999.

15. C. D. Poweleit, A. Gunther, S. Goodnick, and J. Menéndex. "Raman Imaging of Patterned Silicon Using a Solid Immersion Lens," *Applied Physics Letters*, **73**:2275–2277, 1998.

16. T. D. Milster, "Near-Field Optics: A New Tool for Data Storage," *Proceeding of IEEE*, **88**:1480–1490, 2000.

17. N. J. Harrick, "Study of Physics and Chemistry of Surfaces from Frustrated Total Internal Reflections," *Physical Review Letters*, **4**:224–226, 1960.

18. J. Fahrenfort, "Attenuated Total Reflection: A New Principle for the Production of Useful Infra-red Reflection Spectra of Organic Compounds," *Spectrochimica Acta*, **17**:698–709, 1961.

19. N. J. Harrick, M. Milosevic, and S. L. Berets, "Advances in Optical Spectroscopy: The Ultra-Small Sample Analyzer," *Applied Spectroscopy*, **45**(6):944–948, 1991.

20. J. A. Reffner, C. C. Alexay, R. W. Hornlein, "Design of Grazing Incidence and ATR Objectives for FT-IR Microscopy," *Proceeding of SPIE-International Society for Optical Engineering 1575 (8th International Conference on Fourier Transform Spectroscopy)*, 301–302, 1992.

21. T. Nakano, and S. Kawata, "Evanescent Field Microscope for Super-Resolving Infrared Micro-Spectroscopy," *Bunko Kenkyu*, **41**:377–384, 1992.

22. T. Nakano, and S. Kawata, "Evanescent-Field Scanning Microscope with Fourier-Transform Infrared Spectrometer," *Scanning*, **16**:368–371, 1994.

23. Y. Esaki, K. Nakai, and T. Araga, "Development of Attenuated Total Reflection Infrared Microscopy and Some Applications to Microanalysis of Organic Materials," *Toyota Chuo Kenkyusho R&D Rebyu*, **30**(4):57–64, 1995.

24. T. Tajima, S. Takeuchi, Y. Suzuki, T. Tsuchibuchi, and K. Wada, "Mapping Techniques for FTIR Microspectroscopy," *Shimadzu Hyoron*, **53**(1):55–59, 1996.

25. L. Lewis, and A. J. Sommer, "Attenuated Total Internal Reflection Microspectroscopy of Isolated Particles: An Alternative Approach to Current Methods," *Applied Spectroscopy*, **53**(4):375–380, 1999.

26. L. Lewis, and A. J. Sommer, "Attenuated Total Internal Reflection Infrared Mapping Microspectroscopy of Soft Materials," *Applied Spectroscopy*, **54**(2):324–330, 2000.

27. A. J. Sommer, and J. E. Katon, "Diffraction-Induced Stray Light in Infrared Microspectroscopy and its Effect on Spatial Resolution," *Applied Spectroscopy*, **45**(10):1633–1640, 1991.

28. N. E. Lewis, P. J. Treado, R. C. Reeder, G. M. Story, A. E. Dowrey, C. Marcott, and I. W. Levin, "Fourier Transform Spectroscopic Imaging Using an Infrared Focal-Plane Detector," *Analytical Chemistry*, **67**(19):3377–3381, 1995.

29. L. H. Kidder, I. W. Levin, E. N. Lewis, V. D. Kleiman, and E. J. Heilweil, "Mercury Cadmium Telluride Focal-Plane Array Detection for Mid-Infrared Fourier-Transform Spectroscopic Imaging," *Optics Letters*, **22**(10):742–744, 1997.

30. A. J. Sommer, L. Tisinger, G. Story, and C. Marcott, "Attenuated Total Internal Reflection Infrared Microspectroscopy with an Imaging Infrared Microscope," Presented at the Pittsburgh Conference, New Orleans, La., March 2000.

31. A. J. Sommer, L. G. Tisinger, C. Marcott, and G. M. Story, "Attenuated Total Internal Reflection Infrared Mapping Microspectroscopy Using an Imaging Microscope," *Applied Spectroscopy*, **55**(3):252–256, 2001.

32. E. M. Burka, and R. Curbelo, "Imaging ATR Spectrometer," U.S. Patent 6,141,100, October 31, 2000.

33. R. Bhargava, and I. W. Levin, (eds.), *Spectrochemical Analysis Using Infrared Multichannel Detectors*, Blackwell Publishing, Oxford, 2005.

34. B. M. Patterson, and Havrilla, G. J. "Attenuated Total Internal Reflection Infrared Microspectroscopic Imaging Using a Large-Radius Germanium Internal Reflection Element and a Linear Array Detector," *Applied Spectroscopy*, **60**(11):1256–1266, 2006.

35. B. M. Patterson, G. J. Havrilla, C. Marcott, and G. M. Story, "Infrared Microspectroscopic Imaging Using a Large Radius Germanium Internal Reflection Element and a Focal Plane Array Detector," *Applied Spectroscopy*, **61**(11):1147–1152, 2007.

36. L. G. Tisinger, and A. J. Sommer, "Extinction Coefficient Determination of Polymeric Materials for Use in Quantitative Infrared Microspectroscopy," *Applied Spectroscopy*, **56**(11):1397–1402, 2002.

37. M. Born, and E. Wolf, *Principles of Optics: Electromagnetic Theory of Propagation, Interference and Diffraction of Light*, 3rd ed., Pergamon Press, London, 1964.

38. R. K. Dikor, G. M. Story, and C. Marcott, "A Method for Analysis of Clinical Tissue Samples Using FTIR Microspectrocopic Imaging," in *Spectroscopy of Biological Molecules: New Directions*, European Conference on the Spectroscopy of Biological Molecules, 8th, Enschede, Aug. 29–Sept. 2, 1999, Netherlands, Kluwer Academic Publishers, Dordrecht, Netherlands.

39. S. A. Stewart, and A. J. Sommer, "Optical Non-Linearities in Infrared Microspectroscopy: A Preliminary Study into the Effects of Sample Size and Shape on Photometric Accuracy," *Microscopy and Microanalysis*, **3**(Suppl. 2):837–838, 1997.

40. A. J. Sommer, "Mid-Infrared Transmission Microspectroscopy," in *Handbook on Vibrational Spectroscopy*, J. M. Chalmers, and P. R. Griffiths (eds.), John Wiley & Sons, Ltd., England, 2001, Vol. 2, pp. 1369–1385.

41. R. Bhargava, S.-Q. Wang, and J. L. Koenig, "FT-IR Imaging of the Interface in Multicomponent Systems Using Optical Effects Induced by Differences in Refractive Index," *Applied Spectroscopy*, **52**(3):323–328, 1998.

42. M. Romeo, and M. Diem, "Correction of Dispersive Line Shape Artifact Observed in Diffuse Reflection Infrared Spectroscopy an Absorption/Reflection (Transflection) Infrared Micro-spectroscopy," *Vibrational Spectroscopy*, **38**:129–132, 2005.

43. W. Grählert, B. Leupolt, and V. Hopfe, "Optical Modelling vs. FTIR Reflectance Microscopy: Characterization of Laser Treated Ceramic Fibres," *Vibrational Spectroscopy*, **19**(2):353–359, 1999.

44. L. G. Tisinger, "Investigations in Quantitative Infrared Using Attenuated Total Reflectance," Ph.D. Dissertation, Miami University, Oxford, Ohio 2002. http://sc.lib.muohio.edu/dissertaions/AA13043076.

45. J. Anderson, J. Dellomo, A. Sommer, A. Evan, and S. Bledsoe, "A Concerted Protocol for the Analysis of Mineral Deposits in Biopsied Tissue Using Infrared Microanalysis," *Urological Research*, **33**:213–219, 2005.

46. B. M. Patterson, N. D. Danielson, and A. J. Sommer, "Attenuated Total Internal Reflectance Infrared Microspectroscopy as a Detection Technique for Capillary Electrophoresis,". *Analytical Chemistry*, **76**:3826–3832, 2004.

47. A. J. Sommer, and L. G. Tisinger, "Attenuated Total Internal Reflection Infrared Imaging and Spectroscopy of Large Spatial Domains," Presented at the Pittsburgh Conference, New Orleans, La., March 2002, paper 2077.

48. C. Marcott, G. M. Story, L. G. Tisinger, and Sommer, A. J. "Infrared Micro and Macro Spectroscopic Imaging Using Attenuated Total Reflection," Presented at the Pittsburgh Conference, New Orleans, La., March 2002, paper 541.

49. K. L. A. Chan, and S. G. Kazarian, "New Opportunities in Micro- and Macro-Attenuated Total Reflection Infrared Spectroscopic Imaging: Spatial Resolution and Sampling Versatility," *Applied Spectroscopy*, **57**(4):381–389, 2003.

50. T. Takenaka, and H. Fukuzai, "Resonance Raman Spectra of Monolayers Absorbed at the Interface between Carbon Tetrachloride and an Aqueous Solution of a Surfactant and a Dye," *Journal of Physical Chemistry*, **80**(5):475–480, 1976.
51. R. Iwamoto, M. Miya, K. Ohta, and S. Mima, "Total Internal Reflection Raman Spectroscopy," *Journal of Chemical Physics*, **74**(9):4780–4790, 1981.
52. M. Futamata, P. Borthen, J. Thomassen, D. Schumacher, and A. Otto, "Application of an ATR Method in Raman Spectroscopy," *Applied Spectroscopy*, **48**(2):252–260, 1994.
53. M. Yoshikawa, T. Gotoh, Y. Mori, M. Iwamoto, and H. Ishida, "Determination of Anisotropic Refractive Indices of a Single-Crystal Organic Thin Film by Attenuated Total Reflection Raman Spectroscopy," *Applied Physics Letters*, **64**(16):2096–2098, 1994.
54. L. Tisinger, and A. J. Sommer, "Attenuated Total Internal Reflection (ATR) Raman Microspectroscopy," *Microscopy and Microanalysis*, **10**(Suppl. 2):1318–1319, 2004.
55. N. Everall, "The Influence of Out-of-Focus Sample Regions on the Surface Specificity of Confocal Raman Microscopy," *Applied Spectroscopy*, **62**(6):591–598, 2008.
56. A. J. Sommer, "Attenuated Total Internal Reflection Raman Microspectroscopy," Presented at the 35th Annual Conference on the Federation of Analytical Chemistry & Spectroscopy Societies, Memphis, Tenn. October 2007.

sFTIR, Raman, and SERS Imaging of Fungal Cells

Kathleen M. Gough
Department of Chemistry
University of Manitoba
Winnipeg, Manitoba, Canada

Susan G. W. Kaminskyj
Department of Biology
University of Saskatchewan
Saskatoon, Saskatchewan, Canada

5.1 Introduction

In this chapter, we will discuss some of the correlative microscopic techniques that we have brought to bear on the analysis of fungi, including saprotrophs, endophytes, and lichen symbionts. The techniques we will focus on are primarily based on molecular vibrations, including synchrotron-source Fourier transform infrared (sFTIR), Raman, and surface enhanced Raman spectroscopy (SERS). Other chapters in this book also provide introductions to the fundamentals of vibrational spectroscopy. In an effort to reduce duplication, we will describe the fundamentals of each method primarily with respect to the requirements of the fungal samples: appropriate sample preparations, size, shape, lifestyle, and questions posed.

The application of FTIR to imaging of biological samples is now about two decades old; some applications are quite mature but the protocols are not yet routine. Both FTIR and Raman phenomena are well understood; major advances in the latter are bringing it into prominence as a highly sensitive molecular imaging methodology. The term "SERS imaging" is applied to broadly different imaging modalities; there are some very good books and numerous recent review articles to which we refer the interested reader.[1-7] In the 80 years since Raman spectroscopy was first demonstrated, perhaps nothing has received as much fanfare as the potential applications of SERS and surface enhanced resonance Raman spectroscopy (SERRS) for highly sensitive detection down to single molecule level, thus rivaling fluorescence while providing better molecular identification. This chapter presents a brief overview of some of the latest theories, methods, and applications (see Sec. 5.6). This should in no way be taken to imply that the method is now reliable and robust. However, the promise remains bright, and new strategies to achieve the goals are appearing at an ever-increasing rate.

We devote a portion of the chapter to fungi, wonderful subjects to illustrate the power, the promise and the challenges in these techniques. Fungi are exemplary experimental model systems due to their tractable genetics, and their significant impacts on biotechnology as well as on human and plant health. High-spatial resolution chemical analyses are essential to correlate information from other experimental methods with cell genotype and phenotype. Fungal hyphae (tubular cells that form most species) vary tremendously in composition and function over micrometer (μm) spatial scales, so analytical methods such as NMR and mass spectrometry are impractical for understanding fungal biology. FTIR microscopy is an excellent analytical methodology for chemical characterization at high-spatial resolution; however, FTIR analysis of fungal cells is challenging because information on cell walls, through which the hyphae interact with their environment, relates mostly to carbohydrate composition. The most distinctive carbohydrate signatures are at the longer infrared wavelengths (circa 10 μm), where the diffraction limit is 5 to 10 μm at best. Fungal cells are typically rounded and 3 to 10 μm wide (though hundreds of microns long), so in addition to diffraction issues, spectra are prone to scattering artifacts. Also, fungal cells are supported internally by water pressure, so they have relatively low biomass and the signal from individual spores or from hyphal tips is weak. With a synchrotron source to offset these issues, FTIR can assess subtle and cell type specific biochemical differences in fungal cells.[8,9]

Our ongoing research is focused mainly on imaging single fungal cells, with the biological goal of elucidating the biochemistry of fungal lifestyles and the spectroscopic goal of advancing the technology. For the spectroscopists in the audience, we will present some background information on fungi; for the fungal biologists who may be interested in applying these spectroscopic tools to their own research, we present a basic explanation of the fundamentals; finally we present an overview of the ways in which we have been applying these tools in our work. The research presents a curious dilemma, that of the biology and the spectroscopy being equally attractive avenues to pursue; we continue to travel down both roads.

5.2   Introduction to Fungi

Fungi are eukaryotic microbes whose complex internal cellular architecture is comparable to that of animals and plants. Although possibly counterintuitive, fungi are more closely related to animals than plants, despite having carbohydrate cell walls. Apart from mushrooms, the fungi also include single-celled forms called *yeasts* (e.g., baking and brewing yeast, *Saccharomyces cerevisiae*, the first eukaryote to have a completed genome sequence) and multicelled filamentous species that form spreading colonies commonly called *molds* and *mildews*. Most fungi are filamentous; mushrooms are multihyphal assemblages. As described below, these are important organisms that are our experimental systems of choice. Specifically, the filamentous fungal growth habit, which is based on controlled secretion, leads to spatially resolved variation in cellular composition over nano- to micrometer scales.

Fungi exhibit localized cell extension (Fig. 5.1). In filamentous fungi, this occurs at the tips of tubular hyphae.[10] Comparable growth processes are seen in yeasts, which differ in aspects of their cell

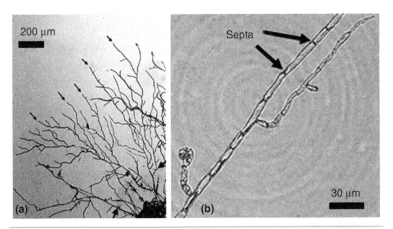

FIGURE 5.1   (a) Photomicrograph of *Fusarium culmorum* hyphae growing out from an inoculated agar block placed on an infrared reflective slide (MirrIR). The hyphae have extended in a typically polarized manner, with some branching, over a 24-hour period. Small arrows show tips of mature hyphae and large arrows (bottom right) show the margin of the medium seeping from the agar medium. (b) Magnification of single hypha showing basal cells separated by septa. Arrows indicate the crosswalls (septa) which are inserted with relatively even spacing. See section on preparation for details.

division. Some fungi (including many human pathogens) can adopt filamentous or yeast-like growth, with morphology depending on environmental conditions. All cells and organisms are *polarized*, that is, they generate and maintain specialized regions that are essential for movement and differentiation. Processes related to cell polarity are fundamentally important in many areas of biology. Hyphal cytoplasm has an extreme level of spatially and temporally predictable polarity. As the hyphal tip extends, the apical cytoplasm secretes wall-building materials and components for nutrient acquisition that were synthesized in basal regions. A saprotrophic fungus like *Aspergillus nidulans*, which forms green mold colonies on bread, has hyphae that are 3 μm in diameter and grow up to 1 μm/min; there are both larger and faster-growing species. The green center of these colonies is due to colored asexual spores generated for survival and dispersal (see Fig. 3 in Ref. 11). Consistent with fungal walls being essential for defining the cell form and as an interface with their environment, about 20 percent of the *Aspergillus* genome is suggested to have wall-related functions.[12] Within the wall, organelle-rich cytoplasm migrates toward the tip, keeping pace with growth, and subapical regions become filled with vacuoles that contribute other metabolic functions.[10] Taken together, fungal hyphae have pronounced structural and functional polarization that unlike many biological systems is relatively predictable and produces cells with simple geometrical forms.

Like animals, fungi acquire their nutrition from other organisms. Saprotrophic fungi consume dead organisms, particularly plants, and have essential roles in recycling. Biotrophic fungi have more or less long-term relationships with living organisms, again particularly with plants. Some saprotrophic and some biotrophic fungi cause disease; however, many others are essential symbionts. Mycorrhizal fungi are associated with the roots of at least 90 percent of plant families, trading minerals for carbohydrates created by photosynthesis. Fossils from 450 million years ago showing these associations are some of the evidence that suggested mycorrhizae might have been necessary for the colonization of land.[13] More recently, endophytic fungi that live within healthy plants have been proposed to be an equally ancient relationship.[14] Some fungal endophytes confer tolerance to environmental stress, and have been shown to be, at the least, very widely distributed.[15]

Fungi are important for ecosystem stability, as threats to the human food supply, as emerging threats to human health, and for their roles in ancient and modern biotechnology. Many fungal species have short life cycles, relatively simple genomes, and are experimentally tractable. Their underlying physiological similarities with animal systems, predictable growth patterns, and myriad ecological and technological impacts, make fungi exemplary systems for scientific inquiry and for assessing the biological relevance of certain analytical methods.

5.2.1   Specimen Preparation

For informative FTIR, Raman, SERS, and micro-SIMS analysis, specimens must be chemically pristine, that is, not chemically fixed or embedded or stained. Our samples are grown in moist chambers across appropriate substrates, nourished from a block of growth medium. Under these conditions, hyphae will extend 1 to 8 mm from the block in 24 hours, depending on the species, and their morphology will be indistinguishable from growth on an agar plate. Extended incubation does not typically lead to ongoing growth, but can lead to tip morphologies suggestive of stress. Migration of medium components by capillary action along hyphal walls (Fig. 5.2a, large arrows) is easily detected. We choose hyphae growing at the colony margin (Fig. 5.2a, small arrows) since they are more likely to be metabolically similar. Spores of some plant pathogenic fungi can germinate on nutrient-free substrates, but this is not generally the case for saprotrophs. Nevertheless, *Aspergillus* and *Neurospora* spores have been shown to germinate at a low frequency without exogenous nutrients, given a humid environment.[8]

Sample harvest and preservation must be rapid to prevent cell degradation or stress-related changes. Our samples are rapidly frozen by placing them sample side up on a −80°C metal plate, so that metabolic processes are arrested within seconds. Frozen samples are freeze-dried or dried at 37°C. We find both methods are effective. Freeze-drying retains cells, three-dimensional structure (good for scanning electron microscopy) but this can lead to scattering artifacts, particularly with sFTIR.[16] Air-drying leads to cell collapse, since fungal hyphae are supported by internal hydrostatic pressure acting against the cell membrane.[17] The membrane is breached by ice crystal damage but the wall restricts the movement of all but small molecules.[18]

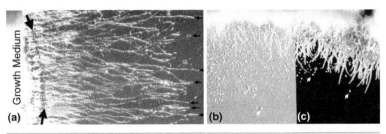

FIGURE 5.2   *Aspergillus nidulans* hyphae grown across pristine gold-coated silicon substrates from an agar-solidified block of medium. (*a*) Large arrows indicate liquid from the medium that extends a short distance along the hyphae. Small arrows indicate hyphae appropriate for sFTIR or Raman analysis. (*b*) *Aspergillus nidulans* hyphae grown across nanopatterned region of Klarite substrate (D3 Technologies Ltd., UK). (*c*) Hyphae in the solid growth medium that nourishes the sample have matured to the point of forming spores that have fallen in clusters across the surface.

Fungal growth is localized at the cell tip, eventually creating a tubular hypha. The hyphae of many fungi (e.g., asexual stages of Ascomycetes) including *A. nidulans* are internally subdivided by cross walls called *septa* into cell-like compartments (Fig. 5.1b). Septa contain additional wall layers as well as a central core that appears (using transmission electron microscopy) to have the same composition as the lateral walls.[19] In vegetative hyphae, septa have a pore that maintains cytoplasmic and pressure continuity between compartments; septum-associated organelles called *Woronin bodies* protect hyphae from excessive loss of cytoplasm following damage.[20]

Using conditional septation-defective strains of *A. nidulans*, septa have been shown to be important for sporulation.[21,22] Isolation of a portion of hyphal cytoplasm is an early stage in spore formation. Notably, the conidiophore foot cell is isolated from the hyphal cytoplasm by an additional wall layer,[23] which is consistent with work in other systems showing that the inner cell wall has important regulatory functions.[24] In fungi such as *Rhizopus*, which have aseptate vegetative hyphae, there is a cross wall that delimits the sporangium from its supporting cell, the sporangiophore. In the oomycete *Saprolegnia*, which also has aseptate vegetative hyphae, septum formation is required for commitment to asexual spore formation.[25] Both of these specialized septa lack a pore, making them substantially similar to the secondary wall of the *Aspergillus* conidiophore foot cell. Taken together, septa are important for both vegetative growth and for sporulation, and may have additional roles that are less well understood. However, because septal walls likely comprise less than 5 percent of the total fungal wall material, studies characterizing their composition have been limited.[26]

5.3   Vibrational Spectroscopy

As this book is entirely devoted to applications of vibrational spectroscopy in biosciences, a complete introduction to the topic, in every chapter, is clearly superfluous. Here we present the basics, with some general tissue examples and then focus on aspects that we have found to be relevant to analyses of fungi.

Vibrational spectroscopy (IR and Raman) is a long-established technique used to identify molecular structures and specific molecular functional groups.[27] Energies are typically reported as inverse wavelength, or *wavenumber*, hence the unit is cm^{-1}. In FTIR, molecules absorb energy (photons in the mid-infrared region of the electromagnetic spectrum) and begin to vibrate: bonds stretch; segments of the molecule twist, rock, or bend. The specific frequencies of light absorbed depend on the number and type of molecular bonds present in the sample.

For any given molecule, there will be a total of $3N-6$ possible vibrational modes, where N is the number of atoms in the molecule.

The *frequency* of each vibration depends on the unique mode of vibration: the atomic masses being displaced and the strength of the connecting bonds. In IR spectroscopy, the *intensity* of the peak depends on the magnitude of the change in the permanent molecular dipole moment accompanying the vibration. A beam of light containing the entire mid-IR radiation band is passed through (or reflected off) a sample. The transmitted (or reflected) light is sent to a detector and displayed as a spectrum (graph) of the amount of light absorbed at each frequency (energy). The mathematics underlying the FT approach is extensively documented in the literature,[28] and described in many of the chapters in this book. Its counterpart, Raman spectroscopy, provides information on the same molecular vibrations, and is similarly well documented.[29,30]

Raman spectroscopy is complementary to IR. Monochromatic laser light is inelastically scattered off the sample. Some of the scattered light is of lower frequency than that of the laser; the missing energy has been deposited in the molecule, in the form of vibrational energy. The physical process of excitation is different from IR; the resultant spectra are different. In Raman scattering, the *intensity*, or differential scattering cross section, recorded for each vibration depends on the induced change in the molecular polarizability accompanying each molecular vibration. Different modes may be excited by IR and Raman; where the same modes are excited, frequencies are the same but relative intensities differ.

The use of vibrational spectroscopy in the study of biological materials was suggested early in the history of the technique and the first experiments on tissues were performed over 60 years ago.[31–33] Widespread use in the biological sciences began with the advent of the interferometer-based instruments (FTIR) in the 1970s, until it became a recognized standard tool for the analysis of model systems in vitro, as well as of components in biotissues in situ.[34–38] For fungi, in situ refers to samples of cells or tissues that have been grown on the imaging substrate or taken from natural systems, and preserved without chemicals. A key aspect is that characteristic absorption peaks from all tissue components (proteins, lipids, carbohydrates, and nucleic acids) may be detected in a single spectrum.

Coupled to a microscope, IR and Raman spectra may be obtained from spot sizes of 1 to 100 μm diameter, permitting subcellular analysis of biological tissues.[39–41] Both techniques may be used to determine molecular component localization: relative amounts and structure of tissue constituents can be ascertained in situ, *without staining*. The two main considerations are (1) spectral differences and (2) spatial resolution. Relevant factors in the success of spectral analysis include:

- Degree to which the components differ from each other chemically (different functional groups, significant conformational differences).

- Presence of strong absorbance peaks arising from these chemical differences.

- Separation between the strong absorbances in the spectrum; this is preferable but not essential, as several numerical tools, carefully applied to sufficiently large datasets, can be employed to overcome some of these issues.
- Physical size and distribution of localized fractions.

5.3.1 Spectral Resolution

Biologically important functional groups produce characteristic, well-separated peaks.[35] Grey and white matter regions of brain tissue are readily distinguished in vibrational spectra, because of the significant difference in the amount of membrane (phospholipid bilayer) present. White matter axons are shielded by a sheath of myelin that is a manyfold thickness of cell membrane. The intense bands at 2800 to 2950 cm^{-1} are characteristic of lipid acyl chains: the dominant peaks at 2848 and 2926 cm^{-1} arise from the symmetric and asymmetric stretch of CH_2, while the weaker peaks at 2875 and 2950 cm^{-1} are from CH_3 stretching of the terminal methyl groups.[42–44] A small absorption due to the stretch of a CH bond on an unsaturated carbon (=C—H) may be observed at 3012 cm^{-1}. Amide groups within an α-helical protein will absorb radiation at 1655 to 1660 cm^{-1}, whereas, when β-sheet amide groups are present, the maximum absorption occurs at lower energy, about ~1630 cm^{-1}.[45]

Our IR data are typically acquired at a nominal spectral resolution of 4 cm^{-1}; absorbance features that are spectrally separated by less than this amount will not be resolved. The absorbance bands of biological molecules are much broader than this limit. The apparently simple bands observed in the IR spectra of tissue (e.g., Fig. 2 in Ref. 45) are really the summation of many, often broad, overlapping absorbance bands centered at nearly identical wavenumbers. Numerical techniques for artificially enhancing the spectral resolution are sometimes employed in these cases; however, to date we have based our analyses on the data as originally recorded. Spectral profiles and sophisticated data analysis algorithms permit classification of large datasets according to small variations in spectral profile.[46–48] These analyses, both supervised and unsupervised clustering algorithms, as well as artificial neural networks, are leading to new applications for rapid recognition of everything from bacteria to cancer.

Spectral resolution in Raman analyses matches the limit in IR, at ~1 to 2 cm^{-1}.[29,30,37] Raman spectra of coronary artery have been used to quantify chemical composition of coronary plaques, in terms of cholesterol, cholesterol esters, triglycerides and phospholipids, and calcium salts.[38,49] Multivariate analysis has been used to improve spectral analysis and decomposition of data according to component fractions. We have used Raman microspectroscopy to verify the identity of crystal depositions as calcium hydroxyapatite in cardiomyopathic hamster heart tissue sections.[50]

5.3.2   Spatial Resolution

Different spectroscopies provide information on different scales: magnetic resonance imaging (MRI) can provide ex vivo images of brain morphology; however, the spatial resolution is on the order of mm^3, much larger than the size of a hippocampal neuron (~20 µm) or a fungal hypha (3 to 10 µm). Scanning electron microscopy (SEM) provides images of molecular organization on the order of nanometers, while x-ray crystallography can reveal accurate molecular structures on the scale of bond lengths (Å). The best spatial resolution possible with Raman and IR microspectroscopy falls between 0.5 and 1 µm, respectively. The physical differences between these two methods result in different issues, which we deal with separately.

Spatial resolution in any IR map ultimately depends on how the individual pixels are apertured or on how an area is imaged onto an array detector; the lower limit is ultimately determined by diffraction. The mid-IR region nominally ranges from 4000 to 400 cm^{-1}; this translates into wavelengths of 2.5 to 25 µm. The diffraction limit becomes an issue because the wavelengths of infrared light impose a physical limit on the spatial resolution that may be achieved. The mid-IR region encompasses most of the fundamental vibrational energies of chemical bonds. If the tissue components of interest are lipids, they may be identified through the CH stretch vibrational modes around 2900 cm^{-1}. In this case, the diffraction limit to spatial resolution would be on the order of 3 µm. However, if the primary interest is sugar composition, the vibrational region of interest is likely the sugar ring vibrational modes around 1000 cm^{-1}, and the associated spatial resolution is about 10 µm. Even if the spatial resolution chosen is on the order of 3 µm, it must be remembered that the thickness of the sample represents the third dimension to the sampling volume, and the IR light may be internally scattered slightly on passage through the sample. Finally, experimental considerations and sample condition can affect spatial resolution.

In Raman microspectroscopy, the wavelength of the laser source is much shorter, typically in the near IR to visible spectrum (though deep UV is also being used); spatial resolution of 1 µm is easily feasible. There is a trade-off between IR and Raman, in that the latter spectra are less intense and, particularly in biosamples, occur in conjunction with broadband fluorescence that gives an intense background signal that must be removed though postprocessing of the data. However, Raman spectroscopy is ideal for analysis of microcrystalline inclusions (see below).

5.4   sFTIR Spectra of Fungi

The information obtained from the spatially resolved sFTIR spectra of fungi relates primarily to differences as cells mature and sporulate.

Major differences have been found as hyphae mature;[8] differences between strains and species may be identified and interpreted in terms of cell morphology and physiology.[51] Since the spatial resolution achievable with sFTIR is typically an order of magnitude poorer than that of Raman, and since the information turns out to be quite different also, we deal with the each separately.

5.4.1   Physical Considerations and Spectral Anomalies in sFTIR Spectra

The walls of fungal hyphae are composed primarily of cross-linked sugars; the hyphae may be hundreds of micrometers in length, but are typically only a few micrometers in diameter. Hence, their physical structure presents some challenges for FTIR imaging. We have acquired all our fungal FTIR spectra described here with synchrotron-source IR light, on instruments setup for single pixel or raster scanning.

The differences between the mature regions of hyphae of three fungi, *A. nidulans*, *Neurospora*, and *Rhizopus*, are readily apparent in their sFTIR spectra, Fig. 5.3; see also Szeghalmi et al.[9] The *A. nidulans* hyphal width is about 3 μm, compared to up to 10 μm for the others. The mature hyphal walls, which are rich in cross-linked sugars, exhibit pronounced differences in number, energy, and relative intensity of bands in the sugar region (900 to 1200 cm^{-1}). The CH-stretch region (2800 to 3000 cm^{-1}) also shows marked differences between species. The peak at 2854 cm^{-1} is a marker for CH$_2$ stretch (characteristic of membrane lipids) and is expected to be present surrounding the cell organelles. The CH bonds in the sugar molecules will also contribute

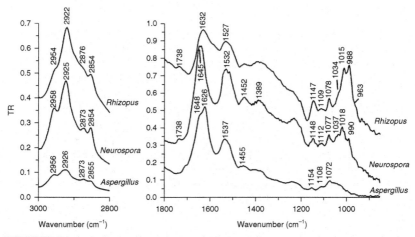

FIGURE 5.3   IR spectra of hyphae from three different filamentous fungi: *Aspergillus nidulans*, *Neurospora*, and *Rhizopus*. Note that the intensity of *Neurospora* spectrum is five-fold increased, for clarity of visual comparison. (*Reproduced from Ref. 9, with permission of Springer Verlag.*)

bands in this region. The variation in the CH profile from *A. nidulans* through *Neurospora* to *Rhizopus* may be attributed to both membranes and walls. This same characteristic feature is also responsible for the increasing slope of the baseline, arising from scattering artifacts associated with the thick, rounded cell structure (see spectral anomalies, below). More recently, we have been using a different preparation, whereby the hyphae are allowed to thaw for 10 seconds at room temperature prior to the drying step. This permits the cells to collapse slightly, reducing the scatter, but is not long enough for enzyme-induced degradation.

Prior to the acquisition of sFTIR data,[51] it had been thought that, within a saprotrophic hypha, the biochemical content would be mainly concentrated toward the growing, organelle-rich, metabolically active tips, since much of the cytoplasm actively migrates to keep up with the extending tip, whereas vacuoles predominate in basal regions.[10] Our sFTIR studies have shown that vacuolate regions are far richer in total biochemical content than had been anticipated. The spectra in Fig. 5.3 not only illustrate the impressive differences in sugar wall composition and structure that exist between different species, they also illustrate that *Aspergillus* hyphae differ biochemically 150 μm from the tip, although atomic force microscopy used for imaging and elastic modulus shows that *Aspergillus* cell wall surfaces have matured by 3 μm from the tip.[52]

The sFTIR spectra have also proved useful in demonstrating drastic differences in cell wall composition in an *A. nidulans* temperature-sensitive mutant, *hypA1*, that had previously been inferred from transmission electron microscopy.[18,51,53] Visible light micrographs of the *hypA1* mutant (Fig. 5.4a) grown at permissive and restrictive temperatures illustrate the morphological differences provoked by growth at slightly elevated temperatures. The sFTIR spectra show significant elevation of absorption across the carbohydrate fingerprint region. The bottom spectrum, showing profiles of *hypA1* tips grown at 28°C, is actually the sum of spectra for five tips that had grown together. Even so, the intensity of the amide I band is <0.1 absorbance unit, much less that that of a single tip grown at 42°C (~0.15 unit). The S/N in the sugar region is only about 3:1, and is heavily affected by diffraction limitations. The restrictive phenotype growth can be seen for the *hypA1* mutant even at 39°C (Fig. 5.4c, top) compared to a wild type *A. nidulans* strain in which long hyphae grow far across the slide. The samples in Fig. 5.4c were grown on the same slide, and thus experienced identical growing conditions.

The greater hyphal diameter of some species presents a different kind of challenge; rounded cell walls can produce huge scattering artifacts, as seen in some much poorer spectra shown in Fig. 5.5 . Similar problems have been identified in the FTIR spectra of cell types, wherein the Mie scatter from the nuclei produced comparable spectral problems.[54]

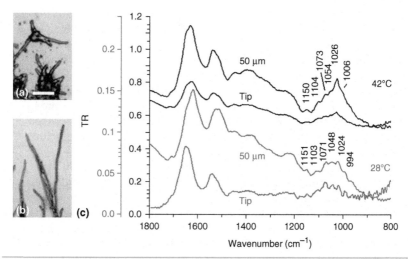

FIGURE 5.4   (a) Photomicrographs of *Aspergillus nidulans* temperature-sensitive *hypA1* morphogenesis mutant cells grown at (a) restrictive (42°C) and (b) permissive (28°C) temperatures. Bar (a) 50 μm, for both. (c) Synchrotron FTIR spectra of *hypA1* hyphae at tip and at 50 μm behind tip for growth at 28°C (red) and 42°C (black). Note the red and black log (1/R) scales for these temperatures. (*Spectra reproduced from Ref. 9, with permission of Springer Verlag.*)

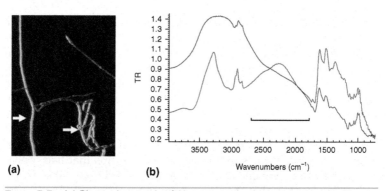

FIGURE 5.5   (a) Photomicrograph of *Neurospora* hyphae grown on a gold-coated Si wafer. Reflected light reveals rounded hyphal walls. (b) sFTIR spectra illustrating scattering artifacts arising from these rounded structures.

Contamination of spectra from materials in the growth medium can be avoided by careful selection of analysis sites. Visual inspection of the prepared samples sometimes reveals a faint halo around the hyphae that are close to the agar block (Fig. 5.4c). Spectra from regions containing medium are compounded of both the mature hypha profile and the protein and sugar content of nonutilized medium. However, where the hyphae have successfully grown far out away from the medium, no such contamination appears; see also the nano-SIMS analyses described later in this chapter.

Other options available on current IR bench instruments are approaching spatial and spectral quality from synchrotron sources. An instrument with a focal plane array (FPA) detector would still require extensive coaddition of scans in order to achieve sufficient signal, but the array collects spectra from an entire map area (e.g., for nominal 5.5 μm pixel, a 64×64 array would cover about 350×350 μm^2) in a single scan. Thus, a 2-3 hour scan would produce excellent S/N and retain spatial resolution of about 7 μm. One problem here is that each individual detector in the array is slightly different, and while one attempts to obtain similar signal at each detector, it is not possible to calibrate the detectors to guarantee uniform absorbance characteristics. A system in which an infrared spectrometer and microscope equipped with an FPA detector has just being commissioned at the Synchrotron Radiation Center, University of Wisconsin at Madison (see Chap. 2).

5.5 Raman Spectroscopy of Fungi

Raman spectroscopy was first shown by C. V. Raman.[55] In this phenomenon, molecular vibrations are excited through inelastic scattering of incident radiation. The magnitude of the scattering cross section for a vibrational mode, counterpart to IR band intensity, depends on the magnitude of the change in the molecular polarizability tensor α during a vibration, where α is a measure of the magnitude of the dipole moment μ induced in a molecule when an electric field ξ is applied:

$$\mu = \alpha\xi \qquad (5.1)$$

The derivative of the polarizability with respect to some normal mode of vibration q is denoted $\partial\alpha/\partial q$. Note that Eq. (5.1), when α is expressed in Å^3, is isomorphic to that defining the dipole moment induced in a perfectly conducting sphere by an external field:

$$\mu = r^3 E \qquad (5.2)$$

This latter relationship is highly important for SERS of nanoparticles (see Sec. 5.6). Raman scattering from molecules provides a molecular fingerprint for every molecule, hence just as for FTIR, Raman offers tremendous potential value for sensitive analyte recognition. The vibrational spectra obtained with Raman can, in principle, be interpreted in terms of molecular composition and in some case, changes in molecular environment. The process of Raman scattering is well understood and is described in many standard texts.[29,30] According to polarizability theory,[56] the differential Raman scattering cross section of a fundamental vibrational band ($\partial\sigma/\partial\Omega$, where Ω = solid angle in this expression) depends on the square of the derivative of the invariants of the molecular polarizability $\bar{\alpha}$ and γ; the trace and anisotropy, respectively, of the polarizability tensor):

$$\left(\frac{\partial\sigma}{\partial\Omega}\right)_i = \frac{\pi^2}{90\varepsilon_0^2}\frac{(\bar{v}_0-\bar{v}_i)^4}{1-e^{-v_i he/kT}}g_i\left[45\left(\frac{\partial\bar{\alpha}}{\partial q_i}\right)^2+7\left(\frac{\partial\gamma}{\partial qi}\right)^2\right] \qquad (5.3)$$

where $\bar{v}_0$ and $\bar{v}_i$ are the wavenumbers of incident light and the molecular vibration, respectively; g_i is the degeneracy of the ith vibrational mode; h, k, T, and ε_0 have the usual meaning (see Refs. 57 to 62). This theory assumes that mechanical and electrical anharmonicity are not significant. Apart from their use as molecular signatures, Raman spectra can provide extremely sensitive information on molecular structure and conformation. See for example the dependence of CH stretching frequency on location within a molecular framework and on orientation to the applied field.[60,61] The Raman trace scattering intensity associated with stretch of individual CH bonds had been calculated at the HF/D95 (d,p) level, for straight chain alkanes to C16 and for the CH bonds in bicyclo-[1.1.1]-pentane. The derivative for the CH bond attached to the terminal methyl group and lying in the plane of the all-trans hydrocarbon chain increases with chain length from $\bar{\alpha}/\partial r$ = 1.05×10^{-30} Cm/V for the CH bond in methane to 1.35×10^{-30} Cm/V for the in-plane CH in hexadecane. The two identical CH lying above and below the chain and the remaining CH methylene bonds located down the length of the chain have derivatives that are smaller (about 0.95×10^{-30} Cm/V), and remain approximately constant regardless of chain length. The calculated and experimental derivatives for the CH methylene and bridgehead CH bonds in bicyclo-[1.1.1]-pentane differ by 25 percent, illustrate the enormous influence of strain and location on individual scattering intensity, even for this simple structural unit. Normal Raman scattering, which is also at the heart of the SERS effect (see Sec. 5.6), depends on the square of this derivative. As is known for SERS experiments, only those molecules in extremely close proximity to the intense localized plasmon field, and usually only those within 10 to 15 nm of the surface, will experience the intense SERS enhancement. The fundamental intensity associated with a particular vibration and the orientation of the functional group relative to the SERS substrate both play a part in the observed SERS intensity, along with enhancement factors that are still being determined.[1-7,63,64] The review by Smith[7] presents a good overview of the state of understanding, without invoking any of the current mathematical models, and serves as an excellent introduction to the phenomenon.

5.5.1   Raman Map from a Hypha, at Growing Tip

Raman maps acquired with the fast acquisition rates now possible permit imaging of fungi in a matter of only a few minutes, for strong spectral features. Where more spectral information is desired, slower scanning can always be chosen to improve the signal to noise. Such maps provide information on the hyphae directly, in the same manner as the sFTIR maps.[51] Different vibrational modes are more prominent, thus complementary information can be obtained.

We have deleted the single-copy UGM sequence (AN3112.4, which we call *ugmA*) from an *A. nidulans nkuAΔ* strain, creating *ugmAΔ*.[11] UgmA catalyzes the conversion between uridine diphosphogalactopyranose (UDP-Gal*p*) and uridine diphosphogalactofuranose (UDP-Gal*f*) in the fungal cell wall. The ugmA gene is nonessential in *A. nidulans*, but the genetic knockout strongly delays asexual sporulation (conidiation) and significantly alters both colony and hyphal morphology. The latter are broader, shorter, and branch more frequently than *A. nidulans* wild type strains. Hyphae of the *ugmAΔ* mutant were grown across a smooth gold-coated substrate at 28°C and pH 6.5 (permissive conditions for normal wild type) and mapped with a Renishaw inVia microscope, 785 nm laser, 50× objective (nominal resolution 2 μm) and laser power 0.5 percent or ~5 mW/cm². A Raman map was taken on a branching hyphal tip growing across the smooth gold surface (Fig. 5.6), at 128 sec/pt, requiring about 2 hours to record (Fig. 5.6). Under these conditions, the main branch produced the characteristic peak at 1050 cm⁻¹ (putative sugar band) with about 4000 total counts relative to a broad fluorescence background of 16,000 counts and an S/N of 6.4. Additional small peaks could just be detected at about 900 and 1004 cm⁻¹ (phenyalanine, hence protein content in the mature hypha).

5.5.2   Raman Map of Spore Branch

A sFTIR map of a spore branch from *Neurospora*, is shown in Fig. 5.7, along with a Raman map recorded on the same region. The spore

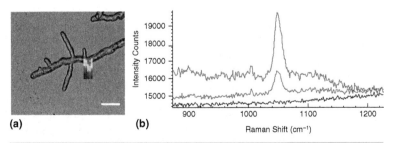

(a)   **(b)**

FIGURE 5.6   (*a*) View of a branching hypha of *Aspergillus nidulans strain ugmAΔ* overlaid with a Raman map, processed on intensity of peak at 1050 cm⁻¹. (*b*) Red spectrum recorded at top center pixel, center of main hypha; blue spectrum from hyphal tip; grey spectrum from bare surface.

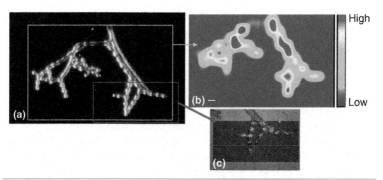

Figure 5.7   (a) Developing spores along a branching structure of *Neurospora*.
(b) The sFTIR map recorded with 8 μm pixel and step size, processed for the
intensity of a CH-stretch band at 2854 cm^{-1}, (c) Raman map acquired on the
lower right branch, 1 μm pixel and step size, 785 nm laser line. [(a) and
(b) reproduced from Ref. 16, with permission of Wiley-Blackwell.]

branch is a fragment that broke during sample acquisition and fell
onto the clean gold surface several mm from the growing colony.
Effectively, the spatial resolution with IR only reveals the general
shape of the branched structure, processing on the CH$_2$ band
indicates lipid within the developing spores. The Raman map has
been processed to show the intensity of the band at 1050 cm^{-1},
typical of the cell wall in this species.

5.5.3   Detection of Crystalline Materials by IR and Raman

In the case of inclusions with fairly high purity, an accurate identity
of the compound can sometimes be made. For example, calcium
hydroxyapatite in calcified tissues,[50] atherosclerotic plaque content,[49]
and silicone in lymph nodes from patients with breast implants[65]
have all been reported. We recently identified elevated deposits of
crystalline creatine in the brain tissue from a transgenic model for
Alzheimer disease[66] using a combination of sFTIR and Raman microscopy. Creatine, with 18 atoms, has a possible 48 vibrational modes,
plus combinations and overtones. The spectral signature is extraordi-
narily distinctive, as each mode or combination appears at a unique
frequency with a unique intensity relative to all the other peaks, and
are easy to detect against the background of normal tissue (Fig. 5.8).

Crystalline compounds can also be associated with fungal
symbionts, such as lichen. For example, usnic acid, a polyketide and
secondary metabolite of certain lichens that has valuable biomedical
properties and also is believed to provide some protective functions
for lichen. More than 800 such natural products have been identified,
most of them are polyketides that have been discovered to possess
multiple useful properties. The lichen *Cladonia* may be found on all
the continents; *C. uncialis* is a slow-growing lichen that can readily
be found in North American forests (Fig. 5.9). The lichen grows over

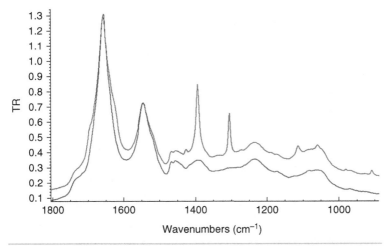

FIGURE 5.8 sFTIR spectrum from grey matter (hippocampus) of transgenic mouse model for Alzheimer disease (blue) and of tissue containing crystalline inclusions of creatine (red). (*Reproduced from Ref. 67.*)

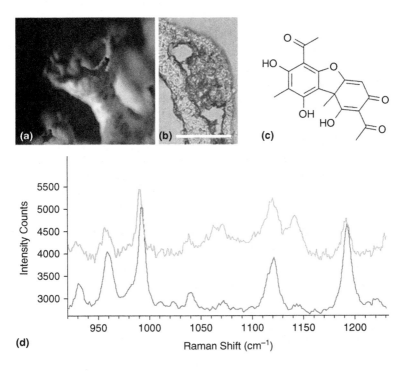

FIGURE 5.9 (*a*) Photomicrograph of the lichen *Cladonia uncialis*. (*b*) Photomicrograph of cryostat section of the lichen, showing the inner and outer walls formed by agglutinated fungal hyphae surrounding an alga, scale bar = 50 μm. (*c*) Molecular structure of usnic acid. (*d*) The Raman spectrum of the sample surface (green) and the Raman spectrum of pure usnic acid crystals (red).

the ground, and may grow over rocks, other plant growth and soils, but does not grow within the soil. Environmental factors may affect gene expression of the polyketide synthase genes of lichen fungi resulting in production of different levels or types of these natural products. Properties of usnic acid include an antibiotic function as part of the lichen's antiherbivore strategy, a strong absorption in the ultraviolet range, allowing it to serve as a lichen "sunblock," as well as anti-inflammatory properties. The compounds are formed from the repeated condensation of carboxylic acids, and are insoluble in water. The material is gradually exuded from the lichen body and may be found as a surface deposit on the exterior of the lichen. Crystalline usnic acid can be detected in Raman spectra of thin sections of *C. uncialis*.[68] Spectra from cryostat sections of the lichen (Fig. 5.9) show a strong signature from the usnic acid exuded by the lichen; the identity is confirmed from spectra of the pure compound, extracted from bulk lichen samples. Spectroscopic tracking of its appearance will assist in identifying the site of production of the polyketide within the thallus tissue.

5.6   SERS Discovery and Development

While spectral signatures in Raman spectroscopy are as unique as those in infrared spectroscopy, the principal drawback to the development of the technique for many decades was the much lower sensitivity. There are many ways of expressing this, for example, only about 1 in 10^7 photons will be Raman scattered; hence, it is far less sensitive technique than IR. This situation was improved by use of lasers as the exciting source nearly 50 years ago but, more importantly, it was changed by the discovery of the potential for surface enhancement of the Raman signal about 30 years ago. The history of that discovery is well-documented, beginning with the experimental observation of immensely and unexpectedly enhanced scattering from pyridine adsorbed onto a roughened silver electrode surface.[69] This experiment was confirmed[70,71] and the first theoretical explanation was presented the following year, wherein the remarkable enhancement was correctly attributed to resonant or preresonant excitation of electrons in the conduction band of the roughened metal surface.[72] The roughened surface was likened to a pseudocolloid, with the bumps on the surface approximating isolated colloidal dots; the resonance condition for absorption of light by the colloid was given by

$$\lambda_R = \lambda_P \left[1 + \varepsilon_b + \frac{(2+q)}{(1-q)} \varepsilon_0 \right]^{1/2} \tag{5.4}$$

in which the resonance wavelength λ_R is defined in terms of the plasmon wavelength, λ_p, ε_b is the interband contribution to the complex dielectric constant, ε_0 is the dielectric constant and q is the volume fraction in the metal particle.[72] This theory was already being developed to explain the colors seen on anodized aluminum, following electrolysis in inorganic salt solutions, and had been shown to be understandable in terms of colloidal particles embedded within the thin dielectric medium.[73]

The intensity of a mode in SERS scatter depends not only on the normal vibrational mode, but also on the orientation of the molecule relative to the surface of the particle, with molecules that are perpendicular to the surface receiving the maximum enhancement. Additional influences in SERS scatter include the possibility of chemical bonding to the surface, charge transfer between the molecule and the surface, and proximity of the exciting laser energy to an electronic absorption of the molecule; the latter would give rise to SERRS, discussed below.

Because of the importance of this discovery, and because the physical explanation for the phenomenon remains incomplete, numerous review articles have appeared in the last 5 years, delineating some of the evolving theories and related practical applications.[1-7] It is not within the scope or goals of the present article to parse these in detail; some of the major points are provided here with references for the interested reader. Before beginning, it is interesting to note the number of papers that self-identify as being concerned with "SERS" and "imaging" has only very recently begun to take off (Fig. 5.10).

The vast majority employ SERS imaging techniques based on nanodots, whether inserted into or taken up by cells whilst alive, or used as tethered tags.[2,6,74,75] The nanodot serves the role of chromophore and reports on the particular site for which it has

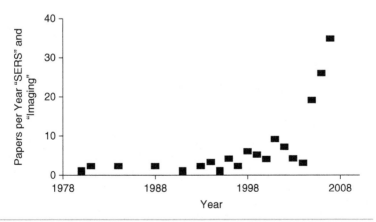

Figure 5.10 Number of papers per year that are self-identified as being concerned with both SERS and imaging from 1978 to 2008.

been manufactured, like a secondary chromophore in fluorescent staining. Only a very few relate to the area map imaging of biotissues mounted onto a regularly patterned SERS active substrate.[5,7,76] We have adopted the latter approach in our study of fungi in order to gain spectroscopic information about the biochemical changes accompanying fungal development from a spore to a mature hypha capable of producing new spores.

As with ordinary Raman spectroscopy, the core of the phenomenon rests on the inelastic scatter from a photon-molecule interaction that produces a second photon, which differs in energy from the incident photon by an amount (either more or less) equal to the vibrational energy of some vibrational mode. The enhancement factor arises from the special interaction between the incident photon and the roughened surface or nanoparticle with which the molecule is in contact.

Electrons in metals can exist in a conduction band, that is, in energy levels that are virtually continuous and high in energy relative to the orbitals of isolated metal atoms. The orbitals of these electrons are delocalized across the metal particle and account for the ease of electrical and thermal conductivity. In the same way that the electrons on a molecule may be polarized by the action of an external field [Eq. (5.1)], the electron distribution in metal nanoparticles may also be polarized. A restoring force acts upon these electrons, due to the strong attraction of the positively charged nuclei which are essentially locked into the metal lattice. When the wavelength of the external electric field is far greater than the dimension of the nanoparticle, the entire particle experiences an effectively homogeneous field, and responds in a uniform manner. For a perfectly uniform metallic sphere, there will be some natural frequency of oscillation of the electrons, acted upon by the nuclear restoring force. If the frequency of incident light matches this resonant frequency, the particle will absorb photons. For gold, silver, and several other coinage metals, the absorption bands of these electrons fall into the visible portion of the spectrum. The electrons in the nanoparticle will now be in a higher energy state in which they oscillate back and forth across the particle, at this resonant frequency: this is the surface plasmon resonance (SPR). The existence of the SPR means that a molecule bound to, or in intimate contact with, the surface will experience a vastly enhanced external field, ξ in Eq. (5.1). The induced dipole moment will be similarly enhanced, as will any derivative of the polarizability, thus the Raman scattering will be enormously enhanced. Not only nanodots, but also hollow gold nanoparticles, silver island films, roughened surfaces, and nanopatterned surfaces have been found to promote the SERS effect. The review articles noted above provide extensive descriptions of the present state of understanding of theory including the various shapes and designs of nanoparticles and nanoparticle aggregates that facilitate the phenomenon. Convincing evidence for the detection of single molecules with SERS has very recently been published.[77] The elegant proof involved detection of a strong Raman scatterer (a dye: Rhodamine 6G) attached to silver aggregate nanoparticles. By ensuring that the number of silver particles greatly exceeded the number of dye molecules, and using two isotopomers which had distinguishable spectra, they were able to demonstrate anticorrelation, that is, the spectra were, in the main, either one isotopomer or the other, but almost never both. The enhancement was aided by the match between the exciting laser line and an electronic transition of the dye molecule, thus the phenomenon was actually SERRS. Discussion continues regarding the possibility of single molecule detection, but the fact of signal enhancement is indisputable, as is the fact that it can be utilized. Ways and means of utilization include tethering chromophores to be used as selective tags, injecting gold nanodots, etc., as discussed in the numerous recent reviews and articles. For fungi, it should be possible to prepare growth media that contains nanoparticles that may interact with the developing hyphae; however, such experiments must be approached with caution, to ensure that abnormal physiological responses are not provoked.

5.6.1   Substrates: The Key to SERS Imaging

In order to obtain further spectroscopic information on fungal development, we have begun a study with SERS imaging that parallels the sFTIR and Raman imaging to date. Nanodots are colloidal particles with a range of diameters; it is now thought that the greatest enhancement may be associated with the small crevices formed within nanodot aggregates. Rather than embark on the development of in-house substrates, for convenience, we elected to use the commercially available Klarite substrates (D3 Technologies Ltd., Glasgow). While nanodots offer much promise for other applications, we have found these regularly nanopatterned substrates provide significant enhancement for the rapid imaging of developing hyphae and possibly of fungal exudates. Other types of regularly patterned substrate are being developed.[76] Many challenges remain; our recent results are described below.

5.6.2   SERS: Applications for Fungi

All fungi secrete compounds during growth; these molecules have roles in substrate adhesion and/or nutrient acquisition. Spores and germination structures of plant pathogens may attach to hydrophobic leaf surfaces by specialized adhesives.[78] Excellent images of hyphal adhesive spore tips and exudates are given in Ref. 78. Much current research is directed toward understanding the fungal "secretome," on the assumption that most interesting molecules are proteins, but spatial resolution and appropriate tools for molecular identification remain challenging issues.

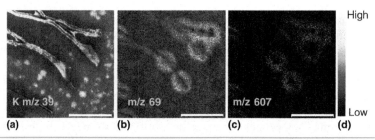

FIGURE 5.11   High-spatial resolution SIMS images of *Aspergillus nidulans*, 256 µm field of view, bar = 100 µm, 1 µm/pixel, positive ions. (*a*) m/z = 39 (likely K⁺); (*b*) m/z = 69 (tentatively C5H9⁺), (*c*) m/z 607 (unknown). Relative intensity from low (dark) to high (white).[79]

Similarly, saprotrophs like *A. nidulans* secrete compounds that likely have a diversity of roles in substrate adhesion and/or nutrient acquisition. We have recently experimented with high-spatial resolution SIMS imaging (Fig. 5.11).

These hyphae were grown over acetone-cleaned soda glass microscope slides from agar plugs, essentially as per sFTIR; frozen to –80°C, freeze dried without interim thaw (three dimensionality retained), gold sputter coated 30 seconds using an Edwards S150B sputter coater (gold coating for SEM typically uses 3 to 6 minute), then shipped under Ar gas for data collection. Sample size was 1 µm × 1 µm × 5 µm thick. Parameters for data collection are similar to those described in Refs. 80 and 81. *Aspergillus nidulans* walls are on the order of 50 nm thick.[11] The K depletion (dark shadow around the hyphae) may be ascribed to sequestration of potassium from hyphal surroundings, and is consistent with the K-rich hyphal walls. In contrast, several organic fragments, such as those for m/z 69 and m/z 607 shown here, appear to be due to exudate secreted by growing hyphae.

SERS can potentially provide valuable information in the study of secreted materials, given that the spectra provide molecular specificity, and the technique provides tremendous sensitivity than can rival fluorescence.[51] The application itself is not yet robust; technical challenges to be met include a much better understanding and control of the spectral blinking phenomenon and the development of appropriate collection parameters that will minimize the danger of photo-oxidation. It is not sensible to use a novel technique for molecular identification when the molecules themselves remain unidentified. Proteins in general have functional group similarities, and a library of specific molecular identifiers does not yet exist. The development of this technique will most likely proceed via advances on both fronts: technical improvements and more standardized procedures on the one hand, better information on the spectral signatures of specific molecular targets on the other. Figure 5.12 shows a SERS map of *Magnaporthe oryzae* (the rice blast fungus) hyphae, grown across the

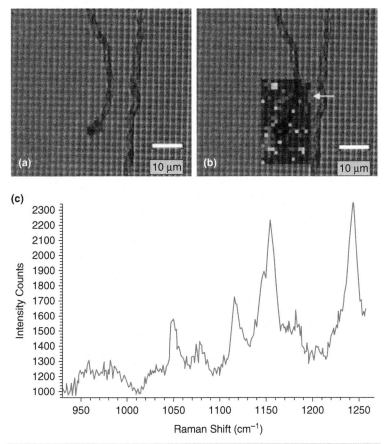

FIGURE 5.12   SERS map of *Magnaporthe oryzae* hyphae. The map was
acquired with the 785 nm linefocus, 1 percent power, 50 × objective, 1200
L/mm grating. Exposure parameter was 10 seconds, 500 points in the map
for a total acquisition time of about 1.5 minutes. (*a*) Photomicrograph of
hyphal tip and a second, longer hypha that has grown past it. (*b*) SERS map
processed for intensity of region at 1244 cm^{-1}. (*c*) Spectrum acquired at
point indicated by yellow arrow in (*b*).

patterned surface of a Klarite substrate. The spectra that appear in the
vicinity of the hypha are quite probably due to fungal exudate.

We have acquired many time lapse spectra on the "hot" spots,
such as that shown in Fig. 5.12*c*. The spectra quite probably are report-
ing a kind of ensemble average SERS, whereby numerous molecules
may be occupying hot spots within the nanopatterned well, as
opposed to single molecule SERS. Time lapse spectra of such hot
spots frequently show considerable strong variation.[51] We are readily
able to observe such variations with any time lapse spectra from hot

spots; however, such data are not reproducible and evidence of sample decomposition often appears after extended dwell times.

The potential usefulness of the method can be seen in the SERS/ non-SERS image of *A. nidulans* hyphae and germinating spores below (Fig. 5.13). The Klarite samples are a 6 × 10 mm gold-coated Si wafer mounted on an ordinary microscope slide (25 × 75 mm), within which a 4 × 4 mm square region has a fabricated nanopatterned surface consisting of a regular array of pyramidal wells that provide SERS enhancement (see Fig. 2D in Ref. 51). SERS samples are prepared in our usual manner: an agar block inoculated with spores was placed

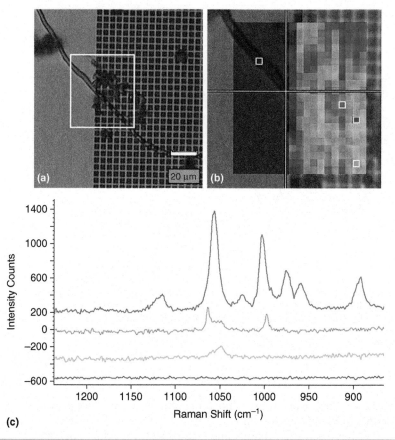

FIGURE 5.13   (a) Photomicrograph of an *Aspergillus nidulans* hypha growing across a Klarite substrate, from nanopatterned surface onto smooth gold surface. Blue box outlines region mapped with Streamline™ option (see text for data collection details). (b) Raman map processed on intensity at the 1050 cm⁻¹ peak, commonly observed in regular Raman spectra of this species. (c) Spectra extracted from the map at pixels indicated by white outline boxes in (b). Spectrum color corresponds to processed pixel color in map (b). Spectra are on matching scale and have been offset for clarity.

adjacent to a Klarite substrate and hyphae were allowed to grow overnight in a moist chamber at 28°C. After freezing and drying, the sample is ready for spectroscopic analysis.

The region shown in Fig. 5.13a contains a hypha that has grown out several millimeters from the agar block and is free of any other material. Several spores from the mature base of the colony have fallen onto the substrate close to the growing hypha, and under the humid growing conditions, have just germinated. Clumps of spores are more likely to germinate in nutrient-free conditions than are isolated spores. Short germ tubes can be seen emerging from several of these spores.

A Raman map of the area was recorded with the 785 nm edge set at 5.0 percent power, 1200 L/mm grating, exposure time 4.00 seconds for an entire column or about 200 ms/point. The SERS enhancement is immediately obvious from the processed image (Fig. 5.13b) that shows the intensity of the peak at 1050 cm^{-1}, characteristic of these hyphae under normal Raman spectroscopy.[51] Four spectra representative of the variety observed in this map are shown in Fig. 5.13c. The weakest intensity (purple) is associated with the hypha growing across the smooth gold surface. The 1050 cm^{-1} peak is not visible; the fact that this region even registers in the mapping image is due to the slight increase in baseline from the natural fluorescence of the biomaterial. The next strongest signal (green) comes from the mature hypha on the SERS patterned surface. Here the 1050 cm^{-1} peak is clearly apparent.

The signal enhancement factor provided by the SERS active substrate cannot be properly quantified from this map, since we have no signal at all from the smooth gold for comparison in this map, but it is clearly several orders of magnitude. An estimate is obtained by comparison between the green spectrum from Fig. 5.13 with a Raman spectrum from a hypha on the smooth gold surface, illustrated in Fig. 5.14. The Raman spectrum was acquired with 640 seconds of exposure at 0.1 percent laser power, compared to 0.045 seconds exposure at 0.5 percent laser power and the same optics for the SERS spectrum. From the signal to noise for these two spectra, scaled for power and exposure time, we get a rough estimate of 4000 as the SERS enhancement factor.

Some additional peaks appear in spectra from regions containing the germinating spores (blue, red) in Fig. 5.13. We interpret these strong spectral peaks as arising from the minute quantities of exudate produced as hyphae extend, from spores as they begin the germination process or from adhesion materials associated with the spore. Such molecules would likely be able to occupy SERS hot spot sites within the pyramidal wells or at the edges of the wells, where the enhanced electric field will provide the expected SERS magnification. Again, neither the molecular identity of these compounds nor the quantity can be established at this point; thus the true SERS enhancement factor cannot be obtained. Likewise, we can only speculate about the possibility that we are detecting spectra from single molecules, or even few molecules.

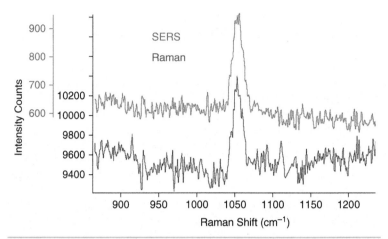

Figure 5.14   Raman (blue) and SERS (red) spectra of hyphae from slide shown in Fig. 5.13. The y axes in blue and red match the corresponding spectra. SERS enhancement is on the order of 4000, see text.

Several variations on SERS are presently being developed. For example, tip enhanced (resonance) Raman spectroscopy [TER(R)S] exploits the molecular sensitivity of SERS, or SERRS, with the tip of a scanning probe microscope.[82] In the latter article, the authors make a strong case for caution, where any of the potential single molecule detection techniques are being applied to complex biological systems. The lessons learned in that article apply equally well to the SERS data presented in this chapter. In particular, we would cite the requirement for distinguishing between genuine single molecule spectra characterized by "blinking" and the rapid fluctuations in spectra resulting from decomposition of material within the intense local energy field. We typically observe the onset of such decomposition as a sudden rise in the fluorescence background, following which the spectrum of carbonaceous material can easily be detected. Control of the experiments to reduce the possibility of these burnouts is not easy but, in general, the use of extremely low laser power and very short dwell times on any one pixel are strongly advised. Thus, these maps serve as tantalizing examples of both the promise and the challenges for regular application of SERS to the understanding of biological problems.

5.7 Conclusions: Lessons Learned, Caveats, Challenges, Promise

The present challenges are a composite of establishing sufficiently robust procedures that the samples may be explored in a simple and easily reproducible manner, and correlating the information from one spectroscopy (and spatial scale) to others. Other imaging tools that we presently employ include the sFTIR and Raman methods described above, several more conventional fluorescence staining, AFM, SEM, TEM, x-ray fluorescence, high-spatial resolution SIMS, as well as the tools of molecular genetics.

Experimental design must be formed around the questions being posed and the methodologies chosen. When considering the desired detection limits in complex biological systems, from desiccated fungal samples described here, to the possibilities of in vivo imaging discussed elsewhere, single molecule detection may be only one of several goals. In the case of fungal analysis with SERS, for example, we must clarify what competes for binding to the most sensitive sites, which compounds might have a more intense signature, and which might interfere. There is now a considerable body of literature on the gross analysis and classification of biomaterials from bacteria[36] and yeasts[47] to cancerous tumors.[83] In the case of fungal identification, it has been shown that different species have different spectroscopic characteristics on which classifications may be based in a statistically robust manner.[84] Our investigations are directed toward understanding fungal lifestyle and the biochemical changes that occur during growth cycles. Spatially resolved molecular information is the goal in this work; single molecule detection may well be unimportant. Multiple copies of the same proteins are likely released, albeit at low levels, to support nutrient acquisition and adhesion activities. Characterization of normal development may not require identification of specific molecules as much as it will require reproducibility of signals. When the latter are achieved, alterations in signals caused by stress, genetic alterations, or antifungal treatments can be more easily detected. The combined application of multiple spectroscopic tools, properly correlated, will continue to be a most fruitful direction of investigation.

Acknowledgments

We are pleased to acknowledge Natural Science and Engineering Research Council (NSERC) Discovery Grant awards to KMG and SGWK, and operating grant to KMG from Canadian Institutes of Health Research (CIHR), and a CIHR-Regional Partnership Program grant to SGWK. Equipment grants were supported by NSERC, Western Economic Diversification Canada, and the University of Manitoba. This work is based in part upon research conducted at the Synchrotron Radiation Center, University of Wisconsin-Madison, which is supported by the National Science Foundation under Award No. DMR-0537588. The technical assistance of Dr. Robert Julian (SRC), Randy Smith and Dr. Lisa Miller (NSLS) is gratefully acknowledged. *Fusarium* was provided by Dr. R. J. Rodriguez (USGS, and University of Washington at Seattle) and Dr. R. S. Redman (University of Washington at Seattle). *Magnaporthe* was provided by Prof. R. A. Dean, North Carolina State University; Profs. M. D. Piercey-Nomore and J. Sorensen (University of Manitoba) provided lichen; all other samples are from SGWK. With the exception of figures reproduced with permissions, all figures are from the Gough and Kaminskyj research groups; we thank the following for their contributions: M. Isenor, K. Jilkine, C. L. Liao, A. M. El-Ganiny, and Dr. S. Gajjeraman.

References

1. R. F. Aroca, *Surface Enhanced Vibrational Spectroscopy*, John Wiley & Sons Ltd., Chichester, 2006, p. 233.
2. K. Kneipp, M. Moskovits, and H. Kneipp, *Surface Enhanced Raman Scattering: Physics and Applications*, Springer-Verlag, Berlin, Heidelberg, 2006, p. 221.
3. E. C. Le Ru, E. Blackie, M. Meyer, and P. G. Etchegoin, "Surface Enhanced Raman Scattering Enhancement Factors: A Comprehensive Study," *Journal of Physical Chemistry C*, 111:13794–13803, 2007.
4. S. Nie, Shuming, and S. R. Emory, "Probing Single Molecules and Single Nanoparticles by Surface-Enhanced Raman Scattering," *Science*, 275:1102–1106, 2997.
5. M. Sackmann, S. Bom, T. Balster, and A. Materny, "Nanostructured Gold Surfaces as Reproducible Substrates for Surface-Enhanced Raman Spectroscopy," *Journal of Raman Spectroscopy*, 38:277–282, 2007.
6. A. M. Schwartzberg and J. Z. Zhang, "Novel Optical Properties and Emerging Applications of Metal Nanostructures," *Journal of Physical Chemistry C*, 112:10323–10337, 2008.
7. W. E. Smith, "Practical Understanding and Use of Surface Enhanced Raman Scattering/Surface Enhanced Resonance Raman Scattering in Chemical and Biological Analysis," *Chemical Society Reviews*, 37:955–964, 2008.
8. K. Jilkine, K. M. Gough, R. Julian, and S. G. W. Kaminskyj, "A Sensitive Method for Examining Whole Cell Biochemical Composition in Single Cells of Filamentous Fungi Using Synchrotron FTIR Spectromicroscopy," *Journal of Inorganic Biochemistry*, 102:540–546, 2008.
9. A. V. Szeghalmi, S. G. W. Kaminskyj, and K. M. Gough, "A Synchrotron FTIR Microspectroscopy Investigation of Fungal Hyphae Grown under Optimal and Stressed Conditions," *Analytical and Bioanalytical Chemistry*, 387:1779–1789, 2007.
10. S. G. W. Kaminskyj and I. B. Heath, "Studies on *Saprolegnia ferax* Suggest the General Importance of the Cytoplasm in Determining Hyphal Morphology," *Mycologia*, 88:20–37, 1996.
11. A. M El-Ganiny, D. A. R. Sanders, and S. G. W. Kaminskyj, "*Aspergillus nidulans* UDP-Galactopyranose Mutase, Encoded by *ugmA* Plays Key Roles in Colony Growth, Hyphal Morphogensis, and Conidiation," *Fungal Genetics and Biology*, 45:1533–1542.
12. de Groot, P. J. W., B. W. Brandt, H. Horiuchi, A. F. J. Ram, C. G. De Koster, and F. M. Klis, "Comprehensive Genomic Analysis of Cell Wall Genes in Aspergillus Nidulans," *Fungal Genetics and Biology*, 46:S72–S81, 2009.
13. K. A. Pirozynski and D. W. Malloch, "The Origin of Land Plants: A Matter of Mycotrophism," *Biosystems*, 6:153–164, 1995.
14. M. Krings, T. Taylor, H. Hass, H. Kerp, N. Dotzler, and E. Hermsen, "Fungal Endophytes in a 400-Million-Yr-Old Land Plant: Infection Pathways, Spatial Distribution, and Host Responses," *New Phytologist*, 174:648–657, 2007.
15. R. J. Rodriguez, J. F. White, A. E. Arnold, and R. S. Redman, "Fungal Endophytes: Diversity and Functional Roles," *New Phytologist*, 182:314–330, 2009.
16. S. G. W. Kaminskyj, A. V. Szeghalmi, K. Jilkine, and K. M. Gough, "High Spatial Resolution Methods for Studying Subcellular Growth and Environmental Responses of Fungal Hyphae," *FEMS Microbiology Letters*, 284:1–8, 2008.

17. S. G. W. Kaminskyj, A. Garrill, and I. B. Heath," The Relation between Turgor and Tip Growth in *Saprolegnia ferax*: Turgor is Necessary, but Not Sufficient to Explain Apical Extension Rates," *Experimental Mycology*, **16**:64–75, 1992.

18. X. Shi, Y. Sha, and S. G. W. Kaminskyj, "*Aspergillus nidulans hypA* Regulates Morphogenesis through the Secretion Pathway," *Fungal Genetics and Biology*, **41**:75–88, 2004.

19. S. G. W. Kaminskyj, "Septum Position Is Marked at the Tip of *Aspergillus nidulans* Hyphae," *Fungal Genetics and Biology*, **31**:105–113, 2000.

20. G. Jedd and N.H. Chua, "A New Self-Assembled Peroxisomal Vesicle Required for Efficient Resealing of the Plasma Membrane," *Nature Cell Biology*, **2**:226–231, 2000.

21. S. D. Harris, J. L. Morrell, and J. E. Hamer, "Identification and Characterization of *Aspergillus nidulans* Mutants Defective in Cytokinesis," *Genetics*, **136**:517–532, 1994.

22. S. D. Harris, "Hyphal morphogenesis in *Aspergillus nidulans*. The Aspergilli: Genomics, Medical Aspects, Biotechnology, and Research Methods," S. A. Osmani and G. H. Goldman, (eds.), CRC Press, Boca Raton FL, pp. 211–222, 2008.

23. C.W. Mims, E. A. Richardson, and W. E. Timberlake, "Ultrastructural Analysis of Conidiophore Development in the Fungus *Aspergillus Nidulans* Using Freeze-Substitution," *Protoplasma*, **144**:132–141, 1988.

24. F. Berger, A. Taylor, and C. Brownlee, "Cell Fate Determination by the Cell Wall in Early Fucus Development," *Science*, **263**:1421–1423, 1994.

25. I. B. Heath, and R. L. Harold," Actin has Multiple Roles in the Formation and Architecture of Zoospores of the Oomycetes, *Saprolegnia ferax* and *Achlya bisexualis*," *Journal of Cell Science*, **102**:611–627, 1992.

26. K. G. A. van Driel, T. Boekhout, H. A. B. Wösten, J. Arie, A. J. Verkleij, and W. H. Müller, "Laser Microdissection of Fungal Septa as Visualized by Scanning Electron Microscopy," *Fungal Genetics and Biology*, **44**:466–473, 2007.

27. E. B. Wilson, J. C. Decius, and P. C. Cross, *Molecular Vibrations: The Theory of Infrared and Raman Vibrational Spectra*," McGraw-Hill Book Co. Inc., New York.

28. P. R. Griffiths and J. de Haseth, *Fourier Transform Infrared Spectroscopy*, 2nd ed., Wiley-Interscience, New Jersey, 2007, p. 529.

29. R. L. McCreery, Raman Spectroscopy for Chemical Analysis, in: *Chemical Analysis*, Vol. 157, J. D. Winefordner (ed.), Wiley Interscience, New York, 2000, p. 420.

30. J. G. Grasselli and B. J. Bulkin, *Analytical Raman Spectroscopy*, Wiley-Interscience, New York, 1992, p. 462.

31. E. R.. Blout and M. Fields, "On the Infrared Spectra of Nucleic Acids and Certain of Their Components," *Science*, **107**:252, 1948.

32. E. R. Blout and R.C. Mellors, "Infrared Spectra of Tissues," *Science*, **110**:137–138, 1940.

33. A. Eliot and E. J. Ambrose, "Structure of Synthetic Polypeptides," *Nature*, **165**:921–922, 1950.

34. R. Bhargava, and I. W. Levin, "Infrared Spectroscopic Imaging Protocols for High-Throughput Histopathology," in: *Vibrational Spectroscopy for Medical Diagnosis*, M. Diem, P. R. Griffiths, and J. M. Chalmers (eds.), J. Wiley & Sons Ltd., Chichester 2008, pp. 155–185.

35. D. Naumann, "The Ultra Rapid Differentiation and Identification of Pathogenic Bacteria Using FT-IR and Multi-Variate Analysis," in: *Fourier and Computerized Spectroscopy*," J. G. Grasselli and D. G. Cameron (eds.), SPIE 553, Bellingham, Wash., 1985, pp. 268–269.

36. D. Naumann, C. P. Schultz, and D. Helm, "What Can Infrared Spectroscopy Tell Us about the Structure and Composition of Intact Bacterial Cells?," in: *Infrared Spectroscopy of Biomolecules*" H. H. Mantsch and D. Chapman (eds.), Wiley-Liss, Toronto, 1996, pp. 279–310.

37. G. J. Puppels, F. F. M. de Mul, C. Otto, J. Greve, M. Robert-Nicoud, D. J. Arndt-Jovin, and T. M. Jovin, "Studying Single Living Cells and Chromosomes by Confocal Raman Microspectroscopy," *Nature*, **347**:301–303, 1990.

38. J. P. Salenius, J. F. Brennan, A. Miller, Y. Wang, T. Aretz, B. Sacks, R. R. Dasari, and M. S. Feld, "Biochemical Composition of Human Peripheral Arteries Examined with Near-Infrared Raman Spectroscopy," *Journal of Vascular Surgery*, **27**:710–719, 1998.

39. G. L. Carr, "Resolution Limits for Infrared Microspectroscopy Explored with Synchrotron Radiation," *Reviews of Scientific Instruments*, **72**:1613–1619, 2001.

40. P. Lasch and D. Naumann, "Spatial Resolution in Infrared Microspectroscopic Imaging of Tissues," *Biochimica et Biophysica Acta*, **1758**:814–829, 2006.

41. L. M. Miller and R. J. Smith, "Synchrotrons versus Globars, Point-Detectors versus Focal Plane Arrays," *Vibrational Spectroscopy*, **38**:237–240, 2005.

42. J. H. Schachtschneider and R. G. Snyder, "Vibrational Analysis of the *n*-Paraffins. II. Normal Co-ordinate Calculations," *Spectrochimica Acta*, **19**:117–168, 1963.

43. S. M. LeVine and D. L. Wetzel, "In situ Chemical Analyses from Frozen Tissue Sections by Fourier Transform Infrared Microspectroscopy: Examination of White Matter Exposed to Extravasated Blood in the Rat Brain," *American Journal of Pathology*, **145**:1041–1047, 1994.

44. K. M. Gough, M. Rak, M. Bookatz, M. R. Del Bigio, S. Mai, and D. Westaway, "Choices for Tissue Visualization with IR Microspectroscopy," *Vibrational Spectroscopy*, **38**:133–141, 2005.

45. M. Rak, M. R. Del Bigio, S. Mai, D. Westaway, and K. M. Gough, "Dense-Core and Diffuse Aβ Plaques in TgCRND8 Mice Studied with sFTIR Microspectroscopy," *Biopolymers*, **87**:207–217, 2007.

46. N. M. Amiali, M. R. Mulvey, B. Berger-Bächi, J. Sedman, A. E. Simor, and A. A. Ismail, "Evaluation of Fourier Transform Infrared Spectroscopy for the Rapid Identification of Glycopeptide-Intermediate *Staphylococcus aureus*," *Journal of Antimicrobial Chemotherapy*, **61**:95–102, 2008.

47. I. Adt, D. Toubas, J. M. Pinon, M. Manfait, and G. D. Sockalingum, "FTIR Spectroscopy as a Potential Tool to Analyze Structural Modifications During Morphogenesis of *Candida albicans*," *Archives of Microbiology*, **185**:277–285, 2006.

48. P. Lasch, W. Haensch, D. Naumann, and M. Diem, "Imaging of Colorectal Cancer Adenocarcinoma Using FT-IR Microspectroscopy and Cluster Analysis," *Biochimica et Biophysica Acta*, **1688**:176–186, 2004.

49. T. J. Römer, J. F. Brennan, G. J. Puppels, A. H. Zwinderman, S. G. van Duinen, A. van der Laarse, A. F. W. van der Steen, N. A. Bom, and A. V. G. Bruschke, "Intravascular Ultrasound Combined with Raman Spectroscopy to Localize and Quantify Cholesterol and Calcium Salts in Atherosclerotic Coronary Arteries," *Arteriosclerosis, Thrombosis, and Vascular Biology*, **20**:478–483, 2000.

50. K. M. Gough, I. M. C. Dixon, D. Zielinski, R. Wiens, and M. Rak, "FTIR Evaluation of Microscopic Scarring in the Cardiomyopathic Heart: Effect of Chronic AT1 Receptor Blockade," *Analytical Biochemistry*, **316**:232–242, 2003.

51. A. Szeghalmi, S. Kaminskyj, P. Rösch, J. Popp, and K. M. Gough, "Time-Fluctuations and Imaging in the SERS Spectra of Fungal Hyphae Grown on Nanostructured Substrates," *Journal of Physical Chemistry B*, **111**:12916–12924, 2007.

52. S. G. W. Kaminskyj and T. E. S. Dahms, "High Spatial Resolution Surface Imaging and Analysis Of Fungal Cells using SEM and AFM," *Micron*, **39**:349–361, 2008.

53. S. Kaminskyj, and M. Boire, "Ultrastructure of the *Aspergillus nidulans hypA1* Restrictive Phenotype Shows Defects in Endomembrane Arrays and Polarized Wall Deposition," *Canadian Journal of Botany*, **82**:807–814, 2004.

54. B. Molenhoff, M. Romeo, M. Diem, and B. R. Wood, "Mie-Type Scattering and Non-Beer-Lambert Absorption Behavior of Human Cells in Infrared Microspectroscopy," *Biophysical Journal*, **88**:3635–3640, 2005.

55. C. V. Raman and K. S. Krishnan, "The Optical Analog of the Compton Effect," *Nature*, **121**:711, 1928.

56. G. Placzek, "U. S. Atomic Energy Commission, UCRL-Trans-524(L); 1962," Translated from *Handbuch der Radiologie*, 2d ed., E. Marx, (ed.), Akademisch, Leipzig, 1934, Vol. 6, Part II, pp. 205–374.

57. K. M. Gough and W. Murphy, "Intensity Parameters for Raman Trace Scattering of Ethane," *Journal of Chemical Physics*, **85**:4290–4296, 1986.
58. K. M. Gough and W. F. Murphy, "Harmonic Force Field and Raman Trace Scattering Parameters in Cyclohexane," *Journal of Chemical Physics*, **87**:1509–1519, 1987.
59. K. M. Gough and W. F. Murphy, "The Raman Scattering Intensity Parameters in Acetylene," *Journal of Molecular Structure*, **224**:73–88, 1990.
60. K. M. Gough, R. Dawes, J. R. Dwyer, and T. Welshman, "QTAIM Analysis of Raman Scattering Intensities: Insights into the Relationship Between Molecular Structure and Electronic Charge Flow," in: *The Quantum Theory of Atoms in Molecules. From Solid State to DNA and Drug Design*, C. Matta and R. Boyd (eds.), Wiley-VCH, Cap. Weinheim, 2007, pp. 95–120.
61. R. Dawes and K.M. Gough, "Absolute Intensities of Raman Trace Scattering from Bicyclo-[1.1.1]-Pentane," *Journal of Chemical Physics*, **121**:1278–1284, 2004.
62. R. Dawes, "Theoretical and Experimental Investigations in Chemistry, Ph.D. Thesis, University of Manitoba, 2004.
63. K. L. Kelly, E. Coronado, L. L. Zhao, and G. C. Schatz, "The Optical Properties of Metal Nanoparticles: The Influence Of Size, Shape, And Dielectric Environment," *Journal of Physical Chemistry B*, **107**:668–677, 2003.
64. K. Hering, D. Cialla, K. Ackermann, T. Dorfer, R. Moller, H. Schneidewind, R. Mattheis, W. Fritzsche, P. Rosch, and J. Popp, "SERS: A Versatile Tool in Chemical and Biochemical Diagnostics," *Analytical and Bioanalytical Chemistry*, **390**:113–124, 2008.
65. W. E. Katzin, J. A. Centeno, L. J. Feng, M. Kiley, and F. G. Mullick, "Pathology of Lymph Nodes from Patients with Breast Implants: A Histologic and Spectroscopic Evaluation," *The American Journal of Surgical Pathology*, **29**:506–511, 2005.
66. M. Gallant, M. Rak, A. Szeghalmi, M. R. Del Bigio, D. Westaway, J. Yang, R. Julian, and K. M. Gough, "Focally Elevated Creatine Detected in Amyloid Precursor Protein (APP) Transgenic Mice and Alzheimer Disease Brain Tissue," *Journal of Biological Chemistry*, **281**:5-8, 2006.
67. M. Rak, "Synchrotron Infrared Microspectroscopy of Biological Tissues: Brain Tissue fromTgCRND8 Alzheimer's Disease Mice and Developing Scar Tissue in Rats." Ph.D. Thesis, University of Manitoba, 2007.
68. M. D. Piercey-Normore, "The genus *Cladonia* in Manitoba: Exploring Taxonomic Trends with Secondary Metabolites," *Mycotaxonomy*, **101**:189–199, 2007.
69. M. Fleischmann, P. J. Hendra, and A. J. McQuillan, "Raman Spectra of Pyridine Adsorbed at a Silver Electrode," *Chemical Physics Letters*, **26**:163–166, 1974.
70. D. L. Jeanmaire and R. P. Van Duyne, "Surface Raman Spectroelectrochemistry. Part I. Heterocyclic, Aromatic, and Aliphatic Amines Adsorbed on the Anodized Silver Electrode," *Journal of Electroanalytical Chemistry and Interfacial Electrochemistry*, **84**:1–20, 1977.
71. M. G. Albrecht and J. A. Creighton, "Anomalously Intense Raman Spectra of Pyridine at a Silver Electrode," *Journal of the American Chemical Society*, **99**:5215–5217, 1977.
72. M. Moskovits, "Surface Roughness and the Enhanced Intensity of Raman Scattering by Molecules Adsorbed on Metals," *Journal of Chemical Physics*, **69**:4159–4161, 1978.
73. J. A. Creighton, C. G. Blatchford, M. G. Albrecht, "Plasma Resonance Enhancement of Raman Scattering by Pyridine Adsorbed on Silver or Gold Sol Particles of Size Comparable to the Excitation Wavelength," *Journal of the Chemical Society, Faraday Transactions 2*, **75**:790–800, 1979.
74. J. Kneipp, H. Kneipp, and K. Kneipp, "SERS —A Single-Molecule and Nanoscale Tool for Bioanalysis," *Chemical Society Reviews*, **37**:1052–1060, 2008.
75. S. Lee, S. Kim, J. Choo, S.Y. Shin, Y.H. Lee, H.Y. Choi, S. Ha, K. Kang, and C. H. Oh, " Biological Imaging of HEK293 Cells Expressing PCLγ1 Using Surface Enhanced Raman Microscopy," *Analytical Chemistry*, **79**:916–922, 2007.
76. D. Cialla, U. Hubner, H. Schneidewind, R. Moller, and J. Popp, "Probing Microfabricated Substrates for Their Reproducible SERS Activity," *ChemPhysChem*, **9**:758–762, 2008.

77. J. A. Dieringer, R. B. Lettan, K. A. Scheidt, and R. P. Van Duyne, "A Frequency-Domain Existence Proof of Single-Molecule Surface-Enhanced Raman Spectroscopy," *Journal of the American Chemical Society*, **129**:16249–16256, 2007.

78. E. D. Braun and R. J. Howard, "Adhesion of Fungal Spores and Germlings to Host Plant Surfaces," *Protoplasma*, **181**:202–212, 1994.

79. E. B. Monroe, S. G. W. Kaminskyj, and J. V. Sweedler, private communication.

80. E. B. Monroe, J. C. Jurchen, J. Lee, S. S. Rubakhin, and J. V. Sweedler, "Vitamin E Imaging and Localization in the Neuronal Membrane," *Journal of the American Chemical Society*, **127**:12152–12153, 2005.

81. S. S. Rubakhin, J. C. Jurchen, E. B. Monroe, and J. V. Sweedler, "Imaging Mass Spectrometry: Fundamentals and Applications to Drug Discovery," *Drug Discovery Today*, **10**:823–837, 2005.

82. K. F. Domke, D. Zhang, and B. Pettinger, "Enhanced Raman Spectroscopy: Single Molecules or Carbon?," *Journal of Physical Chemistry C*, **111**:8611–8616, 2007.

83. B. W. D. de Jong, T. C. Bakker Schut, K. Maquelin, T. van der Kwast, C. H. Bangma, D. J. Kok, and G. J. Puppels, "Discrimination between Nontumor Bladder Tissue and Tumor by Raman Spectroscopy," *Analytical Chemistry*, **78**:7761–7769, 2006.

84. G. Fischer, S. Braun, R. Thissen, and W. Dott, "FT-IR Spectroscopy as a Tool for Rapid Identification and Intra-Species Characterization of Airborne Filamentous Fungi," *Journal of Microbiological Methods*, **64**:63–77, 2006.

CHAPTER 6

Widefield Raman Imaging of Cells and Tissues

Shona Stewart, Janice Panza, Amy Drauch

ChemImage Corporation, Pittsburgh, Pennsylvania, USA

6.1 Introduction

Raman spectroscopy is an analytical tool commonly utilized in research and industrial laboratories of many disciplines. That the Raman scattering phenomenon is highly selective and requires little sample preparation makes it a very promising option for analytical needs. For industries involving high Raman scatterers, such as pharmaceutical and polymer materials, Raman methodologies are routine. Recently there has been a move for many analytical technologies to progress to the imaging paradigm. Images, convenient to convey and comprehend, are becoming more in demand by both practitioners and clients. Chemical images contain, in addition to morphological information, data that describes the molecular composition of a sample. Raman imaging combines Raman spectroscopy with digital imaging technology in order to visualize material chemical composition and molecular structure. There are a variety of ways to generate a Raman image from dispersive spectra; however, these have their own limitations. Widefield Raman imaging provides spectral information of all pixels of an entire field of view at once and thus can be considered to be a more ideal mode of acquisition. This chapter describes widefield Raman imaging, the technological issues involved in the acquisition and preprocessing of data, and the methods that can be employed to analyze the large datasets that result from such experiments. The chapter also describes the state of the technology with respect to the study of cells and tissues.

6.2   Generation of Raman Images

Raman images have evolved over many years, and can be generated using several different modes. The modes in which Raman dispersive spectra are acquired from several adjacent locations within a sample and then stitched together to form an image are called *mapping* experiments. Employment of specialized optical components in order to collect an image without moving the sample is often called *widefield* imaging.

6.2.1   Point Mapping

In point mapping, a Raman image is acquired one pixel at a time by using a point-focused laser beam to collect Raman spectra of adjacent pixels in a grid pattern. This is still the most commonly used mode of generating a Raman image, and has been employed to collect images of a variety of materials, including human cells,[1–4] yeast and plant cells,[5,6] biopsy tissues,[7] food materials,[5] tissue adhesives,[5] emulsions and resins,[8,9] and carbon nanotubes.[10] Point mapping has been used in conjunction with other techniques such as surface enhanced Raman scattering (SERS)[5,11,12] and tip enhanced Raman experiments.[13,14]

The point-mapping approach suffers from several technical drawbacks. Up until now, acquisition of a full complement of spectra in confocal mode to construct an image has taken a long time. The time of acquisition cannot be reduced by increasing laser power because the high laser power density in confocal spectroscopy may damage samples. As a result, generated images have tended to have low-image definition. Furthermore, because of the large time of acquisition, it is commonly suspected that the first few pixels of data acquired may have differing background influences than those at the end of the sample grid. Recent improvements in Raman spectrometers and decreasing time of spectral acquisition, will no doubt have an impact on images generated by point mapping. Other factors which impact the quality of point-mapping images are the limitation of spatial resolution of an image by the size of the laser spot and the microscope stage's mechanical capabilities.

6.2.2   Line Mapping

In line mapping, some of the point-mapping issues are overcome. During excitation, the laser beam is spread into a line with the use of optics, reducing the laser power density and potential for sample damage. At the same time, because a large number of pixels are being illuminated at once, acquisition of the full image tends to be faster. In addition, the optics required to generate line-scanned Raman maps are commercially available. Line mapping has been typically performed by scanning a laser spot with a scanning mirror or using cylindrical optics.[15,16,17] Further specialized optics have been noted to increase beam uniformity across a line.[18,19] Line mapping has been demonstrated to be useful in many applications, including work reported by the Morris group at the University of Michigan,[20,21,22] measurement of chemical composition of plant tissue[23] and investigation of polymer crystallinity.[19]

6.2.3 Other Modes of Generating Raman Images

While point and line mapping techniques are the most typical modes of generating Raman images, other imaging methodologies reported in the literature show promise for progress. Ma and Ben-Amotz described a fiber-bundle image compression technique for acquiring full Raman images with high spatial and spectral resolution in a short period of time.[24] Schulmerich et al. demonstrated that laser power distribution in such a technique not only eliminates thermal damage, but also allows subsurface mapping of polymer samples beneath several millimeters of Teflon.[25] In further work, Schulmerich et al. developed a new system in which Raman maps were collected in concentric circles by a circular fiber-optic array, also enabling the recovery of subsurface Raman data.[26] In another variation of Raman imaging, Ozaki's group describe a Raman imaging probe in which the images at different Raman shifts are achieved by tuning the wavelength of a tunable laser for excitation.[27]

6.2.4 Widefield Imaging

In widefield imaging, also known as chemical imaging, molecular imaging, hyperspectral imaging, direct imaging and global imaging, the entire sample field of view is illuminated and analyzed simultaneously by recording an image at discrete wavenumber increments through the imaging spectrometer. Collected light is measured as a function of both location in the sample field of view and wavelength. In the image domain, the recorded data contains spatial information at each wavelength. The spectral domain describes the molecules contained in each pixel within the field of view. As a result, the widefield image contains both structural and compositional information.[28]

A variety of technologies have been used to achieve widefield Raman imaging. They include dielectric filters in conjunction with tunable lasers,[29] rotating dielectric filters,[30] acousto-optic tunable filters (AOTF)[31] and liquid crystal tunable filters (LCTF).[32] One of the first reported widefield Raman images was collected using AOTF technology. Treado and coworkers demonstrated this Raman imaging approach using a commercially available AOTF to obtain rapidly high-fidelity images to monitor lipid/peptide model systems[31] and identify inclusions in breast tissue.[33]

First to develop an LCTF with a spectral resolution[32] of 7 cm^{-1} and patent the technology for acquisition of widefield Raman images,[34] Treado's group further applied LCTF instrumentation to collect widefield Raman images of chicken breast tissue.[35] At the time, Schaeberle and coworkers claimed that the LCTF is the only demonstrated technology for simultaneously providing high-spatial and spectral resolution.[36] Indeed LCTF-driven widefield imaging has continued to progress in the Raman field.

In general, liquid crystal devices provide diffraction-limited spatial resolution. The spectral resolution of LCTFs is comparable to that provided by a single stage dispersive monochromator. In addition, liquid crystal technology provides high out-of-band rejection, a broad, free spectral range, a moderate transmittance and throughput, and computer-controlled tuning.[37]

Figure 6.1 is a schematic diagram of LCTF-generated imaging. When a field of view of a sample is illuminated by the laser source, the scattered light passes through the LCTF before being imaged onto the charge-coupled device (CCD) camera. The LCTF is electronically tuned to allow throughput of scattered light at individual wavelengths (i.e., Raman shifts), until a whole range of wavelength data has been collected (Fig. 6.1a). Each image frame contains spatial information at a particular Raman shift. Consequently, each frame pixel contains a Raman spectrum in the spectral dimension (Fig. 6.1d). Spectral differences provide image contrast. What might be a virtually featureless field of view (Fig. 6.1b) can be highlighted with Raman imaging using the chemical differences within the sample (Fig. 6.1c).

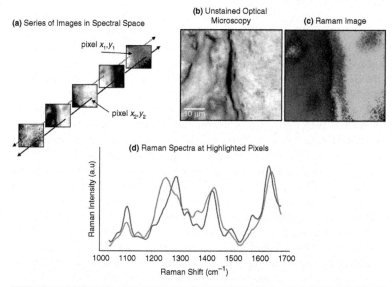

(a) Series of Images in Spectral Space

pixel x_1, y_1

pixel x_2, y_2

(b) Unstained Optical Microscopy

10 μm

(c) Ramam Image

(d) Raman Spectra at Highlighted Pixels

Raman Intensity (a.u)

Raman Shift (cm^{-1})

1000 1100 1200 1300 1400 1500 1600 1700

FIGURE 6.1   Schematic diagram of widefield imaging: (a) tuning the LCTF allows Raman scattered light at different wavelengths to pass to the detector; (b) the original, unstained sample is virtually featureless; (c) the corresponding Raman image, false colored to reflect Raman data, exhibits high contrast; (d) Raman spectra of highlighted pixels x_1, y_1 and x_2, y_2 contain spectral information that indicates different chemical composition at those locations in the sample field of view.

Because both spatial dimensions are collected simultaneously in widefield imaging, the duration of a widefield experiment is proportional to the number of spectral channels and not to the number of pixels. This is particularly advantageous when a limited spectral range provides sufficient chemical and spatial information. By reducing the number of frames required, experimental time decreases without losing spatial resolution. The spatial resolution of the widefield experiment is determined by the combination of diffraction, the CCD pixel size and the microscope magnification at the CCD. Diffraction limited resolution has been reported down to 250 nm.[36]

Used in conjunction with high efficiency solid state lasers, holographic notch rejection filters, and CCD detectors, the quality of widefield Raman images has improved greatly. In addition, the development of microscopic imaging has enabled the acquisition of even higher fidelity information. Widefield Raman imaging has been employed in a variety of applications, including the forensic,[38] pharmaceutical,[39-44] biomedical[45-47] and threat detection[48,49] industries.

6.3 Raman Imaging of Cells and Tissues

Chemical imaging has its roots in biomedical applications, using fluorescent labels to map concentrations of calcium ions and other metabolites.[50] Much of the reported Raman imaging studies on cells and tissues has been carried out by point and line mapping. In contrast, the body of work involving widefield Raman imaging for the analysis of cells and tissues is much less extensive.

Bone is a tissue well suited to Raman techniques. The minerals in bone are high Raman scatterers that have been found to reflect the state of the tissue. Otto and coworkers[51] utilized line scanning technology to show the distribution of hydroxyapatite in bone implants. They were able to ascertain the point in the bone at which there was a transition from crystalline hydroxyapatite of the implant coating to the weaker bone tissue material itself. The Morris group has also employed Raman line mapping to study bone tissue. In their early work, they reported that observation of the mineral and protein bands in bone provide information about maturity[20] and microstructure of human bone.[21] In further investigations, they utilized line mapping to identify and describe the effect of fatigue in bone.[22]

Krafft and coworkers used point mapping to assess the diagnostic potential of Raman images to distinguish between normal brain tissue and the human intracranial tumors, gliomas and meningeomas.[7] They were able to characterize point to point spatial variations in normal brain tissue and in intracranial tumors. In addition, they identified homogeneities in tissue composition, and quantified the relative concentrations of proteins, lipids, and water in the brain tissue.

With the advent of improved spectrometers, Raman microscopes, and faster CCDs, Raman spectroscopy of cells has improved greatly. Uzunbajakava et al. showed Raman mapped images of protein, DNA, RNA, and lipid distribution in live and apoptotic cells, and determined the effect of high laser powers on their samples.[1,2] In the same year Krafft and coworkers observed in their mapped Raman images nucleus, organelles, and membranes of freeze-dried and living cells. They found that the quality of the images was influenced by the hydration state of the molecules in the cells.[3] In 2005, Creely et al. utilized optical trapping technology to collect Raman mapped images of living cells in suspension.[4] The research group of Diem demonstrated label-free detection of drug uptake in cells with Raman mapped images. By staining the Raman mapped cell after acquisition, they were able to prove that mitochondria-rich regions were correctly identified in the Raman image.[52]

Recently, Raman mapping and imaging have been employed in conjunction with labeling or signal enhancement techniques. In work reported by Keren et al., SERS nanoparticles were injected into a live mouse. These were observed in the mouse's liver by collecting SERS mapped images through the mouse's skin.[5] Chao et al. used nanometer-sized diamond particles as detection probes for biolabeling human lung epithelial cells at specific interaction sites.[53] Raman point mapping of labeled cells revealed an easy method for monitoring cell interactions.

The widefield Raman imaging literature is dominated by Treado and coworkers, who were first to demonstrate a novel method for measuring widefield Raman images using an AOTF for wavelength selection.[31] In 1996, Schaeberle et al. demonstrated the first application of widefield Raman imaging (called "Raman chemical imaging" by the group) to pathology by targeting polymer inclusions sometimes found in augmented breast tissues. That same year, Morris et al. introduced the LCTF as an alternative method for collecting widefield Raman images of biological materials.[32] Kline and Treado employed LCTF technology to report Raman images of chicken breast tissue.[35] They demonstrated the capability of this technology to highlight lipid and protein distribution within the tissue.

In more recent tissue work, the results of a preliminary study indicate that widefield Raman imaging can differentiate between two types of kidney cancer commonly difficult to diagnose by pathologists.[46] In this investigation, Raman images, collected from identifiable kidney cancers, were used to create a classification model. The generated model was then utilized to classify image data collected from test biopsy samples. The model exhibited 79 percent sensitivity and 85 percent specificity for the relevant kidney cancers.

Widefield Raman imaging has also been employed at a cellular level. Maier and Treado reported three-dimensional widefield Raman imaging of a single human cell.[54] They discussed the use of deconvolution and show the improvements to the image as a result of the methodology. Panza and Maier demonstrated the employment of widefield Raman imaging to determine the viability of mammalian cells.[45] From Raman image data, they created a classification model to differentiate between normal and apoptotic cells in a prostate cancer cell line.

Stewart and coworkers demonstrated widefield Raman imaging of the waterborne pathogen *Cryptosporidium parvum*.[47] They demonstrated the capability to identify, from a Raman library of microorganism spectra, a single *C. parvum* spore. They also discussed the potential for quantification of samples by Raman imaging and demonstrated that Raman signal increased with increasing *C. parvum* spores. Escoriza et al. also illustrated quantification of waterborne bacteria using widefield Raman imaging.[55] They observed that in higher concentrations, the Raman signal of bacteria directly correlates with the number of cells present in drops of water on sample slides. At very low concentrations (e.g., in drinking water), a filtration step was necessary to concentrate the bacteria. Escoriza's group found that quantification results showed good correlation between Raman measurements and more traditional methods such as turbidity, plate counts, and dry weight.

Kalasinsky and coworkers reported on a comprehensive study of widefield Raman imaging of a suite of pathogens with a view to detecting and identifying them even within a complex background.[48] They noted that for complex, environmental samples, widefield Raman imaging was their preferred method for Raman sensing over others including SERS, UV-excited resonance Raman spectroscopy, and nonlinear Raman spectroscopy. In similar work, Tripathi et al. evaluated widefield Raman imaging technology for waterborne pathogen detection.[49] Taking into account key experimental and background interferences such as laser-induced photodamage threshold, composition of water matrix and organism aging, the group reported the technique's capability of discriminate threat simulants down to the species level.

The information contained in widefield Raman images is immense. In a typical Raman image there may be tens of thousands to hundreds of thousands of pixels, each pixel of which contains spectral information. In addition, because of the heterogeneous nature of their composition, biological materials often have very similar Raman spectra. As a result, differences between biological Raman spectra are usually difficult to see with the eye. In almost all cases, the large datasets generated by Raman imaging require the use of multivariate data analysis methodologies in order to process and interpret the results.

Like all analytical methodology that requires the use of mechanical equipment, several factors influence outputted data—both in quantity and content. Equipment that contains high electronic noise results in Raman data with lower signal-to-noise ratio (SNR). Artifacts that may be produced by instrumentation may also cause misinterpretation of data. For this reason, most imaging technology requires regular implementation of calibration and preprocessing methodologies in order to minimize factors such as instrument response and dark current. These will be discussed in the following sections.

6.4   Background and Image Preprocessing Steps for Widefield Raman Images

Biological samples such as tissues and cells are highly complex and comprise building blocks such as amino acids, proteins, and DNA which exhibit very similar Raman spectra. Consequently, the subtle existing spectral differences between biological samples may be overshadowed by signal attributed to instrument components or background light. In order to eliminate this background signal, a series of preprocessing steps must be completed prior to any data analysis.[56] This way, any differences found in the analysis will be exclusive to the biological content of the tissue only.

Any type of mathematical operation applied to the data prior to analysis is considered preprocessing.[57] In effect, preprocessing separates the Raman signal from the noise, thus removing noninformative data. Contributing factors to the background include instrument response, cosmic events, source illumination intensity variations, and fluorescence. Noise is also present in the data from instrumental components, software computations, and surrounding light.[56]

6.4.1   Fluorescence

Fluorescence is a major source of background in widefield Raman images of biological samples because of the radiation source used and the chemical composition of tissue. Raman scattering occurs as a result of the inelastic scattering of photons. When tissue samples are excited with a 532-nm laser, the Raman signal is often masked by a broad fluorescence emission that occurs simultaneously. This fluorescence is reduced through the process of photobleaching.

Photobleaching is a poorly understood phenomenon that occurs when a fluorophore permanently loses its ability to fluoresce. This loss occurs as a result of photon-induced chemical damage and covalent modification during transitions from excited singlet states to excited triplet states. The number of transitions a fluorophore undergoes prior to photobleaching is dependent upon the molecular structure and the local sample environment. As a result, some fluorophores bleach quickly while others take much longer to bleach due to thousands of transition cycles.[58] For this reason, photobleaching must be completed prior to any Raman image collection.

A method to eliminate the majority of fluorescence masking Raman signal is to monitor the photobleaching process through collection of Raman dispersive spectra prior to image collection. This process is illustrated in Fig. 6.2, where Raman dispersive spectra are collected at 1-second intervals for 30 seconds. The top spectrum collected at 1 second is absent of any Raman signal. Raman peaks become more evident with time as the fluorescence burns down. The last spectrum collected at 30 seconds has peaks evident in both the

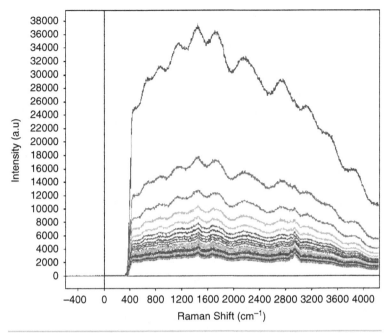

FIGURE 6.2 Photobleaching process monitored through the acquisition of Raman dispersive spectra at 1-second intervals.

fingerprint (400 to 1800 cm⁻¹) and CH-stretching (2600 to 3200 cm⁻¹) regions. When the spectrum no longer changes, the photobleaching process is considered complete. Alternatively, the burn-down rate may be determined through calculation of the SNR in the CH-stretching region. A sample is photobleached once the burn-down rate becomes constant. To photobleach a sample without using Raman dispersive spectroscopy, the sample may be illuminated with laser light for a given time prior to data collection. This process should only be done on samples that have been shown to have consistent fluorescent properties because a major disadvantage is not knowing if the photobleaching process is complete.

The amount of fluorescence a sample exhibits is also dependent on the laser wavelength being used for the excitation source. Shorter excitation wavelengths approach ultraviolet radiation which results in greater fluorescence. Many groups using Raman spectroscopy minimize fluorescence by using laser excitation wavelengths from 700 to 850 nm, approaching the near-infrared region of the electromagnetic spectrum.[59] While this is effective in minimizing background fluorescence, the longer excitation wavelength results in Raman spectra with lower SNR. In addition, silicon-based CCD detectors suffer quantum efficiency losses in the NIR region, also contributing to the low spectral SNR. Consequently, integration times must be increased, resulting in longer experiments. Because of this trade-off between fluorescence and SNR, the laser excitation wavelength chosen should be based on the specific application and goal.

6.4.2   Correction for Dark Current

Cameras containing a CCD sensor are often used as detectors in chemical imaging. Because Raman scattering is a weak phenomenon, a highly sensitive camera containing an electron-multiplying CCD (EMCCD) has become a preference for widefield Raman imaging. For all types of CCDs, the dark current attributed to the sensor must be corrected and thus eliminated from the background.

In the absence of radiation, a constant charge accumulates within the camera wells and on the chip that is the result of the random generation of electrons.[60] This constant response, or dark current, adds to the signal produced. To remove this response, a background image is collected without any source of radiation. This image is then subtracted from each image frame during acquisition. Alternatively, it may also be subtracted from the image stack after data acquisition is complete.

6.4.3   Cosmic Filtering

One important step that is applied to widefield Raman images before other preprocessing steps is a cosmic filter. Random cosmic events often occur while collecting Raman data. These events are seen as bright pixels while scrolling through the wavelength dimension of a widefield Raman image. Applying a cosmic filter eliminates these cosmic spikes. The cosmic filter is a median filter that considers each pixel in the image and checks the nearby neighboring pixels to determine if it is representative of its surroundings. A standard deviation and a window size are set to determine the range and acceptable pixel intensity. An alternative method is to perform a threshold filter to the data. Using this method, upper and lower threshold values are set. Any pixels outside of this threshold value are replaced by the average intensity of the moving window of a selected size of neighboring pixels. This intensity is filtered at the center.

6.4.4   Instrument Response Correction

A major source of background interference in widefield Raman images is signal contributed by the instrument, known as instrument response. Every Raman imaging instrument contains complex optical components. The signal attributed to these components is present in the background and should be removed to reveal the subtle differences in the Raman signal of biological samples. A sample with a known spectral profile, such as calibrated light sources, blackbody radiators, and white light, may be used to correct for instrument response. Recently, the National Institute of Standards and Technology (NIST) has tested and developed standard reference materials (SRM) with known spectral profiles that may be used to correct for instrument response in Raman experiments.

The SRM provides relative intensity corrections for Raman spectrometers. The SRM used is dependent on the laser excitation wavelength. For example, NIST SRM 2242 corrects for instrument response in instruments using a 532-nm laser excitation, while instruments using a laser excitation of 785 nm requires the NIST SRM 2241 for instrument response correction. The standard is a manganese-doped borate matrix glass that emits a broadband luminescence spectrum. This spectrum is described by a polynomial expression that relates the relative spectral intensity to the wavenumber expressed as the Raman shift from the laser wavelength excitation. This polynomial, when compared to the measurement of the luminescence spectrum on the specific Raman instrument used, can determine the spectral intensity-response correction known as instrument response. When corrected with the instrument response correction, Raman spectra are instrument independent.[61]

To correct widefield Raman images, an image of the SRM is acquired immediately after the acquisition of a widefield Raman image of a tissue sample. All experimental parameters are kept the same with the exception of the exposure time which may be decreased to avoid camera overexposure. Upon completion of acquisition, the widefield Raman image of the sample is divided by the instrument response image that was derived from the image collection on the SRM. Evidence of improvement in the spectrum associated with the widefield Raman image can be seen in Fig. 6.3. The image in Fig. 6.3*a* has not yet undergone instrument response correction, whereas the image in Fig. 6.3*b* has been corrected. It is evident that both the spectral features and baseline are smoother and free from instrument-contributed signal in Fig. 6.3*b*.

6.4.5   Flatfielding

The SRM may also be used to flatfield the resultant images. In widefield Raman imaging, the entire field of view is illuminated. This illumination is frequently not uniform across the field of view, and results in images where the intensity is greatest at the center. The flatfield correction improves the image quality and minimizes pixel intensity variations. The SRM image collected is cosmic filtered, blurred, and normalized over the image frames to produce an image that represents the illumination pattern and contains no Raman signal. The sample image is then divided by this image to produce a more evenly distributed data across the field of view. Samples other than the SRM may be used for this procedure, such as silicon; however, it is convenient to use the SRM since an image is already being collected for instrument response correction.

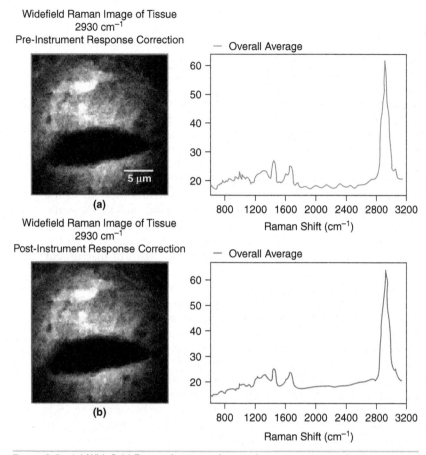

Widefield Raman Image of Tissue
2930 cm^{-1}
Pre-Instrument Response Correction

— Overall Average

(a)

Widefield Raman Image of Tissue
2930 cm^{-1}
Post-Instrument Response Correction

— Overall Average

(b)

Raman Shift (cm^{-1})

Raman Shift (cm^{-1})

FIGURE 6.3   (a) Widefield Raman image and mean image spectrum prior to instrument response correction, (b) the same field of view after correction with instrument response.

6.4.6   Baseline Correction

Once instrument response has been eliminated, the remaining fluorescence is removed from the background. While the photobleaching process eliminates the fluorescence masking the Raman signal, background fluorescence still exists, evident in the sloping baseline. A common method for correcting the sloping baseline is using a polynomial fit.

A polynomial fit may be completed in various ways. Often times the spectral range is selected first upon which the polynomial will be fitted. This is done if the existent errant slope would alter the outcome of the fit. Points are then selected on the spectrum along with a polynomial order. For example, end points and a first order polynomial will result in a linear baseline fit. This example is used in simple baseline fits that correct for spectra containing a few Raman peaks. Often, with biological samples, the spectrum may be more complex, having a curving baseline. Multiple points and higher order polynomials may be chosen. Examples of this type of fit can be seen in Fig. 6.4. Here, a second order polynomial fit was used, and the points chosen for baseline fit included the first two points, the last two points, and a middle range of points encompassing the region from 1810 to 2710 cm^{-1}, where no Raman signal exists. The baseline fit is completed through construction of a polynomial model using a least-squares method. This model is applied to all pixels within the image by subtracting it from each spectrum. Figures 6.4*a* and *b* are the image and the average image spectrum pre-baseline correction, and the spectrum after the baseline fit has been applied is seen in Fig. 6.4*c*.

While this fit is effective and can be generally applied, this does have its disadvantages. It is not usually applied in real time, making Raman in vivo applications difficult, nor does it perform well in low SNR environments.[62] Current research into improving polynomial fits is being done to make the baseline fit procedure automated and account for low SNR data.[62,63]

Polynomial fitting is not the only methodology for baseline correction. Other common techniques utilize the first and second derivatives to correct for background fluorescence. The derivative method is conducted using polynomial fits on small sections of the spectrum, where derivatives of the polynomial in the center of the section are calculated.[56] Similar to polynomial baseline fits; derivative-based techniques do not fit low-SNR data well.

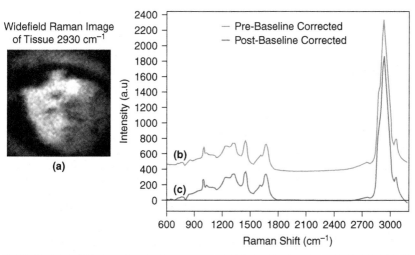

Figure 6.4 (*a*) Widefield Raman image of tissue and the average image spectrum (*b*) pre-baseline correction, (*c*) post-baseline correction.

6.4.7   Normalization

After a baseline is fitted, many spectra associated with each pixel have different baseline intensity levels. Normalization is a process that brings these intensities to the same level thus minimizing any differences attributed to intensity. When normalizing over spectra, the norm, or square root of the sum of the squares of the intensities, is calculated for each pixel and each intensity value in the given spectrum is divided by that value. This process is known as vector normalization. Alternatively, peak normalization may be performed. Using this method, each element or component of a spectrum (or pixel in imaging) is divided by a constant that is determined by the height of a given peak.[56]

6.4.8   Smoothing

In conjunction with these standard preprocessing steps, additional operations may be performed to reduce noise and eliminate other background features through the process of smoothing. While this step is often done, it should be carried out with caution, as it may eliminate minor Raman spectral features that can be utilized in data analysis steps. Benefits of smoothing include improving the aesthetic appearance of the spectra, reducing noise (that already may be present or added during other preprocessing steps), and improving the efficiency of other operations. Savitsky-Golay filtering is a common type of technique that uses polynomials to perform the smoothing operation. The number of data points, polynomial order, and derivative order are taken into account for the smoothing application. An alternative to Savitsky-Golay smoothing is gaussian blurring, which is based on a convolution kernel that is a gaussian function. It is a spatial filter that smoothes sharp contrasts within images. Other convolution filters, such as edge detection or sharpening, may be used to enhance images. These filters have predefined kernel or filter coefficients that are used to perform the filtering. Again, precaution must be taken when applying smoothing and convolution filters to widefield Raman images to prevent the removal of spectral features or the addition of unwanted features.

Once preprocessing is complete, image processing may begin with various multivariate statistical techniques. Performing these functions after preprocessing is complete will illustrate the differences inherent to the sample without interference from the background signal.

6.5   Chemometric Analysis of Widefield Raman Images

The images obtained using widefield Raman imaging are spectrally and spatially resolved images. At every pixel in the image, there is a spectrum to represent the constituent molecular species contained within that pixel. The number of pixels within an image depends on the microchip size of the Raman camera. For example, a typical high resolution EMCCD camera may have 512×512 pixels; therefore, it is possible to have more than 200,000 spectra in an image. However, the number of pixels in an image will decrease when the binning parameters are increased. Nonetheless, there are copious spectra that make up an image and require the use of chemometric techniques to analyze the data. Chemometrics is defined as the use of mathematics and statistics on chemical data in order to extract useful information that can be used for decision making. The following section will describe some of the chemometric techniques used in the analysis of a widefield Raman image. Each analytical technique will be briefly described, but for a more rigorous examination of the techniques, several books on chemometrics are referenced herein.[64–66]

6.5.1 Principal Component Analysis

In the case of a widefield Raman image of a biological tissue, there are abundant associated Raman spectra. Because they consist of the same biomaterials (proteins, lipids, nucleic acids, carbohydrates), they are all very similar; however, within these similar spectra, there should be some finite number of independent variations occurring in the spectral data. Hopefully, the largest variations in the spectral data would be the changes in the spectrum due to different concentrations of biological molecules that comprise the cells or tissue. Other possible variations are due to instrument variation (unless removed by preprocessing steps), environmental conditions, differences in sample preparation, and so on. It is possible to calculate a set of "variation spectra" that represent the changes in the Raman scatter at all wavelengths in the spectra; these variation spectra could be used instead of the spectral data for comparison. There should be fewer common variations amongst the data than the number of spectra; although, since they come from the original data, the variation spectra retain the interrelationship of the original spectra.

The variation spectra are called eigenvectors or principal components (PCs). The method of breaking down a set of spectroscopic data into its most basic variations is called *principal component analysis* or PCA. It is mathematically defined as an orthogonal linear transformation that transforms the data to a new coordinate system such that the greatest variance by any projection of the data comes to lie on the first coordinate (called the first principal component or first PC), the second greatest variance on the second coordinate, and so on.

For example, imagine all the spectra from a widefield Raman image plotted in multi-dimensional space. The first PC would be a vector plotted through the data to find a single axis to capture or span as much of the variance of the data as possible. This is accomplished by a least-squares fit of the data to the new axis. Once the first PC is determined, the second PC is plotted in the same way to capture more variance, but it must be orthogonal to the first PC. This process is continued, where each additional PC must be orthogonal to the other PCs, until all of the variance is adequately described. This reduces the data into a new coordinate system based on the variance.

Once the data is reduced into this new coordinate system, one can visualize the data by plotting different PC scores. The coordinates of data points relative to the PC axes are termed scores. One can plot points from the data along different axes within the PC coordinate system. The resulting plots are called *scatter plots*. The data points within the scatter plot represent the spectra and will cluster in PC space (coordinate system based on PCs) according to similarities in spectral characteristics.

The spectra obtained from widefield Raman images and analyzed using PCA will have been preprocessed according to the steps outlined in the previous section. It is important that all of the data is processed consistently so as to prevent an artifact of variance being introduced to the data from different processing steps. In addition, regions of the spectra where differences are seen can be analyzed without having to evaluate the entire spectrum.

Figure 6.5 illustrates an example of a PC score plot from analysis of Raman spectra derived from widefield Raman imaging of different

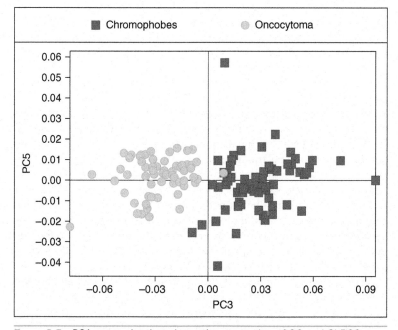

Figure 6.5   PCA score plot that shows the separation of OC and ChRCC spectral data derived from widefield Raman images.

types of kidney tissue.[67] The study evaluated the application of widefield Raman imaging to the differentiation of renal oncocytoma (OC) from chromophobe renal cell carcinoma (ChRCC), which can often have overlapping histological features such that a pathologist is not readily able to distinguish the two types of diseases. A pathologist chose 10 cases of OC and 10 cases of ChRCC. Typical OC and ChRCC regions within each tissue were highlighted by the pathologist. Widefield Raman images were collected from these regions in unstained tissue sections. PCA was employed to reduce the large widefield Raman image datasets to smaller, more comprehensible results. Each point on the scatter plot represents a spectrum derived from the widefield Raman image: the red squares represent spectra obtained from the ChRCC and the turquoise circles represent the spectra obtained from OC. The spectral regions used in this evaluation were within the fingerprint region (1633 to 1850 cm^{-1}). In the plot, the axes are PC5 and PC3. This evaluation shows a clear distinction between ChRCC and OC data, establishing that widefield Raman imaging is an objective method that may provide information to assist the pathologist in a diagnosis.

Similar to the kidney example, distinguishing between invasive ductal carcinoma (IDC) and invasive lobular carcinoma (ILC) in breast tissue is sometimes not possible without the benefit of further staining. Figure 6.6 shows the scatter plot of a PC score of spectra obtained from widefield Raman images of IDC and ILC.[68] There is a slight distinction between the two classes, although the separation is not as good as the kidney tissue. ILC and IDC may not separate well because a subset of ILC cells closely resembles IDC cells in their transcription pattern.[69] These preliminary results indicate that widefield Raman imaging can indicate similarities in different tissue types as well as differences. From the kidney and breast examples, it is apparent that a tool such as widefield Raman imaging can potentially assist pathologists in difficult diagnoses.

6.5.2 Mahalanobis and Euclidean Distance

PCA provides a basis on which models can be created to compare different tissues. As described in the previous section, PCA was useful in highlighting the spectral differences in tissue that are not easily identified visually by a pathologist. The spectral differences in the tissue are due to the collection of molecules that are present within the cell or tissue; and that collection of biomolecules will be dictated by the metabolic state of the cell or tissue. Therefore, widefield Raman imaging and PCA should be useful in showing the spectral differences between tissues in different metabolic states. Different metabolic states will exist when tissue is inflamed, infected, cancerous, or in some other diseased state. Models can be created from spectra derived from widefield Raman images of biological tissue using PCA to differentiate objectively the different metabolic state of cells or tissue.

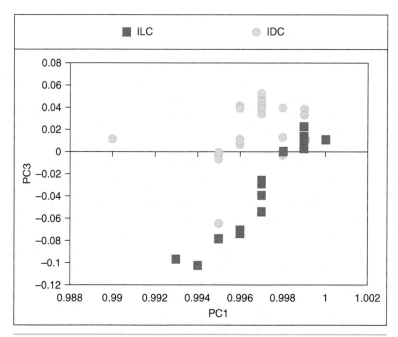

FIGURE 6.6   PCA score plot that shows the separation of IDC and ILC spectral data derived from widefield Raman images.

The PC scores obtained from PCA highlight the clustering or separation of data in PC space. Two methods to measure the clustering of data points in PC space are Euclidean distance (ED) and Mahalanobis distance (MD). These methods are useful in determining the similarity of a set of values from an unknown sample (test point) to an established set of values from known samples. ED only measures a relative distance from the mean point of the clustered data. It does not take into account the distribution of the data points within the cluster. Euclidean distance analysis (EDA) performs spectral similarity assessment by calculating the ED between a reference vector and every spectrum in the image pixels. The ED is simply the sum of the squares of the difference for every spectral dimension for two vectors. The MD, however, does take into account the variability of the data within the cluster. The MD is measured as the distance from one test point to the center of mass of the ellipsoid divided by the width of the ellipsoid in the direction of the test point. Unlike ED, it is based on both the size (determined from standard deviation) and the shape of the ellipse (determined from the covariance within the group).

In order to use MD to classify a test point as belonging to one of N classes, one first estimates the covariance matrix of each class, usually based on samples known to belong to each class. Then, given a test sample, one computes the MD to each class, and classifies the test point as belonging to that class for which the MD is minimal. Using the probabilistic interpretation given above, this is equivalent to selecting the class with the highest probability.

PCA combined with MD calculation not only creates a method to determine patterns within complex datasets, but also provides a way to measure and classify unknown data points to these known groups. Thus a classification system is created for samples that exhibit minor differences spectroscopically.

As an example, the MD is calculated from the PCA of the OC and ChRCC Raman spectra discussed previously (Fig. 6.7).[67] Figure 6.7 is the same as Fig. 6.5, with the addition of the ellipses to indicate the clustering of the data. A J3 criterion is also reported. The J3 criterion is an indication of how well the data separate. It is a quantitative measurement that maximizes the variance between classes and minimizes the variance within a class. The larger the value, then the better the separation is between the groups of data. A J3 criterion of 1 indicates that the groups overlap. The reported J3 criterion for the separation of OC and CHRCC is 3.97, indicating the robustness of the separation. This model can be used to classify other samples as OC or ChRCC.

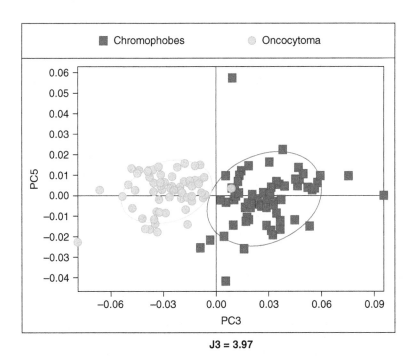

J3 = 3.97

FIGURE 6.7   Mahalanobis distance model of OC and ChRCC spectral data derived from widefield Raman images.

In another example of the MD model approach, a model was constructed using spectra extracted from widefield Raman images, where apoptotic and normal PC3 cells (a prostate cancer cell line) were observed.[45] Figure 6.8 shows a PC score plot (PC6 vs. PC5) of the normal and apoptotic spectral samples obtained from widefield Raman images. The key point from the scatter plot is that the apoptotic and normal spectra are in distinct locations. The model was then used to classify each pixel in a widefield Raman image. The results of this classification are shown in Fig. 6.9. Figure 6.9a is the brightfield reflectance image of a group of PC3 cells. Figure 6.9b shows the widefield Raman image at the Raman shift of 2930 cm⁻¹. Figure 6.9c shows fluorescence labeling of the group of cells with fluorescein isothiocyanate (FITC). The FITC label specifically targets the molecule phosphatidylserine in the plasma membrane of apoptotic cells; therefore, cells that label green with FITC are apoptotic cells and cells that do not fluoresce green are normal. Figure 6.9d shows the Raman image-based classification of each pixel of the widefield Raman image, based on the model shown in Fig. 6.8. Ideally, the green area apoptotic cell region in the classified image (Fig. 6.9d) would overlay with the green in the fluorescently labeled image (Fig. 6.9c) when comparing the two images. This is not completely the case. Reasons for disagreement

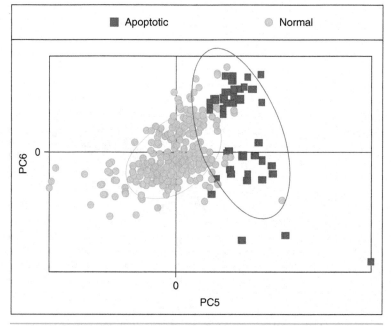

Figure 6.8   The scatter plot of image spectra from normal and apoptotic cells.

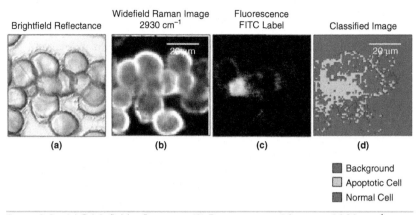

Brightfield Reflectance	Widefield Raman Image 2930 cm^{-1}	Fluorescence FITC Label	Classified Image
(a)	(b)	(c)	(d)

■ Background
□ Apoptotic Cell
■ Normal Cell

Figure 6.9   (a) Brightfield reflectance, (b) Raman spectral frame at 2930 cm^{-1}, (c) fluorescence, and (d) classified images.

include but are not limited to (1) the SNR of the raw spectral image data which can add inaccuracy to classification, and (2) the difference between the measurements. FITC labels phosphatidylserine in the plasma membrane, whereas the Raman measure is not targeted to a specific molecule, but rather the local molecular environment. Nonetheless, the example highlights the usefulness of creating models from spectra obtained from widefield Raman images, and using the models as objective classification mechanisms to determine the metabolic state of cells.

As a further example of ED and MD, Kalasinsky et al. used an optical detection method which combines Raman spectroscopy, fluorescence spectroscopy, and digital imaging to detect and identify pathogens.[48] In this study, widefield fluorescence chemical imaging was used to rapidly screen large surface areas for biological versus nonbiological particulates. Once biological particulates were identified from fluorescence signatures, further identification was performed using widefield Raman imaging. Each Raman spectrum from the widefield Raman image was compared with Raman signatures stored in a library of pathogen spectra. Using this type of detection method, the researchers were able to identify *Bacillus globigii* (*Bg*), a genetic near-neighbor to *Bacillus anthracis* (*Ba*), in an environmental bioaerosol sample using an ED classifier. Furthermore, eleven classes of pathogens were organized into training sets. PCA was performed and a supervised Mahalanobis distance model boundary classifier was constructed for each of the 11 classes. All training set spectra were classified correctly when setting the decision threshold at 99 percent. The results suggest that a robust library was created using the MD classifier.

6.5.3 Spectral Mixture Resolution

As described in the previous section, PCA and MD can be used to create a classification model and classify every pixel within an image. Another method to classify every pixel in an image is the technique of spectral mixture resolution (SMR). SMR uses a set of user-specified reference spectra to find the best linear combination of reference spectra for each pixel in a given spectral image.

The SMR function results in the output of two images: (1) a concentration image with one frame for each spectrum in the reference set and (2) a residual image. The concentration image essentially maps the distribution of reference spectral species in the active image. There will be one concentration map for each reference spectrum. The concentration values for a given frame can vary between 0 and 1, and the set of concentration values for a given pixel (over all frames) sum to 1. A least-squares fit is used with the stipulation that no concentrations can be negative. The residual image contains the spectral information not explained by the reference spectra. If the imaged sample contains only the substances represented by the reference spectra, an average spectrum of the residual image should represent only noise, and have very small intensities compared to an average spectrum of the original image. If there are chemical species present in the sample that are not represented by the set of reference spectra, the residual image should contain spectra that represent the mixture of chemical species remaining after the set of reference spectra has been subtracted.

Figure 6.10 shows an example of an SMR classification scheme. The Raman image classified in the example is that of prostate tissue. The three reference spectra used were (1) an average Raman spectrum of prostate epithelium; (2) an average Raman spectra of prostate stroma; and (3) a glass spectrum. The epithelium and stroma reference spectra were used to segment the different types of cells within the prostate tissue, and the glass spectrum was used to classify where holes exist in the tissue from ducts (no tissue present,

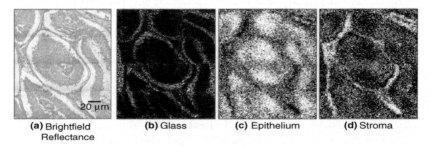

(a) Brightfield
Reflectance

(b) Glass

(c) Epithelium

(d) Stroma

Figure 6.10   (a) Brightfield reflectance image of prostate tissue, (b) the concentration image of glass, (c) the concentration image of epithelium, and (d) the concentration image of stroma.

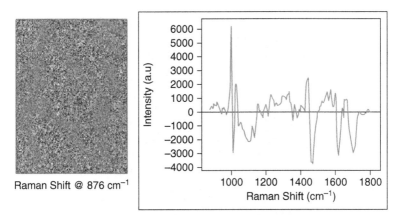

Raman Shift @ 876 cm⁻¹

FIGURE 6.11   Residual image and spectrum from SMR.

just the underlying glass slide). Figure 6.10 shows each frame of the concentration image: glass (Fig. 6.10b), epithelium (Fig. 6.10c), and stroma (Fig. 6.10d). The white areas indicate where the spectral signature is the greatest, whereas the dark areas indicate where there is no spectral signature. Figure 6.11 shows the residual image and the associated spectrum. Both the image and the spectrum show noise only, indicating that the SMR function was able to classify the spectrum at each pixel with the reference spectra, and that it is likely that no other residual spectral species is present in the image. In Fig. 6.12, the concentration images were used to create a colorized image of the prostate tissue to indicate the location of each of the reference species. The blue areas represent epithelium, the red areas represent glass slide, and the green areas represent stroma.

6.5.4   Derivatives

The spectra of biological tissue or cells obtained from widefield Raman images will very similar. Although the molecular composition will be different between different tissues or tissues in different metabolic states, the underlying molecules in each tissue will be the same. Spectra are usually compared by overlaying them to see if they match. If they have multiple points of identification, such as peaks, this is easy to do, especially when the spectra is of a pure component. A match can be reported if the peak position and general shape are the same; however, this is not necessarily true of the spectra of tissue. Raman spectra of tissue have broader peaks due to the mixture of molecules in the tissue sample; therefore when multiple spectra are overlaid, the subtle differences in the spectra of biological samples can be very important. If there is a subtle difference when spectra are overlaid, derivative spectra may be useful to aid comparison. When spectral peaks are broad and featureless, their

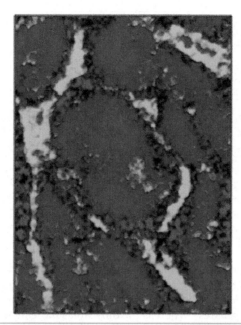

Figure 6.12   Colorized image to show different regions of widefield Raman image based on SMR.

first derivative spectra provide extra points of comparison, and differences in the slope of spectra can be magnified when first derivative spectra are produced. Second derivatives are also useful.

Just like spectra, the first derivatives of spectra can be used in chemometric analysis. Once again as an example, the first derivative spectra of the OC and ChRCC tissues over the range of 1550 to 1800 cm^{-1} were analyzed by PCA and MD.[67] Figure 6.13 shows the results. The OC and ChRCC spectra have distinct locations in PC space with a J3 criterion of 3.11. The J3 criterion of the derivative spectra is lower than that of the original spectra (J3 = 3.79) because although the clusters separate from each other better, there is more scatter within the cluster.

6.6   Chemometrics in the Analysis of Non-Widefield Raman Images

In the previous section, the use of chemometric techniques in the analysis of widefield Raman images of biological samples was discussed. Chemometric techniques can be used in the analysis of Raman spectra extracted from images obtained by other methods, such as point and line scanning. Point and line scanning collect a full spectrum while imaging a point or line, respectively. Like widefield Raman imaging, the resultant image from each of the methods is a

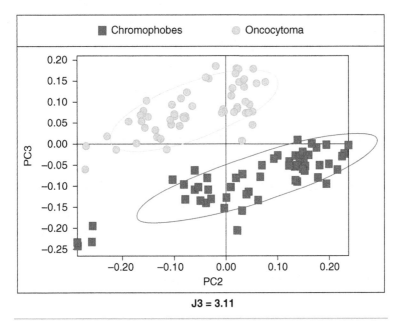

FIGURE 6.13   Mahalanobis distance of first derivative spectral data derived from widefield Raman images of OC and ChRCC.

hypercube of data containing Raman intensity as a function of Raman shift and spatial location. In addition, chemometrics is also used to analyze Raman dispersive spectra. In this section, chemometric analyses of Raman spectra of biological samples, obtained by methods other than widefield Raman imaging, are reviewed.

6.6.1 PCA

In one study, PCA and MD classifiers were used for waterborne pathogen detection.[49] Unsupervised and supervised MD classifiers were able to discriminate Gram-positive organisms (anthrax simulants), Gram-negative organisms (plague simulants), and proteins (toxin simulants). In the unsupervised model, a single MD classifier was constructed from all the spectra based on PCA without any prior knowledge of the spectra. In the supervised model, a priori knowledge provided from the average spectrum of each class was used to construct the supervised MD classifier, resulting in better discrimination. Furthermore, a supervised MD classifier was used to evaluate the effect of tap water and aging on the classifier performance when characterizing *Bacillus thuringiensis* (Bt) spores, Ba spores, and *Escherichia coli* cell preparations. Distilled water, tap water, and residence time in water do not appear to alter significantly the ability of the MD classifier to discriminate between pathogenic species.

Near-infrared Raman spectroscopy was used to map single cells.[3] Thousands of Raman spectra were assembled into a spectral image creating a Raman map of the cell under examination. Raman spectral maps of a freeze-dried human embryonic lung epithelial fibroblast, human osteogenic sarcoma cells in medium, and human astrocytoma cells in medium were created. PCA of the Raman spectral maps of the different cells in the spectral region of 2800-3000 cm^{-1} revealed cellular details like organelles, the membrane, and the nucleus within the image of the cell. The spectral features of the subcellular components were more pronounced for cells in medium than for the freeze-dried cells. It was concluded that the intensity of the spectra strongly depend on the state of hydration of the cell because the conformation of molecules within the cell change upon dehydration.

Similarly, Eliasson et al. also used Raman spectral mapping to identify intracellular components of single living cells, specifically human lymphocytes.[70] In this case, SERS using gold colloids was employed to image an introduced analyte, rhodamine 6G, together with other intracellular components. An inverted confocal Raman microscope was used to map the cells in three dimensions over the spectral range of 450 to 1700 cm^{-1}. Several cells were mapped and the images were analyzed with PCA in order to extract the signal of rhodamine 6G from the image spectra. A score plot of PC1 and PC2 showed that the spectra could be divided into three separate classes: (1) spectra with spectral contributions from rhodamine 6G and native cellular components; (2) spectra with only native cellular component contributions; and (3) spectra without any spectral band information. Therefore, SERS in combination with PCA, makes it possible to measure Raman-active analytes taken up by cells.

PCA of Raman dispersive spectra was used to discriminate normal, inflammatory, premalignant, and malignant conditions in oral tissue.[71] PCA of the spectra from the four different types of tissue showed differentiation between the normal and pathological states; however, there was poor discrimination among the three pathological states. It is likely that the strong presence of proteins in the pathological tissues versus the lipid content in normal tissue accounts for some of the difference. To diagnose correctly the pathological conditions, PCA combined with match/mismatch criteria using three parameters (MD, sum of squared spectral residuals, and scores) was able to discriminate between the different types of tissue. In a similar analytical procedure, discrimination of normal, benign, and malignant breast tissues by Raman spectroscopy was possible.[72] Likewise, the same process was used to discriminate normal and malignant stomach mucosal tissue,[73] normal and malignant mucosal tissues of the colon;[74] and normal and malignant ovarian tissue.[75]

Shafer-Peltier et al. applied morphological modeling to Raman images of human colonic carcinoma cells, human breast tissue, and arterial tissue.[76] Morphological modeling is a technique for analyzing

Raman images, which uses ordinary least squares to fit a set of basis spectra to the data. The basis spectra are acquired from the major morphological features found in a set of representative samples in situ using a Raman confocal microscope. The technique of morphological modeling was compared with other chemometric techniques such as PCA, multivariate curve resolution (MCR) and ED. Point mapping was used to obtain Raman spectral images of biological samples in the study. They found that each technique has advantages and disadvantages. PCA and MCR are excellent techniques when nothing or little is known of the samples beforehand. When some information about the sample is known, ED is useful. From morphological modeling, it is possible to obtain structure and chemical information about subcellular features in a biological sample, although the sample under study must be well understood.

Omberg et al. used principal factor analysis in their study of tumorigenic and nontumorigenic cells.[77] Factor analysis (FA) is a statistical method used to explain variability among observed variables in terms of fewer unobserved variables called factors. FA differs from PCA in the fact that PCA is used to find optimal ways of combining variables into a small number of subsets, while FA may be used to identify the structure underlying such variables and to estimate scores to measure latent factors themselves. FA was used to evaluate the Raman dispersive spectra obtained from two different rat fibroblast cell lines: M1 and MR1. Both cell lines were transfected with the *c-myc* gene which causes immortality; however, the MR1 line was further transfected with the T24Ha-*ras* oncogene which causes tumor formation. Two approaches were used in the study. First, constrained principal factor analysis provided a measure of the relative contribution of protein, lipid, DNA, RNA, and buffer to the Raman spectra in each cell line. When using constrained FA to analyze the spectra of the two cell lines, the increased intensities in specific band assignments were interpreted as increased protein and lipid concentrations in the tumorigenic cells (MR1). Second, FA of the raw spectral data was used to demonstrate that Raman spectra can be used to differentiate the two cell lines. Similar to constrained FA analysis, the spectral differences between M1 and MR1 cells were also attributed to increased protein and lipid relative to DNA; additionally, the ratio of scores of the appropriate factors shows a clear distinction between the M1 and MR1 spectra.

Nijssen et al. used K-means clustering analysis (KCA) in the evaluation of basal cell carcinoma (BCC) from surrounding tissue by Raman spectroscopy.[78] Raman pseudo-color images were constructed from the spectra obtained from each sample by PCA, and then the PCA scores were used as input for KCA. KCA was used to find groups of spectra with similar spectral characteristics (clusters). The main steps of KCA are as follows: (1) the number of clusters is set by the user; (2) for each cluster a spectrum is randomly chosen from the spectra in the data set to act as the initial cluster center; (3) all the spectra in the data set are then compared with the cluster centers and assigned to the center that they most resemble; (4) after all spectra are assigned to a cluster, new cluster centers are calculated by averaging all the spectra assigned to the cluster; and (5) the procedure is repeated until a stable solution is reached. After KCA, a different color was assigned to each cluster and each grid element of the Raman map was assigned the color of the cluster to which it belonged. The pseudo-color Raman maps were compared with hematoxylin and eosin (H&E)-stained sections and closely corresponded with the H&E-stained adjacent section. In all the maps, BCC was separated from its surrounding nontumorous tissue by KCA. In addition, KCA formed separate clusters for dermis in the vicinity of the tumor; dermis further away from the tumor; and a dense inflammatory infiltrate. The difference in the dermal tissue is most likely due to a decrease in collagen in the dermis adjacent to tumors.

Nijssen et al. also created a multivariate statistical classification model based on the cluster means and the histopathological classification of the cluster means.[78] The classification algorithm was based on the method of logistic regression, which is a variation of ordinary regression, useful when the observed outcome is restricted to two values, usually representing the occurrence or nonoccurrence of some outcome event. The result was a two-step classification model where in the first level, dermis is separated from BCC or epidermis, and then in the second level, BCC is separated from epidermis. The performance of the model was tested by a leave-one-out cross validation. The model correctly predicted BCC from dermis, and showed 100 percent and 93 percent sensitivity and specificity, respectively.

Thakur et al. used PCA and discriminant functional analysis (DFA) to evaluate the Raman spectral signatures of mouse mammary tissue and associated lymph nodes.[79] Spectra of tumors and mastitis separated from normal mammary gland tissue and other pathological states.

PCA has been used with other multivariate statistical technique such as classical least squares (CLS) and least-squares analysis (LSA) for the evaluation of pharmaceutical treatment of cells[80] and osteoblast differentiation,[81] respectively; in addition to a quadratic discriminant analysis for the discrimination of cell response to chemical and physical inactivation.[82]

6.6.2   Linear Discriminant Analysis

Linear discriminant analysis (LDA) has proven to be a robust multivariate statistical analytical method. Many studies based on Raman spectra use LDA to create classification models to distinguish between different types of tissue or cells. LDA is used to find the linear combination of features which best separate two or more classes. Unlike PCA, LDA explicitly attempts to model the difference between the classes of data. Typically, a dimensional reduction technique, such as PCA, is used to reduce the variables and identify significant differences in the data (PCs). The identified PCs are then used as input into the LDA model to help create a classification algorithm which minimizes misclassification.

Combining Raman spectroscopy with optical confocal microscopy, called confocal Raman microscopy, adds spatial resolution to spectral information. Taleb et al. used confocal Raman microscopy to map two different prostatic cell lines, PNT1A and LNCaP.[83] In this study, different data preprocessing methods, partial least-squares (PLS) and adjacent band ratios (ABRs) were used as input into LDA to model and classify the two cell lines. The study demonstrated that PLS discriminate analysis and ABRs of Raman data as input into LDA can identify subtle differences between benign (PNT1A) and malignant (LNCaP) prostate cells in vitro.

Mahadevan-Jansen et al. used near-infrared Raman spectroscopy to differentiate cervical precancers from normal tissue.[84] Multivariate statistical algorithms were developed for tissue differentiation using a five-step procedure: (1) data processing; (2) dimension reduction using PCA; (3) selection of diagnostically important PCs using the unpaired Student's t-test; (4) development of a classification algorithm using Fisher's discriminant analysis (FDA); and (5) unbiased evaluation of algorithm performance using cross validation. FDA, similar to LDA, is a statistical technique that uses linear combinations of independent variables to discriminate between two groups such that the misclassification rates are minimized. The algorithms were able to differentiate squamous intraepithelial lesions (SIL) from non-SIL samples.

Similarly, PCA and LDA of Raman spectra of cervical tissue were used to distinguish between normal cervical tissue, cervical intraepithelial neoplasia (CIN), and invasive carcinoma.[85] Based on cross-validation results, sensitivity and specificity values were calculated as 99.5 and 100 percent respectively for normal tissue, 99 and 99.2 percent respectively for CIN, and 98.5 and 99 percent respectively for invasive carcinoma.

LDA was used to create a classification model in another study. The potential of Raman spectroscopy for in vivo classification of normal and dysplastic tissue was investigated by Bakker Schut et al.[86] The Raman dispersive spectra were acquired from normal and dysplastic rat palate tissue in vivo using a fiber optic probe. A procedure similar to that employed by Mahadeven-Jansen et al.[84] was used to create an LDA classification model. PCA was used on the model set to orthogonalize and reduce the number of parameters needed to represent the variance in the spectral data set. PCs that accounted for more than 1 percent of the variance in the data set were retained for the model. A two-sided t-test was used to select those PCs that showed the highest significance in discriminating the different tissue classes. These were used as input for the LDA model. The researchers found that contributions of bone to the spectra influenced the classification model. In a second model, bone signal-corrected PC scores were used as input to the LDA model and yielded slightly better classification results. An advantage of using multivariate statistical analysis classification models is that they become fairly insensitive to noise, which is important for in vivo measurements when the acquisition time must be kept short. However, the number of parameters that can be used as input into the model must be limited to prevent overfitting of the model.

Stone et al. also used LDA in their experiments.[87] Like the previously described work, PCA was used to reduce the number of variables in the analysis. Analysis of variance (ANOVA) was utilized to identify the diagnostically PCs of the spectra. The PCs, used as the variables for entry into an LDA model, maximized the variance between groups and minimized the variance within groups. The classification models were able to separate different types of larynx, bladder, and esophageal tissues; when tested by cross validation, the models exhibited high sensitivities and specificities. Interestingly, the authors noted that instrument variability can decrease the effectiveness of a classification model.

NIR Raman spectroscopy was able to differentiate accurately between parathyroid adenomas and hyperplasia using a PCA fed LDA diagnostic algorithm.[88] Detection sensitivity for parathyroid adenomas was 95 and 93 percent for hyperplasia, indicating Raman spectroscopy is an excellent tool to differentiate between the two types of parathyroid tissue.

In other work, PCA and LDA were applied to Raman spectra of a human osteosarcoma cell line (MG63) in order to classify cells according to the cell cycle phase (G_0/G_1, S, and G_2/M phases).[89] The classification model showed a sensitivity of 93, 67, 60 percent for cells in the cell cycle block of G_0/G_1, S, and G_2/M respectively, and specificity of 87, 80, and 93 percent, respectively, as assessed by a leave-one-out cross-validation.

Chan et al. used PCA and LDA in their experiment that combined optical trapping with Raman spectroscopy.[90] In optical trapping Raman spectroscopy, a tightly focused laser beam functions as both laser tweezers to immobilize a freely diffusing cell and the excitation source for generating Raman spectra of the trapped cell. A PCA-LDA classification model was able to discriminate patient-derived leukemia cells from normal human lymphocytes.

Teh et al. used a PCA-LDA classification approach with Raman spectra to identify dysplasia from normal gastric mucosa tissue.[91] The model yielded the diagnostic sensitivity of 95.2 percent and specificity of 90.9 percent for separating dysplasia from normal gastric tissue.

6.7 Conclusions

Widefield Raman imaging, along with other Raman spectroscopic-based modalities, are effective tools in the study of biological samples. The challenge is to derive useful information from all of the spectral data created by these methods. Chemometrics provides powerful analytic techniques for the evaluation of Raman spectra, as evidenced by the many different types of cells and tissue that can be discriminated from each other based on classification models that were created using one of many multivariate statistical analytical techniques. Each technique has its advantages and limitations; the best choice for analysis depends on each situation and what is known about the sample data beforehand. Ultimately, Raman spectroscopic methods have a promising future in the objective diagnosis of disease and decision making regarding biological samples.

References

1. N. Uzunbajakava, A. Lenferink, Y. Kraan, B. Willekens, G. Vrensen, J. Greve, and C. Otto, "Nonresonant Raman Imaging of Protein Distribution in Single Human Cells," *Biopolymers*, **72**:1–9, 2003.
2. N. Uzunbajakava, A. Lenferink, Y. Krann, E. Volokhina, G. Vrensen, J. Greve, and C. Otto, "Nonresonant Confocal Raman Imaging of DNA and Protein Distribution in Apoptic Cells," *Biophysical Journal*, **84**:3968–3981, 2003.
3. C. Krafft, T Knetschke, A Siegner, R. H. W. Funk, and R. Salzer, "Mapping of Single Cells by Near Infrared Raman Microspectroscopy," *Vibrational Spectroscopy*, **32**:75–83, 2003.
4. C. Creely, B. Volpe, G. Singh, M. Soler, and D. Petrov, "Raman Imaging of Floating Cells," *Optics Express*, **13**(16):6105–6110, 2005.
5. S. Keren, C. Zavaleta, Z. de la Zerda, Z. Cheng, O. Gheysens, and S. Gambhir, "Noninvasive Molecular Imaging of Small Living Subjects Using Raman Spectroscopy," *Proceedings of the National Academy of Sciences*, **105**(15):5844–5849, 2008.
6. N. Gierlinger and M. Schwanninger, "Chemical Imaging of Poplar Wood Cell Walls by Confocal Raman Microscopy," *Plant Physiology*, **140**:1246–1254, 2006.
7. C. Krafft, S. Sobottka, G. Schrackert, and R. Salzer, "Near Infrared Raman Spectroscopic Mapping of Native Brain Tissue and Intracranial Tumors," *Analyst*, **130**:1070–1077, 2005.
8. J. Andrew, M. Browne, I. Clark, T. Hancewicz, and A. Millichope, "Raman Imaging of Emulsion Systems," *Applied Spectroscopy*, **52**(6):790–796, 1998.
9. H. Fenniri, O. Terreau, S. Chun, S. Oh, W. Finney, and M. Morris, "Classification of Spectroscopically Encoded Resins by Raman Mapping and Infrared Hyperspectral Imaging," *Journal of Combinational Chemistry*, **8**(2):192–198, 2006.
10. W. Ren and H. Cheng, "Aligned Double-Walled Carbon Nanotube Long Ropes with a Narrow Diameter Distribution," *Journal of Physical Chemistry B*, **109**:7169–7173, 2005.
11. I. Chourpa, F. Lei, P. Dubois, M. Manfait, and G. Sockalingum, "Intracellular Applications of Analytical SERS Spectroscopy and Multispectral Imaging," *Chemistry Society Reviews*, **37**:993–1000, 2008.
12. V. Guieu, F. Lagugne-Labarthet, L. Servant, D. Talaga, and N. Sojic, "Ultrasharp Optical-Fiber Nanoprobe Array for Raman Local-Enhancement Imaging," *Small*, **4**(1):96–99, 2008.

13. F. Lagugne Labarthet, J. L. Bruneel, T. Buffeteau, and C. Sourisseau, "Chromophore Orientation Upon Irradiation in Gratings Inscribed on Azo-Dye Polymer Films: A Combined AFM and Confocal Raman Microscopic Study," *Journal of Physical Chemistry, B*, **108**:6949–6960, 2004.

14. F. Lagugne-Labarthet, C. Sourisseau, R. Schaller, R. Saykally, and P. Rochon, "Chromophore Orientations in a Nonlinear Optical Azopolymer Diffraction Grating: Even and Odd Order Parameters from Far-Field Raman and Near-Field Second Harmonic Generation Microscopies," *Journal of Physical Chemistry, B*, **108**:17059–17068, 2004.

15. C. de Grauw, C. Otto, and J. Greve, "Line-Scan Raman Microspectrometry for Biological Applications," *Applied Spectroscopy*, **51**(11):1607–1612, 1997.

16. M. Bowden, D. J. Gardiner, G. Rice, D. L. Gerrard, "Line-Scanned Micro Raman Spectroscopy Using a Cooled CCD Imaging Detector," *Journal of Raman Spectroscopy*, **21**:37–41, 1990.

17. C. A. Drumm and M. D. Morris, "Microscopic Raman Line-Imaging with Principal Component Analysis," *Applied Spectroscopy*, **49**:1331–1337, 1995.

18. K. Christensen and M. Morris, "Hyperspectral Raman Microscopic Imaging Using Powell Lens Line Illumination," *Applied Spectroscopy*, **52**(9):1145–1147, 1998.

19. S. L. Zhang, J. A. Pezzuti, M. D. Morris, A. Appadwedula, C. Hsiung, M. Leugers and D. Bank, "Hyperspectral Line Imaging of Syndiotactic Polystyrene Crystallinity," *Applied Spectroscopy*, **52**(10):1264–1268, 1998.

20. J. A. Timlin, A. Carden, M. D. Morris, J. Bonadio, C. Hoeffler II, and K. Kozloff; S. Goldstein, "Spatial Distribution of Phosphate Species in Mature and Newly Generated Mammalian Bone by Hyperspectral Raman Imaging," *Journal of Biomedical Optics*, **4**(1):28–34, 1999.

21. J. Timlin, "Chemical Microstructure of Cortical Bone Probed by Raman Transects," *Applied Spectroscopy*, **53**(11):1429–1435, 1999.

22. J. Timlin, A. Carden, M. Morris, R, Rajachar, and D. Kohn, "Raman Spectroscopic Imaging Markers for Fatigue-Related Microdamage in Bovine Bone," *Analytical Chemistry*, **72**(10):2229–2236, 2000.

23. N. Gierlinger, L. Sapei, and O. Paris, "Insights into the Chemical Composition of Equisetum Hyemale by High Resolution Raman Imaging," *Planta*, **227**:969–980, 2008.

24. J. Ma and D. Ben-Amotz, "Rapid Micro-Raman Imaging Using Fiber-Bundle Image Compression," *Applied Spectroscopy*, **51**(12):1845–1848, 1997.

25. M. V. Schulmerich, W. F. Finney, R. A. Fredericks, and M. D. Morris, "Subsurface Raman Spectroscopy and Mapping Using a Globally Illuminated Non-Confocal Fiber-Optic Array Probe in the Presence of Raman Photon Migration," *Applied Spectroscopy*, **60**(2):109–114, 2006.

26. M. V. Schulmerich, K. Dooley, T. Vanasse, S. Goldstein, and M. Morris, "Subsurface and Transcutaneous Raman Spectroscopy and Mapping Using Concentric Illumination Rings and Collection with a Circular Fiber-Optic Array," *Applied Spectroscopy*, **61**(7):671–678, 2007.

27. H. Sato, T. Tanaka, T. Ikeda, S. Wada, H. Tashiro, and Y. Ozaki, "Biomedical Applications of a New Portable Raman Imaging Probe," *Journal of Molecular Structure*, **598**:93–96, 2001.

28. P. Treado, "Chemical Imaging Reveals More Than the Microscope," *Laser Focus World*, **31**(10):75, 1995.

29. G. Puppels, M. Grond, and J. Greve, "Direct Imaging Raman Microscopy Based on Tunable Wavelength Excitation and Narrow-Band Emission Detection," *Applied Spectroscopy*, **47**:1256–1267, 1993.

30. D. Batchelder, C. Cheng, W. Muller, and B. Smith, "A Compact Raman Microprobe/Microscope: Analysis of Polydiacetylene Langmuir and Langmuir-Blodgett Films," *Makromolekulare Chemie Macromolecular Symposia*, **46**:171–179, 1991.

31. P. Treado, I. Levin, and E. Lewis, "High-Fidelity Raman Imaging Spectrometry: A Rapid Method Using an Acousto-Optic Tunable Filter," *Applied Spectroscopy*, **46**(8):1211–1216, 1992.

32. H. Morris, C. Hoyt, P. Miller, and P. Treado, "Liquid Crystal Tunable Filter Raman Chemical Imaging," *Applied Spectroscopy*, **50**(6):805–811, 1996.

33. M. Schaeberle, V. Kalasinsky, J. Luke, E. Lewis, I. Levin, and P. Treado, "Raman Chemical Imaging: Histopathology of Inclusion in Human Breast Tissue," *Analytical Chemistry*, **68**:1829–1833, 1996.

34. P. Treado, "Chemical Imaging System," US Patent 6002476, 1999.

35. N. Kline and P. Treado, "Raman Chemical Imaging of Breast Tissue," *Journal of Raman Spectroscopy*, **28**:119–124, 1997.

36. M. D. Schaeberle, H. F. Morris, J. F. Turner, and P. Treado, "Raman Chemical Imaging Spectroscopy," *Analytical Chemistry*, **71**:175A–181A, 1999.

37. P. J. Treado and M. P. Nelson, "Raman Imaging," In: I. R. Lewis and H. G. M. Edwards (eds.), *Handbook of Raman Spectroscopy: From the Research Laboratory to the Process Line*. Marcel Dekker, New York, 2001, Chap. 5, pp. 191–249.

38. G. Payne, N. Langlois, C. Lennard, and C. Roux, "Applying Visible Hyperspectral (Chemical) Imaging to Estimate the Age of Bruising," *Medicine, Science and the Law*, **47**(3):225–232, 2007.

39. A. Theophilus, A. Moore, D. Prime, S. Rossomanno, B. Whitcher, and H. Chrystyn, "Co-deposition of Salmeterol and Fluticasone Propionate by a Combination Inhaler," *International Journal of Pharmacutics*, **313**:14–22, 2006.

40. S. Sasic, "Chemical Imaging of Pharmaceutical Granules by Raman Global Illumination and Near-Infrared Mapping Platforms," *Analytica Chimica Acta*, **611**(1):73–79, 2008.

41. S. Sasic, "An In-Depth Analysis of Raman and Near-Infrared Chemical Images of Common Pharmaceutical Tablets," *Applied Spectroscopy*, **61**(3):239–250, 2007.

42. S. Sasic and D. Clark, "Defining a Strategy for Chemical Imaging of Industrial Pharmaceutical Samples on Raman Line-Mapping and Global Illumination Instruments," *Applied Spectroscopy*, **60**(5):494–502, 2006.

43. J. Kauffman, S. Gilliam, and R. Martin, "Chemical Imaging of Pharmaceutical Materials: Fabrication of Micropatterned Resolution Targets," *Analytical Chemistry*, **80**(15):5706–5712, 2008.

44. M. Rios, "New Dimensions in Tablet Imaging," *Pharmacy Technology*, **3**(32):52–62, 2008.

45. J. Panza and J. Maier, "Raman Spectroscopy and Raman Chemical Imaging of Apoptotic Cells," *SPIE Photonics West*, **664108**:1–12, 2007.

46. S. D. Stewart, A. H. Uihlein, J. L. Hammers, J. F. Silverman, J. S. Maier, and J. K. Cohen, "Raman Molecular Imaging—A Potential Tool for Histopathological Tissue Assessment," In: R. Withnall and B. Z. Chowdhry (eds.), *Proceedings of the XXIst International Conference on Raman Spectroscopy*, IM Publications, London, 2008, pp. 219–220.

47. S. Stewart, L. McClelland, and J. Maier, "A Fast Method for Detecting *Cryptosporidium parvum* Oocysts in Real World Samples," *Proceedings of SPIE*, **5692**:341–350, 2005.

48. K. Kalasinsky, T. Hadfield, A. Shea, V. Kalasinsky, M. Nelson, J. Neiss, A. Drauch, G. Vanni, and P. Treado, "Raman Chemical Imaging Spectroscopy Reagentless Detection and Identification of Pathogens: Signature Development and Evaluation," *Analytical Chemistry*, **79**(7):2658–2673, 2007.

49. A. Tripathi, R. E. Jabbour, P. J. Treado, J. Neiss, M. Nelson, J. Jensen, and A. Snyder, "Waterborne Pathogen Detection Using Raman Spectroscopy," *Applied Spectroscopy*, **62**(1):1–9, 2008.

50. P. Treado, "Chemical Imaging Reveals More than the Microscope," *Laser Focus World*, **31**(10):75, 1995.

51. C. Otto, J. de Grauw, and J. Duindam, "Applications of Micro-Raman Imaging in Biomedical Research," *Journal of Raman Spectroscopy*, **28**:143–150, 1997.

52. C. Matthaus, T. Chemenko, J. Newmark, C. Warner, and M. Diem, "Label-Free Detection of Mitochondrial Distribution in Cells by Nonresonant Raman Microspectroscopy," *Biophysical Journal*, **93**:668–673, 2007.

53. J. Chao, E. Perevedentseva, P. Chung, K. Liu; C. Cheng, C. Chang, and C. Cheng, "Nanometer-Sized Diamond Particle as a Probe for Biolabeling," *Biophysical Journal*, **93**:2199–2208, 2007.

54. J. S. Maier and P. J. Treado, "Raman Molecular Chemical Imaging: 3D Raman Using Deconvolution," In: B. M. Cullum (ed.), *Smart Medical and Biomedical Sensor Technology II. Proceedings of SPIE*, Bellingham, Wash, **5588**:98–105, 2004.

55. M. F. Escoriza, J. M. Vanbriesen, S. Stewart, J. Maier, and P. J. Treado, "Raman Spectroscopy and Chemical Imaging for Quantification of Filtered Waterborne Bacteria," *Journal of Microbiological Methods*, **66**(1):63–72, 2006.
56. N. Afseth, V. Segtnan, and J. Wold, "Raman Spectra of Biological Samples: A Study of Preprocessing Methods," *Applied Spectroscopy*, **60**:1358–1367, 2006.
57. M. Pelletier, "Quantitive Analysis Using Raman Spectroscopy," *Applied Spectroscopy*, **57**:20A–39A, 2003.
58. "Specialized Techniques," Molecular Expressions, http://micro.magnet.fsu.edu/primer/techniques/fluorescence/fluorointrohome.html., 2008.
59. J. F. Brennan III, Y. Wang, R. R. Dasari, and M. S. Feld, "Near-Infrared Raman Spectrometer Systems for Human Tissue Studies," *Applied Spectroscopy*, **51**:201–208, 1997.
60. "Glossary of Terms," Princeton Instruments, http://www.princetoninstruments.com/library/glossary.aspx#d. (*last accessed* on March 2010)
61. National Institute of Standards & Technology, "Certificate of Analysis. Standard Reference Material 2242," Gaitersburg, Md., 2004.
62. J. Zhao, H. Lui, D. McLean, and H. Zeng, "Automated Autofluorescence Background Subtraction Algorithm for Biomedical Raman Spectroscopy," *Applied Spectroscopy*, **61**:1225–1232, 2007.
63. C. Lieber and A. Mahadevan-Jansen, "Automated Methods for Subtraction of Fluorescence from Biological Raman Spectra," *Applied Spectroscopy*, **57**:1363–1367, 2003.
64. K. R. Beebe, R. J. Pell, and M. B. Seasholtz, *Chemometrics A Practical Guide*, John Wiley & Sons, New York, N.Y. 1998.
65. R. Kramer, *Chemometric Techniques for Quantitative Analysis*, CTC Press, Boca Raton, Fla., 1998.
66. H. Martens and T. Naes, *Multivariate Calibration*, John Wiley & Sons, New York , N.Y., 1989.
67. A. H. Uihlein, S. Stewart, J. Maier, J. Cohen, Y. L. Liu, and J. F. Silverman, "Diagnostic Utility of Raman Molecular Imaging (RMI) in Separating Renal Oncocytoma from Chromophobe Renal Cell Carcinoma," in *97th Annual Meeting United States and Canadian Academy of Pathology*, USCAP Denver, Colo., 2008.
68. J. L. Lindner, S. Stewart, J. Maier, J. Cohen; J. F. Silverman, "Separation of Invasive Lobular from Ductal Carcinoma of the Breast Using Raman Molecular Imaging," in *97th Annual Meeting United States and Canadian Academy of Pathology*, USCAP, Denver, Colo., 2008.
69. H. Zhao, A. Langerod, Y. Ji, K. Nowles, J. Nesland, R. Tibshirani, I. Bukholm, R. Karesen, D. Botstein, A. Borresen-Dale, and S. Jeffrey, "Different Gene Expression Patterns in Invasive Lobular and Ductal Carcinomas of the Breast," *Molecular Biology of the Cell*, **15**(6):2523–2536, 2004.
70. C. Eliasson, A. Loren, J. Engelbrektsson, M. Josefson, J. Abrahamsson, and K. Abrahamsson, "Surface-Enhanced Raman Scattering Imagin of Single Living Lymphocytes with Multivariate Evaluation," *Spectrochimica Acta, Part A*, **61**:755–760, 2005.
71. R. Malini, K. Venkatakrishna, J. Kurian, K. Pai, L. Rao, V. Kartha, and C. M. Krishna, "Discrimination of Normal, Inflammatory, Premalignant, and Malignant Oral Tissue: A Raman Spectroscopy Study," *Biopolymers*, **81**: 179–193, 2006.
72. M. V. P. Chowdary, K. K. Kumar, J. Kurien, S. Mathew, and C. Murali Krishna. "Discrimination of Normal, Benign, and Malignant Breast Tissues by Raman Spectroscopy," *Biopolymers*, **83**:556–569, 2006.
73. K. K. Kumar, A. Anand, M. V. P. Chowdary, J. K. Keerthi, C. Murali Krishna, and S. Mathew, "Discrimination of Normal and Malignant Stomach Mucosal Tissue by Raman Spectroscopy: A Pilot Study," *Vibrational Spectroscopy*, **44**:382–387, 2007.
74. M. V. P. Chowdary, K. K. Kumar, K. Thakur, A. Anand, J. Kurien, C. Murali Krishna, and S. Mathew, "Discrimination of Normal and Malignant Mucosal Tissues of the Colon by Raman Spectroscopy," *Photomedicine and Laser Surgery*, **25**:269–274, 2007.

75. K. Maheedhar, R. A. Bhat, R. Malini, N. Prathima, P. Keerthi, P. Kushtagi, and C. Murali Krishna, "Diagnosis of Ovarian Cancer by Raman Spectroscopy: A Pilot Study," *Photomedicine and Laser Surgery*, **26**:83–90, 2008.

76. K. Shafer-Peltier, A. Haka, J. Motz, M. Fitzmaurice, R. Dasari, and M. Feld, "Model-Based Biological Raman Spectral Imaging," *Journal of Cellular Biochemistry Supplement*, **39**:125–137, 2002.

77. K. Omberg, J. Osborn, S. Zhang, J. Freyer, J. Mourant, and J. Schoonover, "Raman Spectroscopy and Factor Analysis of Tumorigenic and Non-Tumorigenic Cells," *Applied Spectroscopy*, **56**(7):813–819, 2002.

78. A. Nijssen, T. C. Schut, F. Heule, P. Caspers, D. Hayes, M. Neumann, and G. Puppels, "Discriminating Basal Cell Carcinoma from Its Surrounding Tissue by Raman Spectroscopy," *The Journal of Investigative Dermatology*, **119**(1):64–69, 2002.

79. J. S. Thakur, H. Dai, G. K. Serhatkulu, R. Naik, V. Naik, A. Cao, A. Pandya, G. Auner, R. Rabah, M. Klein, and C. Freeman, "Raman Spectral Signatures of Mouse Mammary Tissue and Associated Lymph Nodes: Normal, Tumor, and Mastitis," *Journal of Raman Spectroscopy*, **38**:127–134, 2007.

80. C. Owen, J. Selvakumaran, I. Notingher, G. Jell, L. Hench, and M. Stevens, "In Vitro Toxicology Evaluation of Pharmaceuticals Using Raman Micro-Spectroscopy," *Journal of Cellular Biochemistry*, **99**:178–186, 2006.

81. G. Jell, I. Notingher, O. Tsigkou, P. Notingher, J. Polak, L. Hench, and M. Stevens, "Bioactive Glass-Induced Osteoblast Differentiation: A Noninvasive Spectroscopic Study," *Journal of Biomedical Materials Research Part A*, **86A**:31–40, 2008.

82. M. Escoriza, J. van Briesen, S. Stewart, and J. Maier, "Raman Spectroscopic Discrimination of Cell Response to Chemical and Physical Inactivation," *Applied Spectroscopy*, **61**(8):812–823, 2007.

83. A. Taleb, J. Diamond, J. McGarvey, J. Renwich Beattie, C. Toland, and P. Hamilton, "Raman Microscopy for the Chemometric Analysis of Tumor Cells," *Journal of Physical Chemistry B*, **110**:19625–19631, 2006.

84. A. Mahadevan-Jansen, M. F. Mitchell, N. Ramanujam, A. Malpica, S. Thomsen, U. Utzinger, and R. Richards-Kortum, "Near-Infrared Raman Spectroscopy for In Vitro Detection of Cervical Precancers," *Photochemistry Photobiology*, **68**(1):123–132, 1998.

85. F. Lyng, E. O. Faolain, J. Conroy, A. Meade, P. Knief, B. Duffy, M. Hunter, J. Byrne, P. Kelehan, and H. Byrne, "Vibrational Spectroscopy for Cervical Cancer Pathology, from Biochemical Analysis to Diagnostic Tool," *Experimental Molecular and Pathology*, **82**:121–129, 2007.

86. T. Bakker Schut, M. J. Witjes, H. J. Sterenborg, O. Speelman, J. Roodenburg, E. Marple, H. Bruining, and G. Puppels, "In Vivo Detection of Dysplastic Tissue by Raman Spectroscopy," *Analytical Chemistry*, **72**:6010–6018, 2000.

87. N. Stone, C. Kendall, N. Shepherd, P. Crow, and H. Barr, "Near-Infrared Raman Spectroscopy for the Classification of Epithelial Pre-Cancers and Cancers," *Journal of Raman Spectroscopy*, **33**:564–573, 2002.

88. K. Das, N. Stone, C. Kendall, C. Fowler, and J. Christie-Brown, "Raman Spectroscopy of Parathyroid Tissue Pathology," *Lasers in Medical Science*, **21**:192–197, 2006.

89. R. Swain, G. Jell, and M. Stevens, "Non-Invasive Analysis of Cell Cycle Dynamics in Single Living Cells with Raman Micro-Spectroscopy," *Journal of Cellular Biochemistry*, **104**:1427–1438, 2008.

90. J. Chan, D. Taylor, S. Lane, T. Zwerdling, J. Tuscano, and T. Huser, "Nondestructive Identification of Individual Leukemia Cells by Laser Trapping Raman Spectroscopy," *Analytical Chemistry*, **80**:2180–2187, 2008.

91. S. Teh, W. Zheng, K. Ho, M. The, K. Yeoh, and Z. Huang, "Diagnostic Potential of Near-Infrared Raman Spectroscopy in the Stomach: Differentiating Dysplasia from Normal Tissue," *British Journal of Cancer*, **98**:457–465, 2008.

CHAPTER 7

Resonance Raman Imaging and Quantification of Carotenoid Antioxidants in the Human Retina and Skin

Mohsen Sharifzadeh, Igor V. Ermakov

Department of Physics
University of Utah
Salt Lake City, Utah

Paul S. Bernstein

Moran Eye Center
University of Utah
Salt Lake City, Utah

Werner Gellermann*

Department of Physics
University of Utah
Salt Lake City, Utah

7.1 Introduction

Motivated by the growing importance of carotenoid antioxidants in health and disease, we investigate resonance Raman scattering (RRS) as a novel approach for the noninvasive optical detection of carotenoids

*Corresponding author.

in living human tissue. Raman spectroscopy is a well-known, highly molecule specific form of vibrational spectroscopy that is commonly used to identify molecular compounds through their spectrally narrow "Raman fingerprint" responses. Most frequently, off-resonance Raman techniques are used for this purpose since they avoid the strong intrinsic electronic fluorescence transitions typically encountered in complex molecules. Carotenoid molecules, however, possess a unique energy level structure and associated optical pumping cycle. While easily excited from the ground state into a higher excited state within a strong, electric dipole-allowed absorption transition, they relax quickly into a new, lower-lying excited state, from which fluorescence transitions back to the ground state are forbidden. This offers the opportunity to use fluorescence-background-free resonant excitation of the carotenoids in their visible absorption bands, which results in a resonance enhancement of the carotenoid Raman response by about five orders of magnitude relative to nonresonant Raman scattering.[1] It becomes possible, therefore, to explore RRS not only for the identification of carotenoids in biological tissue environments, but also, through the intensity of their RRS response, for the quantification of their tissue concentrations. The tissue environment can be expected to have only a minor effect on the molecule's vibrational energy. This should cause the Raman signature to be virtually identical for the isolated carotenoid molecule, the molecule in solution, or the molecule in a cell environment. However, the applicability of the method can be expected to depend heavily on potentially confounding tissue properties, such as a saturation of the carotenoid Raman response at high concentrations, or the existence of other molecules with potentially interfering scattering, absorption and/or fluorescence contributions. A crucial task therefore is the validation of the RRS detection method for the particular tissue environment. If successful, RRS could be used as a novel optical diagnostic method for the measurement of tissue carotenoid levels, potentially allowing one to measure large populations in clinical and field settings, and to track their changes occurring over time as a consequence of developing pathology and/or tissue uptake.

A tissue site that is particularly interesting for the application of the Raman method is the *macula lutea*. It is the well-known "yellow spot" in the human retina that contains the highest concentration of carotenoids in the human body. Of the approximately 10 carotenoid species identified in human serum, only two carotenoids, lutein and zeaxanthin, are selectively taken up at this retinal tissue site. Their concentrations can be as high as several 10 ng/g of tissue, however, in the healthy human retina. Due to their strong absorption in the blue-green spectral range, the macular carotenoids, also termed *macular pigment* (MP) impart a yellow coloration to the macula. When viewing the retinal in cross section, MP is located anterior to the photoreceptor outer segments and the retinal pigment epithelium.[2,3] It is therefore thought to shield these vulnerable tissues from light-induced oxidative damage by blocking phototoxic short-wavelength visible light. Also, MP is thought to directly protect the cells in this area through the antioxidant function of the carotenoids, lutein and zeaxanthin, which are well-known scavengers of reactive oxygen species. There is increasing evidence that MP may help mediate protection of the cone photoreceptor cells in the macula against visual loss from age-related macular degeneration (AMD)[4–8] the leading cause of irreversible blindness affecting a large portion of the elderly population. Since the MP compounds are taken up through the diet, there is a chance that early age screening of MP concentrations can be used to identify individuals with low levels of MP. If indicated, dietary interventions such as increased consumption of lutein and zeaxanthin containing vegetables or nutritional supplements could be used to help prevent or delay the onset of the disease.

MP concentrations in the healthy human retina have been assumed to be highest in the very center of the macula, the foveola, and to drop off rapidly with increasing eccentricity.[9] Recently emerging high-resolution optical imaging techniques based on lipofuscin fluorescence ("autofluorescence") excitation and reflection methods, however, are beginning to reveal a much more complex behavior of MP distributions in the living human retina, such as patterns with locally depleted MP levels and MP distributions with ring-shaped patterns.[10–13] It is important to confirm these interesting features with an imaging Raman method, which in comparison to the other methods would be a much more direct, carotenoid-specific detection. Furthermore, it could provide a valuable new method that is useful to track MP distributions, their local MP levels, and potential changes occurring in them upon the occurrence of vision problems, macular pathology, and subsequent dietary modifications or supplementation.

In addition to ophthalmology, RRS spectroscopy is interesting also for the detection of carotenoids in human skin. In this tissue, which constitutes the largest organ of the human body, the predominant carotenoid species are lycopene and β-carotene, followed in order of tissue concentrations by β-cryptoxanthin, lutein, and zeaxanthin. All are thought to play an important protective role as antioxidants, like in the protection of skin from premature aging caused by ultraviolet and short-wavelength visible radiation. The carotenoids, lutein and lycopene, may also have protective functions for cardiovascular health, and lycopene may play a role in the prevention of prostate cancer. It is conceivable that skin levels of these carotenoids are correlated with corresponding levels in internal tissues.

Objective measurements of skin carotenoid levels would be of interest also for improving dietary data collected in epidemiological studies, which in turn are used in developing public health guidelines that promote healthier diets. The protective effects of diets rich in fruits and vegetables has been observed for many disease outcomes, including various cancers[14,15] and cardiovascular disease.[16] Since carotenoids are a good biomarker for fruit and vegetable intake, Raman measurements of skin carotenoid levels could be used as rapid optical method to assess fruit and vegetable consumption in large populations.

For many decades, the standard technique for measuring carotenoids has been high-pressure liquid chromatography (HPLC). This chemical method works well for the measurement of carotenoids in serum, but it is time consuming, expensive, and highly invasive for human tissue, since it requires biopsies of relatively large tissue volumes. Additionally, serum antioxidant measurements are more indicative of short-term dietary intakes of antioxidants rather than steady-state accumulations in body tissues exposed to external oxidative stress factors such as smoking and UV-light exposure.

7.2   Optical Properties and Resonance Raman Scattering of Carotenoids

Carotenoids are π-electron conjugated carbon-chain molecules and are similar to polyenes with regard to their molecular structure and optical properties. Distinguishing features are the number of conjugated carbon double bonds (C=C), the number of attached methyl side groups, and the presence or absence and composition of attached end groups. The molecular structure of β-carotene is shown as an example in Fig. 7.1, which consists of a linear, conjugated, carbon backbone with alternating carbon single (C—C) and double bonds (C=C), two ione end groups, and 4 methyl side groups. The structure is very similar to all other carotenoids of interest in this chapter. Resonance Raman spectroscopy detects the vibrational stretch frequencies of the carbon bonds as well as the rocking motion of the attached methyl side groups. Also shown is a configuration coordinate diagram for the three lowest lying energy states, and an indication of all optical and nonradiative transitions connecting the states. The configuration coordinate represents the displacement of a normal coordinate of the molecule's atoms its equilibrium position. Absorption, fluorescence, and Raman transitions occur at fixed nuclear positions of the molecule's atoms (fixed configuration coordinate) as vertical transitions. A strong, electric-dipole allowed absorption transition occurs between the molecule's delocalized π-orbital from the 1^1A_g singlet ground state to the 1^1B_u singlet excited state. As illustrated in Fig. 7.2a, this gives rise to a broad absorption band in the blue-green spectral region (peak at ~460 nm, spectral width ~100 nm) that features clearly resolved vibronic substructures (with spectral spacing of ~1400 cm^{-1}). After excitation to the 1^1B_u state, the carotenoid molecule relaxes very rapidly, within ~200 to 250 fs,[17] via nonradiative transitions, to the lower-lying 2^1A_g excited state. Since this is a state with gerade parity, a radiative transition to the ground state is parity-forbidden. As a consequence, the luminescence transitions from the 1^1B_u and 2^1A_g states to the ground states are very weak (10^{-5} to 10^{-4} efficiency). It is this de facto absence of intrinsic luminescence of carotenoids that allows one to

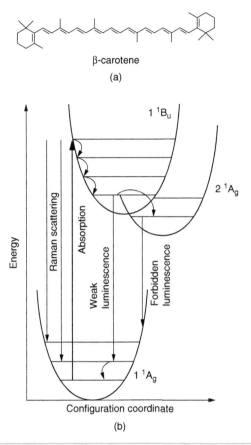

β-carotene

(a)

Configuration coordinate

(b)

FIGURE 7.1   (a) Molecular structure of β-carotene; (b) configuration coordinate diagram for three lowest lying energy levels of carotenoids, with indication of optical and nonradiative transitions between all levels.

detect the relatively weak RRS response of their molecular vibrations virtually free of otherwise overwhelming luminescence signals.

For tetrahydrofuran solutions of β-carotene, zeaxanthin, lycopene, lutein, and phytofluene, we obtain the RRS spectra displayed in Fig. 7.2b. All reveal strong and clearly resolved Raman signals which are comparable or even stronger than the intrinsic fluorescence background. Three prominent Raman Stokes lines, characteristic for all carotenoids, appear at ~1525 cm^{-1} (C=C stretch), 1159 cm^{-1} (C—C stretch), and 1008 cm^{-1} (C—CH$_3$ rocking motions).[1] In the shorter-chain phytofluene molecule, only the C=C stretch appears, and it is shifted significantly to higher frequencies (by ~40 cm^{-1}), which is a consequence of a shorter conjugation length in this molecule.

Raman scattering is a linear spectroscopy, in principle, meaning that the Raman scattering intensity I_S scales linearly with the intensity

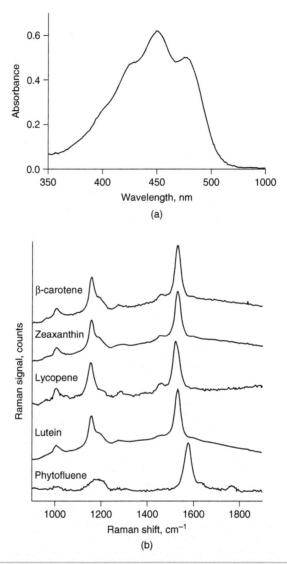

Figure 7.2 (a) Absorption spectrum of a β-carotene solution corresponding to the molecule's $1^1A_g \rightarrow 1^1B_u$ transition, showing the characteristic broad absorption with vibronic substructure in the blue-green spectral range; (b) resonance Raman spectra of β-carotene, zeaxanthin, lycopene, lutein, and phytofluene solutions.

of the excitation light I_L. This assumes that the scattering compound can be considered as optically thin. At fixed excitation light intensity I_L, the Raman response scales with the population density of the scatterers $N(E_i)$ according to

$$I_s = N(E_i) \times \sigma_R \times I_L$$

Here, σ_R is the Raman scattering cross section, a constant whose magnitude depends only on the excitation and collection geometry. However, in optically thick media, such as in geometrically thin but optically dense tissue, a deviation from the linear Raman response of I_s versus concentration N can be expected. This could occur, for example, due to self-absorption of the Stokes Raman signal by the strong electronic absorption, or due to insufficient excitation light penetration. In these cases, a nonlinear calibration between RRS response and molecule concentration may be required using suitable tissue phantoms.

7.3   Spatially Integrated Resonance Raman Measurements of Macular Pigment

The human retina can be optically accessed relatively easily. Targeting of the macular region with excitation light is simply accomplished by having the subject fixate on an aiming beam. The excitation and Raman light have to traverse cornea, aqueous humor, lens, and vitreous, as sketched in Fig. 7.3a. All these ocular media are generally highly transparent. However, confounding light absorption and scattering can be expected in cases of lens cataracts and in cases of scattering from the vitreous. The macula is essentially free of blood vessels. When containing a healthy concentration of lutein and zeaxanthin concentrations, it appears as a gray-shaded area in black-and-white reflection images of the retina, or in autofluorescence images taken with blue light, as shown in Fig. 7.3b.

In vivo spectroscopy of the macula can take advantage of favorable anatomical features of the tissue structures encountered in the excitation and Raman light scattering pathways. A cross section of the retinal tissue layers in the macular region, shown in Fig. 7.4, helps to illustrate the concept. First, the major sites of macular carotenoid deposition are the Henle fiber layer, which has a thickness of only about hundred micrometers, and to a lesser extent the plexiform layer (both layers are shown in Fig. 7.4 together with the outer nuclear layer as a single layer, HPN). Considering that the optical density of MP in the peak of the absorption band is typically lower than 1, as determined from direct absorption measurements of MP in excised eyecups, these tissue properties provide essentially an optically thin film with minimal self-absorption for both the excitation light and the Raman scattered light, provided the carotenoids/MP are properly excited on their long-wavelength absorption shoulder. Second, the Raman scattering approach relies only on the backscattered, single-path Raman response from the lutein and zeaxanthin-containing MP layers. Since these retinal layers are located anteriorly to other retinal layer structures in the optical pathway through the retina, absorption and fluorescence effects originating from chromophores such as rhodopsin in the photoreceptor layer, PhR, and melanin and lipofuscin

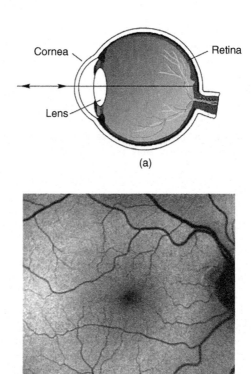

(a)

(b)

FIGURE 7.3   (a) Cross section of human eye with indication of optical beam paths propagating back and forth to the macular region of the retina; (b) autofluorescence image of healthy human retina, showing macular region in center with dark shading caused by macular pigment. Part of the optic nerve head can be seen as dark spot at center right.

in the retinal pigment epithelial layer, RPE, respectively, can be ignored or subtracted from the Raman spectra.

Our initial "proof of principle" studies of ocular carotenoid RRS employed a laboratory-grade high-resolution Raman spectrometer and flat mounted human cadaver retinas and eyecups. We were able to record characteristic carotenoid RRS spectra from these tissues with a spatial resolution of approximately 100 μm, and we were able to confirm linearity of response by extracting and analyzing tissue carotenoids by HPLC after completion of the Raman measurements.[18] For in vivo experiments and clinical use, we developed Raman instruments with lower spectral resolution but highly improved light throughput.[19,20] A current version that is combined with a fundus camera to permit independent operator targeting of the subject's macula[21] is shown in Fig. 7.5a. The instrument's Raman module, containing a

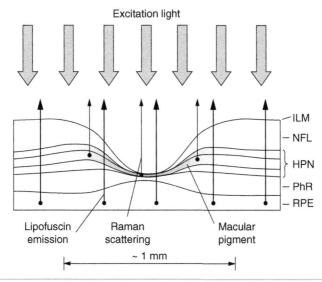

Excitation light

ILM
NFL
HPN
PhR
RPE

Lipofuscin / Raman   Macular
emission   scattering   pigment

~ 1 mm

FIGURE 7.4 Schematics of retinal layers participating in light absorption, transmission and scattering of excitation and emission light. ILM: inner limiting membrane; NFL: nerve fiber layer; HPN: Henle fiber, Plexiform and Nuclear layers; PhR: photoreceptor layer; RPE: retinal pigment epithelium. (*Reprint with permission from Ref. 27.*)

488-nm laser excitation source, a spectrograph, and a CCD array detector, is optically connected with the fundus camera using a beam splitter that is mounted between the front-end optics of the fundus camera and the eye of the subject. Once alignment is established, an approximately 1 mm diameter, 1.0 mW, light excitation disk is projected onto the subject's macula for 0.25 second through the pharmacologically dilated pupil, and the backscattered light is routed to the Raman module for detection. Retinal light exposure levels of the instrument are in compliance with ANSI safety regulations, since ocular exposure levels are a factor of 19 below the thermal limit, and a factor of 480 below the photochemical limit for retinal injury.[21]

Typical RRS spectra, measured from the macula of a healthy human volunteer through a dilated pupil are displayed, in near real time, on the instrument's computer monitor, as shown in Fig. 7.5*b*. The left panel shows the raw spectrum obtained from a single measurement, and clearly reveals the three characteristic carotenoid Raman signals, which are superimposed on a steep, spectrally broad fluorescence background. The background is caused partially by the weak intrinsic fluorescence of lutein and zeaxanthin, and partially by the short-wavelength emission tail of lipofuscin, which is present in the retinal pigment epithelial layer, and which is excited by the portion of the excitation light that is transmitted through the MP-containing Henle fiber and plexiform layers. The ratio between the intensities of

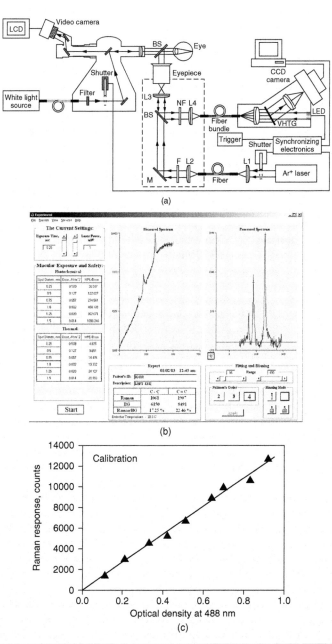

(a)

(b)

(c)

FIGURE 7.5   (a) Schematics of fundus-camera-interfaced RRS instrument for measurement of integral MP concentrations in human clinical studies; (b) computer monitor display showing raw Raman spectrum obtained after single measurement (left panel) and processed, scaled spectrum obtained after subtraction of fluorescence background (right panel); (c) calibration curve for RRS response of tissue phantom with nine lutein and zeaxanthin concentration levels.

the carotenoid C=C Raman response and the fluorescence background is high enough (up to ~0.25) that it is easily possible to quantify the amplitudes of the C=C peak after digital background subtraction. This step is automatically accomplished by the instrument's data processing software, which approximates the background with a fourth-order polynomial, subtracts the background from the raw spectrum, and displays the final result as processed, scaled spectrum in the right panel of the computer monitor, as shown in Fig. 7.5b. MP carotenoid RRS spectra measured for the living human macula were indistinguishable from corresponding spectra of pure lutein or zeaxanthin solutions measured with the same instrument.

While the fundus-camera-interfaced Raman instrument is well suited for measurements of elderly subjects, subjects with macular pathologies, and also research animals, we found that simplified instrument versions can be used for healthy human subjects, provided they have good visual acuity (with or without correcting lenses) and are able to self-align on a fixation target prior to a Raman measurement. An example for a particularly simple self-alignment instrument configuration is a version in which the CCD/spectrograph combination is replaced with a single photomultiplier/filter combination.[22]

In order to cross calibrate different instrument versions, we constructed a simple tissue phantom consisting of a lens and a thin, 1 mm path length, cuvette placed in the focal plane of the lens, and measured the RRS response for preset lutein and zeaxanthin solutions with optical densities in the range 0.1 to 1.0, a range that at the higher end exceeds typically encountered physiological concentration levels. An example of a calibration curve for a particular instrument version is shown in Fig. 7.5c. It demonstrates a linear RRS response up to a relatively high-optical density of 0.8. Besides for cross calibration purposes, this calibration method can be used to correlate the RRS response of a subject's MP with its corresponding optical density value.

An example for clinical RRS measurements of a relatively young subgroup (33 eyes), ranging in age from 21 to 29 years, is shown in Fig. 7.6a. A striking observation is the fact, that the RRS-measured MP levels can vary drastically between individuals (up to ~10 fold difference). Since the ocular transmission properties of all anterior optical media in this age group can be assumed to be very similar, the variations must be attributed to strongly varying MP levels. Subjects with extremely low-carotenoid levels may be at higher risk of developing macular degeneration later in life.

When measuring a large population of normal subjects, none of whom were consuming nutritional supplements with containing substantial amounts of lutein or zeaxanthin, we found a striking decline of average macular carotenoid levels with age,[20,23] as shown in Fig. 7.6b. Part of this decline can be explained by "yellowing" of the crystalline lens with age, which would attenuate some of the illuminating and

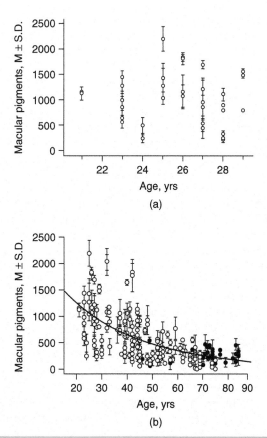

FIGURE 7.6 (a) RRS MP measurements of 33 normal eyes for a young group of subjects ranging in age from 21 to 29 years. Note the large (up to ~10-fold) variation of RRS levels that can exist between individuals; (b) RRS measurements of 212 normal eyes as a function of subject age, revealing statistically significant decrease of MP concentration with age. (*Panel a adapted from Ref. 26, panel b reprinted with permission from Ref. 20.*)

backscattered light, but we also found consistently low MP levels even in patients who had previously had cataract surgery with implantation of optically clear prosthetic intraocular lenses (pseudophakia). Also, we have noted that patients with unilateral cataracts after trauma or retinal detachment repair typically have very similar RRS carotenoid levels in the normal and in the pseudophakic eye. Thus, we have concluded that there is a decline of macular carotenoids that reaches a low steady state just at the time when the incidence and prevalence of AMD begins to rise dramatically. The conclusion is further confirmed by our spatially resolved MP detection methods, as outlined below. These results emphasize the importance of assuring that populations are properly age-matched when using RRS spectroscopy in case-control studies.

7.4 Spatially Resolved Resonance Raman Imaging of Macular Pigment—Methodology and Validation Experiments

MP distributions are often assumed to have strict rotational symmetry, with high-central pigment levels and a monotonous decline with increasing eccentricity toward the peripheral retina. In order to gain more insight into spatial distribution aspects of MP, we developed resonance Raman imaging (RRI) of carotenoids.[24] The experimental setup for this purpose is shown in Fig. 7.7. Blue light from a solid-state 488-nm laser is routed onto a subject's retina via optical fiber,

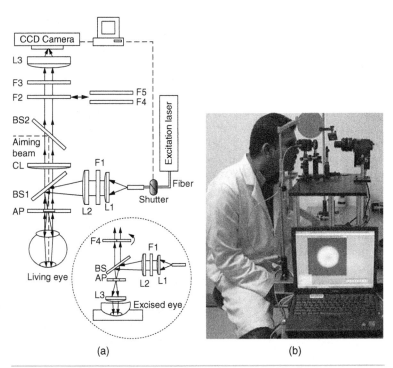

(a)   (b)

FIGURE 7.7 (a) Schematics of experimental setup used for in vivo resonance Raman imaging, RRI, of MP distributions. Light from a blue laser source is projected onto the macula as a ~3.5-mm-diameter excitation disk. The backscattered light is collimated by the eye lens and imaged with a two-dimensional CCD camera array detector. Two sets of filters are used sequentially to selectively image light at the C=C Raman wavelength ("Raman image") and at a slightly longer wavelength ("offset image"). The two images are digitally subtracted and displayed as topographic or three-dimensional pseudocolor images of the spatial MP concentrations. L1-3: lenses; F1: laser line filter; BS: dichroic beam splitters; F2: tunable filter; F3: bandpass filter. Inset shows modifications for use with excised tissue; (b) photograph of subject measured with instrument. RRI images are recorded with 0.2 second exposure time for dilated or nondilated pupils.

collimating lens, laser line filter, achromatic beam expander, dichroic beam splitter, and aperture. The laser excitation disk projected onto the retina has a diameter of ~3.5 mm, and has an intermediate focus at the position of the aperture. Laser speckle was effectively removed by mechanically shaking the light delivery fiber. This generates a spatially homogeneous laser excitation spot via fiber mode mixing. A red aiming light, coupled into the setup serves as a fixation target for the subject during a measurement. The optical shutter is designed such that it allows a small portion of the excitation light to be transmitted even when it is closed, so the subject can view both the red fixation target and the superimposed excitation disk for optimum head alignment. The light emitted from the excited disk of the retina is collimated by the subject's eye lens and imaged onto a CCD camera which was kept at a temperature of $-10°C$ during measurements. The instrument is interfaced to a computer that controls a mechanical shutter and acquires data for export into appropriate image processing software.

For each measurement two separate images are recorded. In the first image, the light returned from the retina under 488 nm excitation is filtered to transmit only 528 nm light, which is the spectral position of the resonance Raman response of the 1525 cm^{-1} carbon-carbon double bond stretch frequency of the MP carotenoids. This is achieved with a combination of a narrow-bandpass filter (transmission range 528 ± 1 nm), a broader bandpass filter (transmission window 525 ± 15 nm), and a notch filter. The obtained image contains the Raman response of MP and overlapping background fluorescence components. In the second image, "fluorescence image," the light returned from the retina is filtered with a bandpass filter to only transmit background fluorescence components slightly above the Raman wavelength, in the 530 to 550 nm range. As further explained below, this image is subtracted from the first image to retrieve the pure RRI image of the MP distribution.

For measurements of excised eyecups, the setup is modified as sketched in inset of Fig. 7.7. In this case, the Raman images are derived from two measurements taken with a narrow-band tunable filter that is angle-tuned to "on" and "off" Raman spectral positions.

Human subjects were recruited from an eye clinic and had their eyes either dilated or undilated depending on their prior eye examination. To exclude pupil size effects, subjects with dilated eyes were chosen for intersubject comparisons involving absolute MP levels. Other measurements, such as the identification of a specific type of spatial MP pattern in a subject's eye, or the monitoring of the MP distribution in a subject's eye over time, were carried out with undilated eyes.

Measurements were carried out in a semi-darkened room. Laser power levels at the cornea were 4 mW during a measurement; exposure times were 100 ms for background fluorescence measurements, and 300 ms for resonance Raman imaging. At a retinal spot size of 3.5 mm diameter, the area of the exposed retinal field is 0.096 cm^2, and the retinal radiant exposure is $(4\,mW)(400\,ms)/0.096\,cm^2 = 16.7\,mJ/cm^2$, assuming all radiant power enters the pupil. This level is considered safe according to limits set by the ANSI standard.[25] The laser light exposures caused after-images that typically disappeared within a few minutes. During this time, the filters were changed to switch from Raman to fluorescence imaging.

Raman and fluorescence signals produced within the ocular media and retinal layers upon optical excitation with intensity $I(\lambda_{exc})$ are illustrated in Fig. 7.8. Under laser excitation with spectral overlap of the MP absorption, a resonance Raman response is obtained only from the MP containing retinal layer.[26] Fluorescence signals of MP can be neglected since the corresponding fluorescence quantum efficiency is extremely weak. The influence of the photoreceptors in the detector response can be neglected as well since the used light intensities are high enough to ensure temporary bleaching see below, and reference.[27]

Under these assumptions, the light intensity I_{Det} that is detected in each pixel of the CCD detector array at the spectral position of the Raman wavelength λ_R can be written as

$$I_{Det}(\lambda_R) = \alpha_{OM}(\lambda_{exc}) \cdot \eta_{OM}(\lambda_R) \cdot I(\lambda_{exc})$$
$$+ T_{OM}(\lambda_{exc}) \cdot T_{OM}(\lambda_R) \cdot N_{MP} \cdot \sigma_{MP}(\lambda_R) \cdot I(\lambda_{exc})$$
$$+ T_{OM}(\lambda_{exc}) \cdot T_{MP}(\lambda_{exc}) \cdot \alpha_{LP}(\lambda_{exc}) \cdot \eta_{LP}(\lambda_R) \cdot I(\lambda_{exc}) \cdot T_{MP}(\lambda_R) \cdot T_{OM}(\lambda_R)$$
$$(7.1)$$

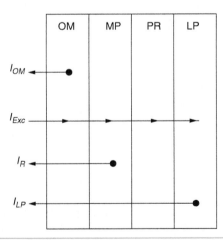

FIGURE 7.8 Schematics of retinal layer system with indication of excitation and emission light intensities encountered in resonant Raman imaging of macular pigment. OM: anterior optical media, MP: macular pigment containing layer; OS: photoreceptor outer segment layer; LP: lipofuscin containing layer. I_{OM}: light intensity originating from optical media fluorescence; I_R: Raman response from MP; I_{LP}: light intensity due to lipofuscin fluorescence (see text). (*Reprinted with permission from Ref. 24.*)

The first term describes fluorescence from the ocular media (OM), the second term the Raman response from macular pigment (MP), and the third term fluorescence from lipofuscin (LP). In Eq. (7.1), α_{OM} and α_{LP} are the absorption coefficients of the optical media and lipofuscin, respectively; T_{OM} and T_{MP} are the percentage transmissions of the optical media and MP at the indicated absorption and emission wavelengths; N_{MP} is the concentration of the MP molecules, σ_{MP} their Raman scattering cross section, and η_{OM} and η_{LP} describe the quantum efficiencies for optical media and lipofuscin fluorescence transitions.

The detector signal at the Raman wavelength is a superposition of a weak, spectrally narrow Raman signal, and typically 100-fold stronger and spectrally broad background fluorescence (mostly from the lens and lipofuscin). To retrieve the MP Raman response, the superimposed fluorescence intensities need to be measured and subtracted from the total detector signal. The contribution of the broad fluorescence at the Raman wavelength is approximately the same as at a wavelength slightly offset to a longer wavelength position, λ_{offset}. It can be shown that the Raman component, $I_R(\lambda_R)$, for each image pixel is approximately

$$I_R(\lambda_R) \approx T_{OM}^{-1}(\lambda_{exc}) \cdot T_{OM}^{-1}(\lambda_R)(I_{Det}(\lambda_R) \, / \, T_R - I_{Det}(\lambda_{offset}) / (T_{offset})$$

where $I_{Det}(\lambda_R)$ and $I_{Det}(\lambda_{offset})$ = detector intensities

T_R and T_{offset} = filter transmissions at the respective wavelengths

T_{OM} = unknown transmission of the ocular media

The RRI image of a MP distribution can thus be derived with a digital image subtraction routine, where the intensities obtained for each pixel of the two images are divided by the appropriate filter transmission coefficient.

In order to determine the correct position of the angle tunable transmission filter, to check for the linearity between resonance Raman response and optical density of MP, and to determine the dynamic range of the method, we first measured a number of dried lutein drops with widely differing concentration levels. All drops were spotted onto polyvinyl pyrolidine difluoride (PVDF) substrates. In Fig. 7.9 we show the RRI image obtained for the lutein drop with the highest concentration level in the center (OD $\approx$ 0.8). The intensity of each pixel is color coded according to the linear intensity scale shown at right. Similar imaging measurements were carried out with half a dozen drops of varying lutein concentrations. The integrated response was found to be roughly linear in concentration up to the highest level.

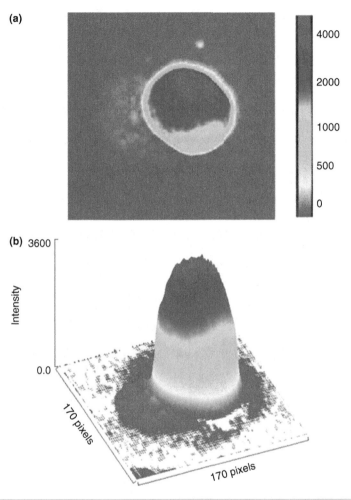

FIGURE 7.9   En-face (*a*) and three-dimensional (*b*) pseudo-color-scaled RRI images of a dried drop of lutein spotted onto a PVDF substrate. The lutein solution had a physiological carotenoid concentration OD ≈ 0.8. The intensity of each pixel is color coded according to the scale shown in (*a*) and is displayed as a function of pixel position in the camera CCD array (1 pixel = 20 μm). (*Reprinted with permission from Ref. 24.*)

In a second step, to further validate the RRI setup with physiologically encountered human pigment distributions, we imaged 11 excised human donor eyecups, again widely varying in MP levels, and compared RRI-derived MP levels with HPLC-derived levels.[28] Postmortem human eyes were obtained from the Utah Lions Eye Bank within 24 to 48 hours after corneas had been removed for surgical transplantation. All eyes measured were from donors free of macular and retinal diseases. None of the donors had a history of significant lutein supplementation premortem. The eyes were placed on a glass holder, the anterior segment was carefully cut away, and the posterior segments with the macula were left in situ for RRI imaging. At the conclusion of the imaging experiments, the macula was punched out with a 5-mm diameter trephine, and macular carotenoids were extracted and analyzed by well-known HPLC methods.

Two-dimensional and three-dimensional pseudo-color Raman images are shown for two representative eyecups in Fig. 7.10. The first eyecup features a distribution with a relatively strong central peak that has a small depression in it and that is surrounded by a broken-up ring structure (Fig. 7.10a and b); the other eyecup displays a strongly elongated asymmetrical distribution with high-central levels and relatively smooth decline toward increasing eccentricities (Fig. 7.10c and d). In Fig. 7.10e we

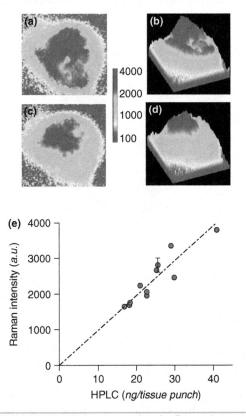

FIGURE 7.10   (a), (c): two-dimensional, pseudo-color Raman images of two of 11 donor eyecups imaged to establish a correlation between Raman- and HPLC-derived carotenoid levels; (b), (d): corresponding three-dimensional images. The color scale bar indicates the color coding of light intensities. The graph in (e) shows the correlation between optical intensities integrated over the macular regions of the eyecups and subsequently derived HPLC levels obtained for 8-mm diameter tissue punches centered on the macula (correlation coefficient $R = 0.92$; $p < 0.0001$). (Reprinted with permission from Ref. 24.)

plot the spatially integrated Raman intensities obtained from the MP RRI images of all eyecups, and compare these optically derived intensities with HPLC-derived MP concentration levels. A high correlation is obtained between optical and biochemical method ($R = 0.92; p = 0.0001$).

To further test the RRI imaging method, we compared it with a recently developed, nonmydriatic version of the lipofuscin fluorescence imaging ("autofluorescence imaging") method.[27] Autofluorescence imaging, AFI, is a less specific detection methods since it detects the light emitted from a compound other than MP, and thus derives the concentration of MP only indirectly. The method has to take into account light traversal through deeper retinal layers, has to carefully eliminate image contrast diminishing fluorescence and scattering from the optical media like the lens (via confocal detection techniques, filtering, etc.), has to bleach the photoreceptors, and has to use a location in the peripheral retina as a reference point. The peripheral reference could potentially lead to an underestimation of the MP density, especially in individuals regularly consuming high-dose lutein supplements, which can cause substantial increases in even peripheral carotenoid levels.[29] AFI has an advantage, however, since the peripheral reference location allows one to eliminate, in first order, any potentially confounding attenuation arising from anterior optical media, and the instrumentation is less complex compared to RRI imaging.

7.5 Spatially Resolved Resonance Raman Imaging of Macular Pigment in Human Subjects

For RRI imaging of MP distributions in human subjects we recruited 17 healthy volunteers. All procedures were performed with Institutional Review Board (IRB) approval and were in accordance with the tenets of the Declaration of Helsinki. Imaging details for one of the subjects are shown in Fig. 7.11. Figure 7.11a shows the gray-scale image obtained after subtraction of the fluorescence background, Fig. 7.11b a pseudo-color, three-dimensional representation of the distribution, and Fig. 7.11c two-line plots along the nasal-temporal and inferior-superior meridians, respectively. The MP distribution in this subject features a wide, axially slightly asymmetric distribution with relatively high-central MP levels, and a monotonous decline of MP levels toward the peripheral retina. The spatial resolution obtainable with the instrument is approximately sub-50-μm as can be determined from the size of small blood vessels discernable in the gray-scale images.

To check whether the obtained imaging results depend on the degree of bleaching of the photoreceptors, we compared two images of a subject's retina, one obtained after exposing it to blue bleaching light for 2 minutes, and the other obtained after no bleaching. The resulting images were seen to be identical. Evaluating the MP distributions of all subjects, distinctly different categories are clearly revealed. Representative distributions are shown in Fig. 7.12a. One is

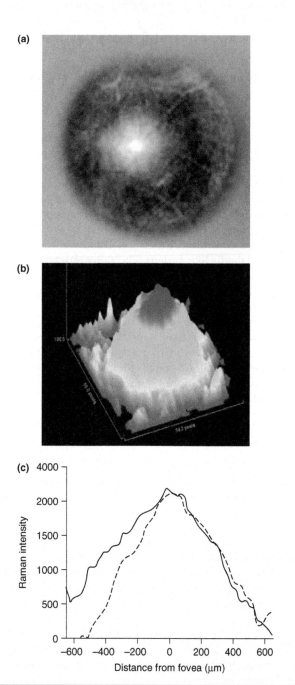

FIGURE 7.11 RRI results for MP distribution in living human eye. (*a*) Typical gray-scale image obtained after subtraction of fluorescence background from pixel intensity map containing Raman response and superimposed fluorescence background. (*b*) Pseudo-color-scaled, three-dimensional representation of gray-scale image; (*c*) line plots along nasal-temporal (solid line) and inferior-superior meridians (dashed line), both running through the center of the MP distribution. (*Reprinted with permission from Ref. 24.*)

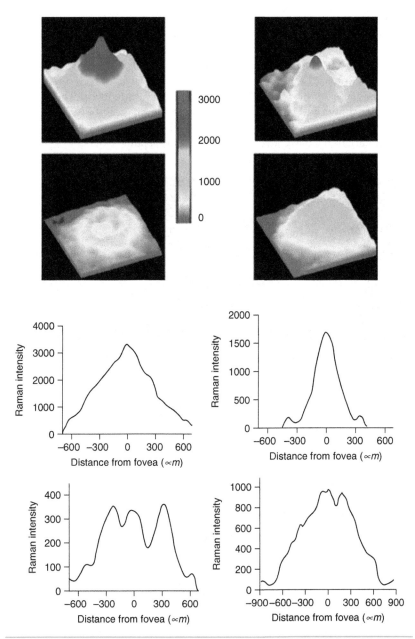

a category displaying relatively wide spatial distributions with a high-central level, a second one again with high-central levels but narrower spatial extent, a category with a ring-like MP distribution surrounding a central MP peak, and a category with relatively wide but overall low levels. Intensities in the four distributions are all color coded with the same intensity bar. Also, the spatial dimensions are identical. The line plots shown in Fig. 7.12*b* correspond to the images in Fig. 7.12*a*. They run along primary meridians and highlight the significant inter-subject variations in MP levels, symmetries, and spatial extent.

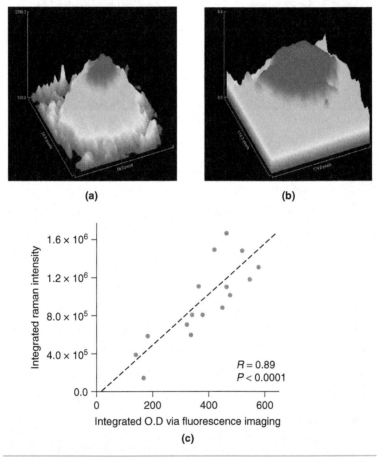

(a)   (b)

(c)

Figure 7.13   Images of MP distributions obtained for a same subject with (*a*) RRI and (*b*) fluorescence based imaging. (*c*) Comparison of integrated MP densities obtained for 16 subjects with Raman- and fluorescence-based imaging methods. Vertical scale shows integrated MP densities derived from RRI images by integrating intensities over the whole macular region; horizontal scale shows corresponding densities derived via fluorescence imaging. A high-correlation coefficient of $R = 0.89$ is obtained for both methods, using a best fit that is not forced through zero. If forcing the fit through zero (not shown), one obtains a correlation coefficient $R = 0.80$. (*Reprinted with permission from Ref. 24.*)

In Fig. 7.13, we summarize the main results of a comparison of MP distributions and concentrations obtained with Raman imaging and nonmydriatic lipofuscin fluorescence excitation,[24] respectively. Figure 7.13*a* and *b* compare images of both methods, obtained for the same subject. Compared to the RRI image, the fluorescence-based image is nearly identical, with the exception of a smoother appearance of the distribution. This is due to the derivation of the MP density map as the logarithm of a ratio between perifoveal and foveal fluorescence intensities, which tends to slightly compress the "dynamic range" of the density map amplitudes, and smoothen out the resulting MP distribution. For a subgroup of 16 subjects, we integrated the MP levels of images obtained with both methods for each individual over the whole macula region, and plot the results in Fig. 7.13*c*. Using a best fit that is not forced through zero, we obtain a high-correlation coefficient of $R = 0.89$ between both methods. Forcing the fit through zero, the correlation coefficient drops slightly to $R = 0.80$. The high correlation is remarkable in view of the completely different optical beam paths and derivation methods used to calculate MP densities in both methods.

7.6 Raman Detection of Carotenoids in Living Human Skin

Levels of carotenoids are much lower in the skin relative to the macula of the human eye, but higher light excitation intensities and longer acquisition times can be used in Raman detection approaches to compensate for this drawback. Since the bulk of the skin carotenoids are in the superficial layers of the dermis, and since the concentrations are relatively low, the thin-film Raman equation given above should still be a good approximation.

A cross section of excised human skin, histologically stained, is shown in Fig. 7.14. It shows a layer structure of the tissue with an increased homogeneity in the bloodless stratum corneum layer, where the cell nuclei are absent, and where the potentially confounding melanin concentrations are minimal as well. The penetration depth of visible light into the stratum corneum is approximately 400 μm and therefore is confined to this outermost layer, as sketched in Fig. 7.14 for a hemispherical beam penetration into the tissue. Using skin tissue sites with thick stratum corneum layer in RRS measurements, such as the palm of the hand or the sole of the foot, one therefore realizes measuring conditions of a fairly homogeneous uniform tissue layer with well-defined absorption and scattering conditions.

To check the spatial uniformity of skin carotenoids, we Raman imaged the carotenoid distribution of an excised skin tissue sample. The resulting RRI image of an approximately 40-μm diameter skin area is shown in Fig. 7.15 along with a line plot through the center of the carotenoid distribution. These and similar results with large

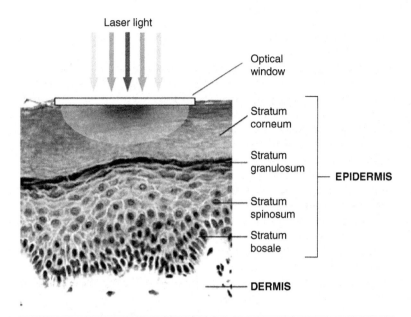

Laser light

Optical window

Stratum corneum

Stratum granulosum

EPIDERMIS

Stratum spinosum

Stratum bosale

DERMIS

FIGURE 7.14   (a) Layer structure of human skin as seen in a microscope after staining, showing the morphology of dermis, basal layer, stratum spinosum, stratum granulosum, and stratum corneum. Cells of the stratum corneum have no nucleus (missing dark stain spots), and form a relatively homogeneous optical medium well suited for Raman measurements. For visible wavelengths, the excitation light has a penetration depth of about 400 μm, and stays within the 0.7 to 2 mm-thick stratum corneum, as indicated.

diameter spots show that skin carotenoid concentrations vary significantly only on a microscopic scale, but that they are rather uniform over mm-scale spot sizes. For Raman detection of average skin carotenoid levels, imaging is therefore not needed. Instead, a mm-scale beam spot size is adequate to integrate over these microscopic spatial concentration changes, and thus to obtain a reproducible noninvasive Raman measure of skin carotenoid content in selected tissue sites.

A recent field-usable instrument configuration that evolved out of the development of RRS for in vivo skin carotenoid measurements[30] is shown in Fig. 7.16. It is based on a miniaturized, fiber-based, computer-interfaced spectrograph with high light throughput, and hand held probe module, as shown in Fig. 7.16a.[31] For a RRS skin carotenoid measurement, the palm of the hand is held against the window of the probe head module and the tissue is exposed for about 10 seconds with 488-nm laser light at laser intensities of ~10 mW in a 2-mm-diameter spot. Carotenoid RRS responses are detected with a CCD array integrated into the spectrograph. Typical skin carotenoid RRS spectra measured in vivo, are shown in Fig. 7.16b. The raw spectrum shown at the top of the panel (trace 1) was obtained directly after laser exposure, and reveals a broad, featureless, strong

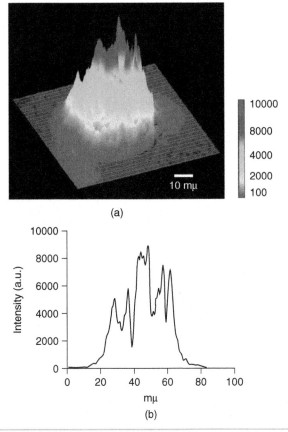

(a)

(b)

FIGURE 7.15   Pseudo-color-scaled microscopic RRI image of excised palm tissue sample (*a*) and intensity plot (*b*) along a line running through middle of distribution. Results show strong spatial variation of skin carotenoids on microscopic scale.

"autofluorescence" background of skin, with three superimposed Raman peaks characteristic for the carotenoid molecules at 1008, 1159, and 1524 cm^{-1}. Even though the intensity of the skin fluorescence background is about 100 times higher than the carotenoid signals, it is possible to measure the skin carotenoid RRS responses with high accuracy by using a detector with high-dynamic range. Approximation of the fluorescence background with a higher order polynomial and subsequent subtraction from the raw spectrum yields an isolated Raman spectrum of the skin carotenoids (trace 2), that is virtually undistinguishable from a solution of pure β-carotene, shown for comparison (trace 3). The skin carotenoid RRS response originates from contributions of all skin carotenoid species absorbing in the visible spectral range. Since all individual C=C stretch positions and bandwidths are indistinguishable at the instrument's spectral resolution,

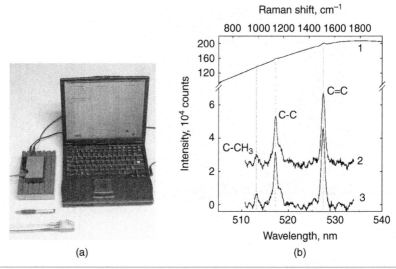

(a)   (b)

Figure 7.16   Image of clinic- and field-usable, computer-interfaced, skin carotenoid RRS instrument, showing solid state laser, spectrograph, and light delivery/collection module; (b) typical skin carotenoid Raman spectra measured in vivo. Spectrum (1) is obtained directly after exposure, and reveals a strong, spectrally broad, skin autofluorescence background with superimposed weak, but recognizable Raman-peaks characteristic for carotenoids. Spectrum (2) is obtained after fitting the fluorescence background with a fourth-order polynomial, subtraction from (1), and scaling of the spectrum. Spectrum (2) is indistinguishable from a spectrum of a β-carotene solution, shown as (3) for comparison.

our RRS approach allows us to use the absolute peak height of the C=C signal at 1524 cm^{-1} as a measure for the overall carotenoid concentration in human skin.

Experiments with varying light excitation intensities showed that the skin carotenoid RRS response is stable up to the highest intensities permissible for skin applications.[31] To validate the skin carotenoid RRS detection approach, we initially carried out an indirect validation experiment that compared HPLC-derived carotenoid levels of fasting serum with RRS derived carotenoid levels for inner palm tissue sites. Measuring a large group of 104 healthy male and female human volunteers, we obtained a significant correlation ($p < 0.001$) with a correlation coefficient of 0.78.[32] Recently, we carried out a direct validation study, in which we compared in vivo RRS carotenoid skin responses with HPLC-derived results, using the thick stratum corneum layer of heel skin tissue sites. Following RRS measurements of the sites in eight volunteer subjects, the subjects removed thin skin slivers of 10 to 50 mg weight around the optically measured area with a razor blade for subsequent HPLC analysis. In Fig. 7.17a, the comparison of RRS skin carotenoid responses is shown for all subjects with corresponding HPLC-derived carotenoid content. The latter is a

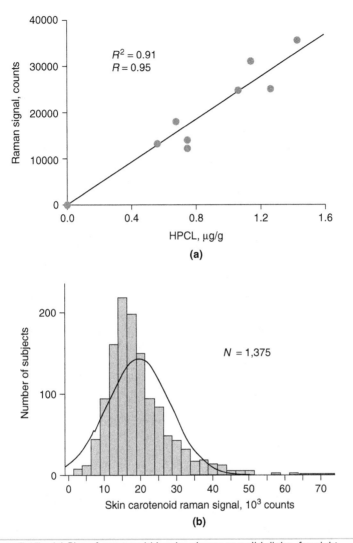

FIGURE 7.17   (a) Plot of carotenoid levels, shown as solid disks, for eight samples of human tissue measured with RRS technique in vivo, and subsequently, after tissue excision, with HPLC methods. The solid line is the resulting linear regression crossing the origin, and reveals a correlation coefficient R equal to 0.95; (b) histogram of skin carotenoid RRS response measured in the palm of 1375 subjects, showing wide distribution of skin carotenoid levels in a large population.

sum of individual concentrations determined for each excised sample for the main skin carotenoids lutein, zeaxanthin, cis-lutein/zeaxanthin, α-cryptoxanthin, β-cryptoxanthin, trans-lycopene, cis-lycopene, α-carotene, trans-β-carotene, cis-β-carotene, and canthaxanthin. Using a regression line fit that passes through the origin we obtained a near perfect correlation between Raman and HPLC data, as evidenced by a correlation coefficient R as high as 0.95.[33] The results show excellent linearity of RRS-derived carotenoid levels over a wide range of physiological skin carotenoid concentrations, and provide a direct validation of the skin carotenoid RRS detection approach.

Measurements of large populations with the Raman device reveal a bell-shaped distribution of carotenoid levels, as shown in Fig. 7.17b for a group of 1375 healthy volunteer subjects that could be screened with the RRS method within a period of a few weeks.[26,32] Analysis of the data confirmed a pronounced positive relationship between self-reported fruit and vegetable intake (a source of carotenoids) and skin Raman response. Furthermore, the study showed that people with habitual high-sunlight exposure have significantly lower skin carotenoid levels than people with little sunlight exposure, independent of their carotenoid intake or dietary habits, and that smokers had dramatically lower levels of skin carotenoids as compared to nonsmokers.[26] Importantly, is also showed that RRS detection can track the increase of skin carotenoid levels occurring in subjects with low-skin carotenoid levels within a relat ively short time frame of weeks as a result of dietary supplementation with carotenoid-containing multi-vitamins.

Based on these capabilities, the RRS detection method has already found commercial application in the nutritional supplement industry, (BioPhotonic Scanner™, Pharmanex LLC, Provo, Utah), which has placed thousands of portable instruments with their customers for rapid optical measurements of dermal carotenoid levels, and which has further developed the instrumentation for rugged field use.[34] For medical applications, the method had found initial interest in dermatology, where a tentative correlation was demonstrated between certain types of cancerous lesions and depleted carotenoid levels.[35] Quantitative RRS measurements in these tissues, however, which extend to layers beyond the stratum corneum, are more complicated due to additional absorption and scattering caused by other layers and chromophores, and need to be further refined for future studies. In the field of epidemiology, the RRS method has recently been applied to subjects with increased bitter taste sensitivities. Measuring the stratum corneum layer of palm tissue, an inverse relationship was observed between taste sensitivity and fruit and vegetable uptake,[36] a finding that may be helpful to promote healthy behavioral patterns of dietary change in large populations. In neonatology, skin carotenoid RRS measurements are in progress to investigate correlations of carotenoid levels with retinopathy of prematurely born infants.[37] Measuring the sole of the foot, it could be shown that retinopathy is influenced by the carotenoids in human milk-fed infants, and that it appears likely that carotenoids are important nutrients in decreasing the severity of the disease.

In all previous RRS measurements of dermal carotenoids we measured the total concentration of all long-chain carotenoid species since the method only detects the chain's carbon double-bond vibration, which is identical in all species. Lycopene has an increased conjugation length compared to the other carotenoids in skin and therefore features a small (~10 nm) but distinguishable red shift of the absorption. This shift can be explored to measure skin lycopene levels independently of the other carotenoid concentrations,[38] and to investigate specific correlations between lycopene tissue levels and tissue protection/diseases in future clinical studies.

7.7   Conclusions

In ocular applications, Raman spectroscopy can quickly and objectively assess composite lutein and zeaxanthin concentrations of macular pigment using spatially averaged, integral, Raman measurements or Raman images that quantify and map the complete MP distribution with high-spatial resolution. Importantly, both methods can be validated with HPLC methods in excised human eyecups and in animal models.

Both integral and spatially resolved MP Raman methods use the backscattered, single-path, Raman response from lutein and zeaxanthin in the MP containing retinal layer, and largely avoid light traversal through the deeper retinal layers. Since they do not rely on any reflection of light at the sclera, the overlapping fluorescence signals from the ocular media can be subtracted from the overall light response. Importantly, the Raman methods have no assumptions other than approximating the spectrally broad background fluorescence with the fluorescence response at a wavelength that is slightly offset from the MP Raman response. MP Raman measurements are a measure of absolute MP concentration levels since the method does not use a reference point in the peripheral retina. Attenuation effects caused by the optical media are therefore fully effective, and have to be avoided or minimized, particularly when comparing MP levels between subjects. Optical losses from the lens can be neglected in longitudinal studies provided these are carried out over a time span in which lens absorptions can be considered to stay constant (1 to 2 years), or in any other studies involving subjects with lens implants. Therefore, the Raman method would be well suited, for example, in important nutritional supplementation trials, studies in which significant increases in individual MP levels have been demonstrated to be achievable in a time span of 12 months.[39]

Raman imaging clearly reveals the existence of spatially complex MP distribution patterns throughout the subject population. The distributions vary strongly regarding widths, axial and rotational asymmetries, locally depleted areas, and integrated concentration levels. RRI-derived results provide highly molecule-specific evidence for a more complicated nature of MP distributions in human subjects than previously thought. It is therefore a valuable method to explore correlations between MP patterns and retinal health and pathology in future clinical studies.

In dermal applications, the Raman method can rapidly assess dermal carotenoid content in large populations. Spatially resolved RRI is not needed due to the rather smooth spatial change of skin carotenoid levels. Instead, composite skin carotenoid levels can be measured with simple, spatially integrating Raman detection. Measurements have to be limited to tissue sites with a thick stratum corneum. In this case the probed tissue is thicker than the penetration depth of the excitation light, thus avoiding absorption of hemoglobin. Furthermore, the stratum corneum tissue in the palm of the hand or the sole of the foot is free of melanin. A correlation of our Raman-derived carotenoid data with HPLC-derived serum levels again confirms the validity of the carotenoid Raman detection technique in the physiologically relevant concentration range under these measuring conditions.

Carotenoid RRS detection has exciting potential for in-vivo applications. In the nutritional supplement industry it is already established as an objective, portable device for the monitoring of the effect of carotenoid-containing supplements on skin tissue carotenoid levels. In ophthalmology, it may become a fast screening method for MP levels in the general population. In epidemiology, it may serve as a noninvasive novel biomarker for fruit and vegetable intake, replacing costly plasma carotenoid measurements with inexpensive and rapid skin Raman measurements. In neonatology, it may serve as a noninvasive method to assess carotenoid levels in prematurely born infants to investigate their correlation with oxidative stress related degenerative diseases. Lastly, due to its capability of selectively detecting lycopene, the technology may be useful to investigate a specific role of lycopene in the prevention of prostate cancer and other diseases.

References

1. Y. Koyama, "Resonance Raman spectroscopy," in: *Carotenoids*, Vol. 1B, *Spectroscopy*, G. Britton Liaaen-Jensen, and H. Pfander (eds), Birkhäuser, Basel, 1995, pp. 135–146.
2. D. M. Snodderly, P. K. Brown, F. C. Delori, and J. D. Auran, "The Macular Pigment. I. Absorbance Spectra, Localization, and Discrimination from Other Yellow Pigments in Primate Retinas," *Investigative Ophthalmology & Visual Science*, 25:660–673, 1984.
3. D. M. Snodderly, J. D. Auran, and F. C. Delori, "The Macular Pigment. II. Spatial Distribution in Primate Retinas," *Investigative Ophthalmology & Visual Science*, 25:674–685, 1984.
4. Age-Related Eye Disease Study Research Group, "The Relationship of Dietary Carotenoid and Vitamin A, E, and C Intake with Age-Related Macular Degeneration in a Case-Control Study," AREDS Report No. 22, *Archives of Ophthalmology*, 125:1225–1232, 2007.
5. J. T. Landrum and R. A. Bone, "Lutein, Zeaxanthin, and the Macular Pigment," *Archives of Biochemistry and Biophysics*, 385:28–40, 2001.

6. N. I. Krinsky, J. T. Landrum, and R. A. Bone, "Biologic Mechanisms of the Protective Role of Lutein and Zeaxanthin in the Eye," *Annual Review of Nutrition*, **23**:171–201, 2003.

7. N. I. Krinsky and E. J. Johnson, "Carotenoid Actions and their Relation to Health and Disease," *Molecular Aspects of Medicine*, **26**:459–516, 2005.

8. J. M. Seddon, U. A. Ajani, R. D. Sperduto, R. Hiller, N. Blair, T. C. Burton, M. D. Farber, E. S. Gragoudas, J. Haller, and D. T. Miller, "Dietary Carotenoids, Vitamins A, C, and E, and Advanced Age-Related Macular Degeneration," *Journal of the American Medical Association*, **272**:1413–1420, 1994.

9. D. M. Snodderly, J. A. Mares, B. R. Wooten, L. Oxton, M. Gruber, and T. Ficek, "Macular Pigment Measurement by Heterochromatic Flicker Photometry in Older Subjects: The Carotenoids and Age-Related Eye Disease Study," *Investigative Ophthalmology & Visual Science*, **45**:531–538, 2004.

10. A. G. Robson, J. D. Moreland, D. Pauleikoff, T. Morrissey, G. E. Holder, F. W. Fitzke, A. D. Bird, and F. J. G. M. D. van Kuijk, "Macular Pigment Density and Distribution: Comparison of Fundus Autofluorescence with Minimum Motion Photometry," *Vision Research*, **43**:1765–1775, 2003.

11. M. Trieschmann, G. Spittal, A. Lommartzsch, E. van Kuijk, F. Fitzke, A. C. Bird, and D. Pauleikoff, "Macular Pigment: Quantitative Analysis on Autofluorescence Images," *Graefe's Archieve for Clinical Experimental Ophthalmology*, **241**:1006–1012, 2003.

12. F. C. Delori, "Autofluorescence Method to Measure Macular Pigment Optical Densities: Fluorometry and Autofluorescence Imaging," *Archieves of Biochemistry and Biophysics*, **430**:156–162, 2004.

13. T. T. J. M. Berendschot and D. van Norren, "Macular Pigment Shows Ringlike Structures," *Investigative Ophthalmology & Visual Science*, **47**:709–714, 2006.

14. D. S. Michaud, D. D. Feskanich, E. B. Rimm, G. A. Colditz, F. E. Speizer, W. C. Willett, and E. Giovannucciet, "Intake of Specific Carotenoids and Risk of Lung Cancer in 2 Prospective US Cohorts," *American Journal of Clinical Nutrition*, **92**:990–997, 2000.

15. L. N. Kolonel, J. H. Hankin, A. S. Whittemore, A. H. Wu, R. P. Gallagher, L. R. Wilkens, E. M. John, G. R. Howe, D. M. Dreon, D. West and R. S. Paffenbarger, Jr. "Vegetables, Fruits, Legumes, and Prostate Cancer: A Multiethnic Case-Control Study," *Cancer Epidemiology Biomarkers and Prevention*, **9**:795–804, 2000.

16. S. Liu, J. E. Manson, I. M. Lee, S.R. Cole, C. H. Hennekens, W. C Willett and J. E Buringet, "Fruit and Vegetable Intake and Risk of Cardiovascular Disease: The Women's Health Study," *American Journal of Clinical Nutrition*, **72**:922–928, 2000.

17. A. P. Shreve, J. K. Trautman, T. G. Owens, and A. C. Albrecht, "Determination of the S2 Lifetime of β-Carotene," *Chemical Physics Letters*, **178**:89, 1991.

18. P. S. B. Bernstein, M. D. Yoshida, N. B. Katz, R. W. McClane, and W. Gellermann, "Raman Detection of Macular Carotenoid Pigments in Intact Human Retina," *Investigative Ophthalmology & Visual Science*, **39**:2003–2011, 1998.

19. I. V. Ermakov, R. W. McClane, W. Gellermann, and P. S. Bernstein, "Resonant Raman Detection of Macular Pigment Levels in the Living Human Retina," *Optics Letters*, **26**:202–204, 2001.

20. W. Gellermann, I. V. Ermakov, M. R. Ermakova, R. W. McClane, D. Y. Zhao, and P. S. Bernstein, "In Vivo Resonant Raman Measurement of Macular Carotenoid Pigments in the Young and the Aging Human Retina," *Journal of Optical Society of America A*, **19**:1172–1186, 2002.

21. I. V. Ermakov, M. R. Ermakova, P. S. Bernstein, and W. Gellermann, "Macular Pigment Raman Detector for Clinical Applications," *Journal of Biomedical Optics*, **9**:139–148, 2004.

22. I. V. Ermakov, M. R. Ermakova, W. Gellermann, "Simple Raman Instrument for in Vivo Detection of Macular Pigments," *Applied Spectroscopy*, **59**:861–867, 2005.

23. P. S. Bernstein, D. Y. Zhao, S. W. Wintch, I. V. Ermakov, and W. Gellermann, "Resonance Raman Measurement of Macular Carotenoids in Normal Subjects and in Age-Related Macular Degeneration Patients," *Ophthalmology*, **109**:1780–1787, 2002.

24. M. Sharifzadeh, D.-Y. Zhao, P. S. Bernstein, and W. Gellermann, "Resonance Raman Imaging of Macular Pigment Distributions in the Human Retina," *Journal of Optical Society of America A*, **25**:947–957, 2008.
25. American National Standards Institute, "American National Standard for Safe Use of Lasers," ANSI Z136.1-2000, Laser Institute of America, Orlando, Fla., 2000, Sec. 8.3.
26. I. V. Ermakov, M. Sharifzadeh, M. R. Ermakova, and W. Gellermann, "Resonance Raman Detection of Carotenoid Antioxidants in Living Human Tissue," *Journal of Biomedical Optics*, **10**(6):064028-1–064028-18, 2005.
27. M. Sharifzadeh, P. S. Bernstein, and W. Gellermann, "Non-Mydriatic Fluorescence-Based Quantitative Imaging of Human Macular Pigment Distributions," *Journal of Optical Society of America A*, **23**:2373–2387, 2006.
28. W. Gellermann, I. V. Ermakov, R. W. McClane, and P. S. Bernstein, "Raman Imaging of Human Macular Pigments," *Optics Letters*, **27**:833–835, 2002.
29. P. Bhosale, Da You Zhao, and P. S. Bernstein, "HPLC Measurement of Ocular Carotenoid Levels in Human Donor Eyes in the Lutein Supplementation Era," *Investigative Ophthalmology & Visual Science*, **48**:543–549, 2007.
30. W. Gellermann, R. W. McClane, N. B. Katz, and P. S. Bernstein, "Method and Apparatus for Non-Invasive Measurement of Carotenoids and Related Chemical Substances in Biological Tissue," U.S. patent # 6,205,354 B1, 2001.
31. I. V. Ermakov, M. R. Ermakova, R. W. McClane, and W. Gellermann, "Resonance Raman Detection of Carotenoid Antioxidants in Living Human Skin," *Optics Letters*, **26**:1179–1181, 2001.
32. C. R. Smidt, W. Gellermann, and J. A. Zidichouski, "Non-Invasive Raman Spectroscopy Measurement of Human Carotenoid Status," *Federation of American Society for Experimental Biology Journal*, **18**(4):A480, 2004.
33. I. V. Ermakov and W. Gellermann, unpublished results.
34. D. Bergeson, J. B. Peatross, N. J. Eyring, J. F. Fralick, D. N. Stevenson, and S. B. Ferguson, "Resonance Raman Measurements of Carotenoids Using Light Emitting Diodes," *Journal of Biomedical Optics*, **13**(4):2008.
35. T. R. Hata, T. A. Scholz, I. V. Ermakov, R. W. McClane, F. Khachik, W. Gellermann, and L. K. Pershing, "Non-Invasive Raman Spectroscopic Detection of Carotenoids in Human Skin," *Journal of Investigative Dermatology*, **115**: 441–448, 2000.
36. S. N. Scarmo, B. Cartmel, W. Gellermann, I. V. Ermakov, D. J. Leffell, H. Lin, and S. T. Mayne, "Perceived Bitter Taste and Fruit and Vegetable Intake Measured by Self-Report and an Objective Indicator," in print.
37. G. M. Chan, C. Rau, W. Gellermann, and M. Ermakova, "Retinopathy of Prematurity and Carotenoids in Human Milk Fed Infants," Abstract, 2006 Meeting of the American Academy of Pediatrics, Washington, D.C.
38. I. V. Ermakov, M. R. Ermakova, W. Gellermann, and J. Lademann, "Non-Invasive Selective Detection of Lycopene and β-Carotene in Human Skin Using Raman Spectroscopy," *Journal of Biomedical Optics*, **9**:332–338, 2004.
39. S. Richer, W. Stiles, L. Statkute, J. Pulido, J. Frankowski, D. Rudy, K. Pei, M. Tsipursky, and J. Nyland, "Double-Masked, Placebo-Controlled, Randomized Trial of Lutein and Antioxidant Supplementation in the Intervention of Atrophic Age-Related Macular Degeneration: The Veterans Last Study (Lutein Antioxidant Supplementation Trial)," *Optometry*, **75**:216–230, 2004.

CHAPTER 8

Raman Microscopy for Biomedical Applications: Toward an Efficient Diagnosis of Tissues, Cells, and Bacteria

Christoph Krafft, Ute Neugebauer

Institute of Photonic Technology
Jena, Germany

Jürgen Popp

Institute of Physical Chemistry
University Jena
Jena, Germany

The advantages of Raman spectroscopy to diagnose tissue, cells, and bacteria include that the method provides a wealth of information about the molecular structure and biochemical composition without labeling. The disadvantages such as low sensitivity and overlapping autofluorescence have been overcome by better instrumentation and excitation with near-infrared lasers. This contribution summarizes recent research results on fiber-optic Raman spectroscopy of tissue, Raman imaging of tissue and cells, and Raman spectroscopy of bacteria. The sections are organized from low-spatial resolution which was obtained using multimode optical fiber probes to high-spatial resolution which was obtained in tip-enhanced Raman spectroscopy (TERS) using functionalized AFM tips.

8.1   Introduction

The expression "biophotonic" is composed of the Greek syllables "bios" for life and "phos" for light. The term denotes the scientific discipline which applies light-based techniques to problems in medicine and life sciences. Biophotonic methods can provide information about inherent optical properties of cells and tissues, and the presence or absence of endogenous or exogenous fluorophores. The optical detection offers advantages in (1) spatial resolution in the sub-micrometer range whereby Abbe's detection limit can even be overcome by super resolution approaches such as stimulated emission depletion in fluorescence spectroscopy,[1] (2) nondestructivity because photons with wavelengths from visible to the infrared do not harm cells and tissues, (3) speed because data acquisition usually takes seconds, and (4) costs because most instruments are less expensive compared to clinical devices. These advantages are utilized in various areas in life sciences. Absorption, fluorescence, reflection, and bioluminescence belong to the most prominent techniques. A general drawback of all optical methods for diagnostic purposes is the low penetration depth in tissue which depends on absorption and scattering properties. High absorbance exists in the visible range (350 to 700 nm) mainly due to hemoglobin of blood and in the infrared range (>900 nm) mainly due to water and lipids. In the near-infrared region all biomolecules show minimal absorption so techniques using this spectral window take advantage of maximum penetration in the order of 1 to 2 cm. However, this value is still too low, e.g., to localize brain tumors inside the skull using optical procedures. Therefore, clinical applications require microscopes or handheld fiber-optic probes during open surgery, or miniaturized fiber optic probes during minimal invasive endoscopy.

Raman spectroscopy has been recognized as another powerful technique for biomedical applications. Its main advantage is that the method yields a wealth of information about the molecular structure and biochemical composition without labeling or any other sample preparation. The last decade brought a tremendous progress in Raman instrumentation. Improved detection sensitivity and excitation with near-infrared or ultraviolet lasers enabled to record Raman spectra from biological samples. More bands are resolved in Raman spectra than in other optical spectra because numerous bands in biomolecules can simultaneously be excited. This fingerprint capability of Raman spectroscopy is applied to characterize and identify tissue, cells, and bacteria. As the spectral contributions of all constituents such as proteins, lipids, nucleic acids, and carbohydrates overlap to complex signatures, features in Raman spectra from tissue, cells, and bacteria are distributed over a broad region and multivariate algorithms are required for analysis and classification. The principle of supervised classification is that a model is trained by spectra of known class membership and subsequently spectra of unknown origin can be assigned to one of these classes.

This contribution summarizes selected research results about fiber-optic Raman spectroscopy of tissue, Raman imaging of tissue and cells, and Raman spectroscopy of bacteria. Recent reviews described more comprehensively biomedical applications of Raman and infrared spectroscopy to diagnose tissues,[2] metabolic fingerprinting in disease diagnosis,[3] disease recognition by Raman and infrared spectroscopy,[4] noninvasive studies on individual cells,[5,6] and Raman and coherent anti-Stokes Raman spectroscopy of cells and tissues.[7]

8.2 Raman Imaging of Tissue

In principle, diseases and other pathological anomalies lead to chemical and structural changes on the molecular level which also change the Raman spectra and which can be used as sensitive, phenotypic markers of the disease. As these spectral changes are very specific and unique, they can be considered as a fingerprint. Early reports in the literature regarding the utility of Raman spectroscopy to biomedical problems were based on macroscopic acquisition of spectral data only at single points which required an a priori knowledge of the location or a preselection of the probed position. Since the inhomogeneous nature of tissue on the microscopic level was not considered in these early studies, an accurate correlation between the histopathology of the sampled area and the corresponding spectra was only possible for homogenous tissue on the macroscopic level. Considerable progress was made when high throughput and more sensitive instruments became available for Raman microspectroscopic imaging. They enable to microscopically collect larger number of spectra from larger sample populations in less time, improving statistical significance and spatial specificity.

Besides microscopic imaging, another promising medical application of Raman spectroscopy is the combination with endoscopy. The basic components consist of a light source, a light guide, and an endoscope which can be rigid or flexible. The imaging contrast is usually provided by white light illumination and registration of the scattered light by an optical system. To validate a medical diagnosis micromechanical instruments are inserted into the working channel and a tissue biopsy is collected which is subsequently analyzed using other methods such as light microscopy. However, the selection of a biopsy and the localization of early or small cancer lesions based on visual inspection are often difficult. Tissue changes should be detected as early as possible because the prospects for cure are at maximum in an early stage of tumor genesis.

The accuracy of biopsies can be optimized by coupling with spectroscopic techniques. Spectroscopic diagnosis of tissue is also called optical biopsy. The capability to perform optical biopsies is highly desired in medical procedures, e.g., to guide microsurgery, to identify suspicious tissue more sensitive and specific, and to collect tissue biopsies more accurate. Unnecessary tissue excisions and the biopsy related costs and risks could be avoided. Furthermore, these developments would enable in the future to monitor the effects of therapies under real-time conditions and to optimize the treatment using this information.

8.2.1 Mouse Brains

An important step in the development of new diagnostic methods is their application in animal models. A mouse model was used to test the detection of brain tumors by fiber optic Raman spectroscopic imaging.[8] Injection of tumor cells in the carotid artery induces tumors in mouse brains. Primary brain tumors that originate from cells inside the brain (see Sec. 8.2.2) as well as secondary brain tumors that are metastases from tumors outside the brain can be induced using selected tumor cells. After injection the tumor develops in one hemisphere whereas the other hemisphere remains tumor free and serves as an internal control. This constitutes an advantage compared with human brain tumor specimens because normal human brain tissue is preserved as much as possible during neurosurgery and consequently, no control specimen is usually available. Figure 8.1a shows the experimental setup to study nondried mouse brains with and without tumors. Before data acquisition, two tissue sections were prepared from the mouse brains. The first tissue section was stained by hematoxylin and eosin (H&E) for histopathological inspection. The second tissue section was transferred onto a calcium-fluoride slide for acquisition of FTIR images. The remaining tissue was covered by a window to prevent the surface from drying. Such a subsequent preparation enables to compare the Raman, FTIR, and histopathological results. A Raman spectrometer (Kaiser Optical System, United States) was coupled to a commercial fiber optic probe (Inphotonics, United States) which focused the laser to a spot of approximately 60 μm diameter. This setup is more flexible, more compact, and less expensive than a microscope. The sample was placed on a motorized stage, a raster was defined, and Raman spectra were collected from each raster point with 8 seconds exposure time. The laser intensity of 100 mW at 785 nm was found to be nondestructive. This is an important prerequisite to apply the method to living animal or even patients in the future. The obtained Raman image was segmented into groups of similar spectra using k-means cluster analysis. Cluster analysis belongs to the unsupervised algorithms by which objects are grouped into subsets, called clusters, so that the differences between data within each cluster are minimized and the differences between clusters are maximized according to some defined distance measure. The k-means algorithm assigns each point to one of k clusters whose center is nearest. The results of the

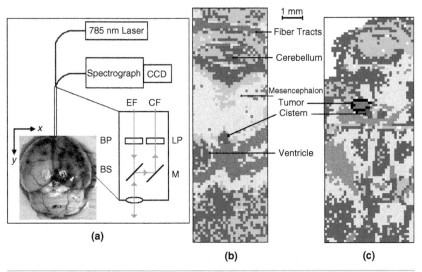

(a)

(b)   (c)

FIGURE 8.1 (a) Schematic diagram to study tumors in mouse brains by fiber-optic Raman imaging. (b) Raman images of mouse brain without tumor and (c) with tumor next to the cistern. Color codes represent segmentation of spectra by k-means cluster analysis. EF: excitation fiber, CF: collection fiber, BP: bandpass filter, BS: beam splitter, M: mirror, LP: long pass filter.

k-means cluster analysis are the centers of each cluster that represent an average spectrum and the cluster membership map. The clusters in the color-coded membership map can be assigned to anatomical features. Fiber tracts corresponding to white matter could be identified in the cerebellum and around a ventricle (Fig. 8.1b). A cistern was resolved between the left- and the right-brain hemisphere and below the mesencephalon. In a second specimen, brain metastases of malignant melanomas were found near the cistern (Fig. 8.1c). Further details could be visualized in FTIR images with a spatial resolution of 25 μm which were recorded from consecutive tissue sections.[8]

Mean cluster spectra are shown in Fig. 8.2. Raman spectra of a protein mixture comprising the all-alpha helical protein serum albumin and the all-beta sheet protein concanavalin A, and a phospholipid are included for comparison. Bands in the Raman spectrum of phosphatidylcholin which represents a phospholipid are assigned to the hydrophilic head group choline (717, 875 cm^{-1}), the phosphate group (1089 cm^{-1}), the ester group (1738 cm^{-1}), and the hydrophobic unsaturated fatty acid moiety (1066, 1130, 1270, 1301, 1442, 1660, 2852, 2886 cm^{-1}). The Raman spectra of other brain lipids have previously been reported.[9] Bands in the Raman spectrum of a protein mixture are assigned to the peptide backbone (940, 1243, 1665 cm^{-1}), disulfide bonds between the amino acid cysteine (508 cm^{-1}) the aromatic amino acids phenylalanine (621, 1004 cm^{-1}), tyrosine (643, 828, 853 cm^{-1}), tryptophan (760, 1340 cm^{-1}), and aliphatic amino acids (1447, 2932 cm^{-1}).

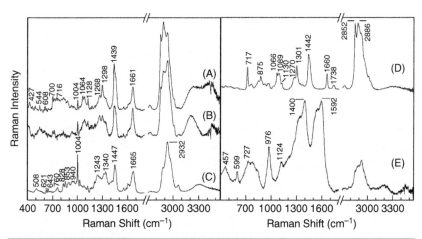

FIGURE 8.2   Raman spectra from 400 to 1800 cm^{-1} and 2700 to 3550 cm^{-1} from fiber tracts (A, orange cluster in Fig. 8.1), normal tissue (B, blue/cyan cluster in Fig. 8.1) of mouse brain, a protein mixture (C), the lipid phosphatidylcholin (D) and metastasis of malignant melanoma in mouse brain (E, black cluster in Fig. 8.1).

The Raman spectrum of fiber tract shows the typical signature of white matter with maximum spectral contributions of lipids and cholesterol. Spectral contributions of cholesterol are evident at 427, 544, 608, and 700 cm^{-1}. The intensity ratio of lipid-to-protein bands decreases in Raman spectra of gray matter which is represented by the blue and cyan cluster in normal mouse brain tissue. Raman spectra of brain metastases can easily be distinguished by intense spectral contributions of the pigment melanin at 457, 599, 727, 976, 1124, 1400, and 1592 cm^{-1}. The pigment partially absorbs the excitation wavelength 785 nm which gives a preresonance effect in the spectral range 400 to 1800 cm^{-1}. It is important to note that the expression of the pigment melanin is a molecular property of the primary tumor cells. That means the secondary tumor contains the molecular information of the primary tumor. As Raman spectroscopy probes the molecular fingerprint of tissues and cells, this opens the exciting perspective to develop a Raman-based approach for identification of the primary tumor of metastases. Classification models have already been reported for the related vibrational spectroscopic technique infrared spectroscopy. They assigned the primary tumors of the four most frequent human brain metastases based on the infrared spectroscopic fingerprint.[10,11] Unknown primary tumors constitute a severe problem in patients with brain metastases as an organ specific therapy cannot be performed.

The fiber optic probe which was used in this proof-of-concept study combines several attractive features. Integrated filters suppress the elastic Rayleigh scattering and background signals that are generated by the intense excitation laser inside the fiber. A lens focuses the laser light onto the sample and enables optimized collection geometry for the inelastic scattered light. An unfiltered probe which consisted of a single fiber of 300 µm diameter was recently applied to collect Raman spectra from porcine brain tissue.[12] Whereas the small diameter allows inserting the probe into the working channel of an endoscope, the missing collimating optic gives lower lateral resolution in the range of 300 µm and reduced collection efficiency. Another disadvantage is the smaller wavelength range from 2000 to 3900 cm^{-1}. As the Raman bands of tumor tissue (Fig. 8.2, spectrum E) are very similar to the Raman bands of normal brain tissue (Fig. 8.2, spectrum B) in the high-wavenumber region from 2700 to 3550 cm^{-1}, detection of tumor cells would be more difficult in this spectral interval.

8.2.2   Human Brain Tumors

Tumors that originate from brain tissue are called primary brain tumors. According to the recommendation of the World Health Organization (WHO) they are classified to the cell types from which they are derived.[13] Among them are glial cells, meninges, pituitary cells, and periphery nerves. Gliomas from different glial cells constitute the largest group of primary brain tumors (50 percent). The histopathological assessment of H&E-stained tissue sections is the so called golden standard for the distinction of gliomas into astrocytomas, oligodendrogliomas, ependymomas, and oligoastrocytomas as mixed gliomas. The malignancy of the most frequent astrocytomas is divided into grade II (low-grade astrocytoma), grade III (anaplastic astrocytoma), and grade IV (glioblastoma multiforme, GBM). Tumor staging is difficult as a tumor mass may encompass different tumor grades. The highest tumor grade is an indicator of the prognosis which is in the order of 12 months survival for GBM patients. In spite of the medical progress in tumor treatment in the last years, this value did not increase significantly. GBMs are characterized by fast, diffuse, and infiltrative growth making them to one of the greatest challenge in tumor research.

Meningiomas are tumors of the meninges. The meninges are a system of connective tissues that consist of the dura mater, the arachnoid mater, and the pia mater. The arachnoid and pia mater are sometimes together called the leptomeninges. Meningiomas form the second largest group of primary brain tumors (20 percent). Most of them are benign tumors (grade I) that means they show slow growth, do not infiltrate brain tissue and do not form metastasis. Schwannomas are the third largest group of primary brain tumors. They originate from Schwann cells that produce myelin which is an electrically insulating phospholipid layer that surrounds brain or periphery nerves. Most of them are acoustic neurinomas of grade I. Raman imaging can contribute to brain tumor classification and staging. First, Raman-based classification can complement on-site diagnosis of tissue sections. Second, tumor margins can be determined during surgery which is important to maximize tumor resection upon minimum damage of brain functions. Raman images of tissue sections were reported for GBMs,[14] meningiomas,[15] and GBM, meningiomas and schwannomas.[16] The results were compared with FTIR images[16] and have been transferred to nondried brain tumors.[17]

Dried Tissue Sections

Figure 8.3 shows three photomicrographs and Raman images of unstained, dried brain tissue sections. These samples were prepared from snap frozen specimens using a cryotome. It is important that the specimens that should be studied by Raman spectroscopy were not embedded in paraffin. Such a preparation requires dehydration of tissue in solvents, treatment with paraffin and subsequent removal of paraffin by washing with solvents. Virtually all lipids in the tissue section are removed as well. As brain tissue belongs to the tissue class with high-lipid content and brain tumors are characterized by significant changes in the composition and amount of lipids, an important fraction of the molecular information would be lost during these procedures. Raman images were collected using a 785-nm excitation laser and a microscope objective with a high-magnification and high-numerical aperture (here 100×/NA 0.9). Such a microscopic setup simultaneously maximizes the photon flux within the focus spot of approximately 5 µm and the collection efficiency of the scattered radiation. The step size in the mapping registration mode was increased to 100 µm which gave an undersampling ratio of $100^2/5^2 = 400$. That means sample regions between the raster points were omitted. A step size of 5 µm in the x and y dimension would increase the number of spectra and the total exposure time by a factor of $20 \times 20 = 400$. The

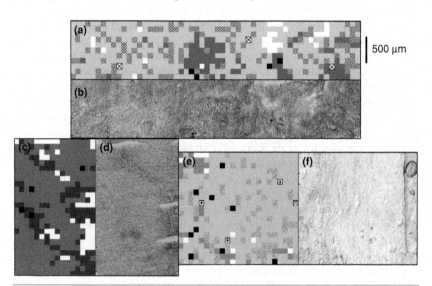

FIGURE 8.3   Raman images (a, c, e) and photomicrographs (b, d, f) of dried tissue sections of brain tumors: meningioma (a, b), glioblastoma (c, d) and schwannoma (e, f).

larger step size was sufficient to give an overview of the Raman microspectroscopic features of each sample.

K-means clustering segmented the spectra of the Raman image (Fig. 8.3a) into five groups. The Raman spectra of the blue and cyan clusters are overlaid in Fig. 8.4. The main differences at 857, 939, and 1246 cm^{-1} are assigned to higher collagen content in the blue cluster. Collagen can be distinguished from other proteins due to the primary structure with a high fraction of amino acids glycin, prolin, and hydroxyprolin, and due to the secondary structure with three-stranded triple helices. The Raman spectra of gray-shaded cluster within the cyan cluster are characterized by increased spectral contributions of nucleic acids (not shown). This is consistent with a higher proliferation rate which is an inherent property of tumor cells. The Raman spectra of the black cluster contain spectral contributions of hydroxyapatite at 960 cm^{-1}. Calcification is not an unusual phenomenon in pathological tissue. Calcium salts are formed in tumors as a consequence of functional irregularities and can be used to distinguish the pathological state and tumor type. Calcification was also detected in Raman images of GBM (black cluster in Fig. 8.3c) and schwannoma (black cluster in Fig. 8.3e). A Raman spectrum with spectral contributions of calcium oxalate at 969 cm^{-1} is displayed in Fig. 8.4. As this spectrum was obtained from the Raman image Fig. 8.3e of schwannoma, it was compared with a Raman spectrum representing this tumor type (green). The drying process of the tissue section induces crystallization of hydrophobic material. Two types of microcrystals could be identified in Raman image Fig. 8.3e: cholesterol (box with black dot) and cholesterol ester (box with black cross). The Raman spectrum of the cholesterol ester microcrystal in Fig. 8.4 is

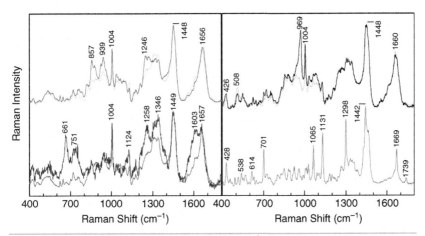

FIGURE 8.4   Raman spectra from 400 to 1800 cm^{-1} of meningioma (blue, cyan), glioblastoma (red, vine), schwannoma (green), calcifications (black) and cholesterol ester microcrystals (gray). Colors represent clusters in Fig. 8.3.

characterized by spectral contributions due to unsaturated fatty acids (1065, 1131, 1298, 1669 cm^{-1}), the ester group (1739 cm^{-1}) and the cholesterol moiety (701 cm^{-1}). Further Raman bands which distinguished cholesterol ester and cholesterol can be found[9] at 428, 538, and 614 cm^{-1}. Interestingly, the cholesterol ester content in normal brain tissue is very low and significantly increases (up to 100 fold) in brain tumors. Therefore, the detection of cholesterol ester can be considered as a molecular marker of brain tumors. This observation has recently been confirmed in mass spectra of lipid extracts of brain tumors.[18] Raman spectra with higher and lower spectral contributions of hemoglobin were found to be the two main clusters in GBM. Hemoglobin in dried tissue sections contributed to Raman bands at 661, 751, 1004, 1124, 1258, 1346, and 1603 cm^{-1}. These bands significantly deviate from hemoglobin in fresh, nondried brain tumors (see Fig. 8.5) and in red blood cells (see Fig. 8.9 in Sec. 8.3.2). The Raman signature of hemoglobin in Fig. 8.4 (vine spectrum) is typical for the heme group in an aggregated and denatured state. Spectral contributions of hemoglobin in GBM are consistent with high blood perfusion rate which is a typical property of malignant brain tumors.

Nondried Brain Tissue

The investigations of thin, dried tissue sections have been transferred to ex vivo studies of nondried brain tissue.[17] These samples do not show artifacts due to drying such as crystallization of hydrophobic constituents, denaturation of biomolecule structures or shrinkage after dehydration. Analogous to the investigation of mouse brains, the specimens were covered by a window to prevent them from drying. Instead of a fiber-optic probe, the sample was placed under a microscope with a 10×/NA 0.25 objective. The lower magnification objective focuses the laser to a spot of approximately 50 µm diameter. Raman images of 2×2 mm^2 areas were collected with a step size of 100 µm. Cluster analyses were performed in the spectral interval 1200 to 1800 cm^{-1} because it contains spectral contributions of the main constituents water (1635 cm^{-1}), proteins (1660 cm^{-1}) including collagen (1247 cm^{-1}), hemoglobin (1566 cm^{-1}), and lipids (1440 cm^{-1}) including cholesterol (1670 cm^{-1}). A Raman image of GBM and the spectra of the main clusters are shown in Fig. 8.5. Both spectra are dominated by protein bands near 1005, 1447, and 1662 cm^{-1}. The spectra differ by spectral contributions of hemoglobin at 755, 1004, 1214, 1547, 1564, 1607, and 1621 cm^{-1}. The intense hemoglobin bands are attributed to a resonance effect of the chromophor heme which is described in more detail in the context of red blood cells (Sec. 8.3.2). Chemical analysis of brain tumors and Raman spectra of brain lipid extracts confirmed[18] that GBM are characterized by increased water content and decreased lipid content. More subtle differences are increased cholesterol ester and phosphatidylcholine contents. These differences constitute the basis for a Raman-based classification

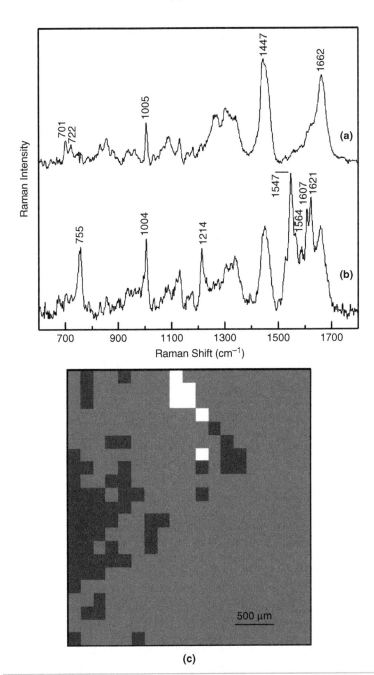

FIGURE 8.5   (a) Raman spectra from 600 to 1800 cm⁻¹ of the red cluster and (b) the vine cluster (c) of the Raman image of nondried glioblastoma brain tumor tissue.

model of gliomas. A classification model based on infrared spectroscopy has already been developed and validated using magnetic resonance tomography and histopathology.[19]

8.2.3   Human Colon Tissue

Raman spectroscopy was applied to investigate epithelial cancers of the gastrointestinal tract, including colon, esophageal, and stomach disease because of its relative accessibility for the in vivo application as an adjunct to current diagnostics using endoscopic techniques. A review has detailed the application of optical techniques to colon cancer.[20] Human colon tissue was also selected for a comparative Raman and FTIR imaging study. A prerequisite was that the sample thickness and the substrate material were properly chosen. Ten-micrometer tissue sections absorb approximately 90 percent of mid-infrared radiation which yields an absorbance equals 1. Substrates made of pure calciumfluoride (CaF_2, vacuum ultraviolet grade) are transparent for mid-infrared radiation with wavenumbers above 900 cm^{-1}. In addition, they show an extremely low Raman background except for a single band near 322 cm^{-1}. Furthermore, they have a low solubility in water (1.5×10^{-3} g per €100 g H_2O). Although quite expensive (approximately :100 per slide) CaF_2 is a better choice for biological samples than other alkaline halogenated salts such as NaCl, KBr, or BaF_2 that are more soluble in water. Less expensive alternatives are glass slides that are coated with a thin metal film. Selecting an appropriate thickness, these slides are transparent enough for visible light microscopy, they reflect mid-infrared radiation, and they suppress spectral contributions from the glass substrate that are very intense relative to the Raman signals of tissue using 785 nm excitation. Raman spectra can be detected in the relative wavenumber range (= Raman shift) from 200 to 3550 cm^{-1} using 785 nm excitation. In particular, the low-wavenumber range is extended for samples on CaF_2 substrates compared with FTIR spectroscopy. The low-wavenumber limit in modern Raman systems is determined by the transmission properties of the Notch filter which suppresses the elastic Rayleigh scattering. A number of informative vibrations occur in the wavenumber interval between 400 and 1030 cm^{-1}. They include well resolved bands of cholesterol (700 cm^{-1}), hydrophilic groups in lipids such as choline (717 cm^{-1}), amino acids bands of phenylalanine (621, 1004, 1030 cm^{-1}), tyrosine (643, 828, 853 cm^{-1}) and tryptophan (760, 880, 1011 cm^{-1}) and nucleic acids (see next paragraph). The high-wavenumber limit using 785 nm is determined by the quantum efficiency of the silicon-based charge-coupled-device (CCD) detector which drops toward zero for Raman shifts above 3400 cm^{-1} (1070 nm). The Raman spectra in Fig. 8.2 were normalized for the different detector sensitivity throughout the wavenumber interval. However, the normalization procedure also amplifies the noise above 3400 cm^{-1}.

A photomicrograph and a Raman image of mucosa in a colon tissue section are displayed in Fig. 8.6*b* and *c*, respectively. The color code indicates the segmentation of the Raman spectra by *k*-means cluster analysis: mucus beyond the top tissue margin (pink/magenta/violet), connective tissue of submucosa (red) and epithelium (blue/cyan/black). The averaged Raman spectra of the cyan and black clusters are overlaid in Fig. 8.6*a*. Details and further Raman spectra have previously been published.[21] A difference spectrum was calculated to inspect the spectral changes. Whereas the protein and lipid bands that are labeled in the Raman spectra did not change significantly, all positive difference bands can be assigned to DNA. A Raman spectrum of calf thymus DNA was inserted for comparison. The labeled bands in the reference spectrum can be assigned to thymine (669, 747, 781, 1368, 1670 cm^{-1}), guanine (680, 1482, 1575 cm^{-1}), adenine (726, 1303, 1336, 1482, 1575 cm^{-1}), cytosine (781, 1253 cm^{-1}), the phosphate backbone (833, 1094 cm^{-1}), and CH$_2$ groups of the backbone (1420 cm^{-1}). Note that some bands contain spectral contributions from two nucleotides. Negative difference bands at 833 and 875 cm^{-1} were correlated with the expression of secretion products in glands. The 833 cm^{-1} band overlaps with the DNA backbone band.

A disadvantage of Raman imaging in the serial registration mode is longer exposure times compared with FTIR imaging in the parallel registration mode using focal plane array (FPA) detectors. Collecting 39 × 71 = 2769 Raman spectra at a step size of 10 μm took more than 24 hours using the Raman system Envia (Renishaw, UK). Collecting eight FTIR images of 64 × 64 = 4096 FTIR spectra per image could be finished within 15 minutes. Each image covered an area of 170 × 170 μm^2 as indicated by the boxes in Fig. 8.6*b*. Even shorter exposure times can be obtained using coherent anti-Stokes Raman scattering (CARS) and rapid laser scanning microscopes. CARS microscopy is a nonlinear variant of Raman spectroscopy. Details of theoretical and technical consideration of CARS microscopy are outside the scope of this contribution and can be found in Chap. 11 and elsewhere.[22–24] Its advantages include (1) signal enhancement of several orders of magnitude for single vibrational modes, (2) coherent emission in a direction determined by phase matching which allows efficient signal detection, (3) narrow spectral bandwidth which allows registration without monochromators, and (4) absence of an autofluorescence background as the signal is recorded on the anti-Stokes side of the excitation light. CARS images of colon tissue sections could be obtained from approximately 600 × 600 μm^2 regions within 5 seconds.[25] CARS images were compared with Raman images. The information content of the both techniques was found to be complementary. Whereas variations within tissue sections could be visualized more rapidly using CARS microscopy than Raman microscopy, identification of tissue types and characterization of variances between different tissue sections can be achieved better in Raman images because the information from an extended spectral interval is simultaneously recorded.

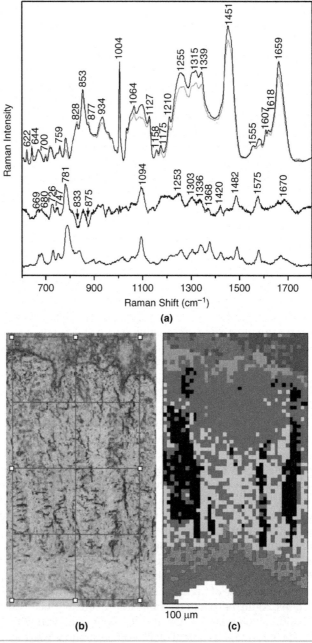

FIGURE 8.6   (a) Raman spectra from 600 to 1800 cm^{-1} of mucosa in a colon tissue section representing the cyan (black) and black (gray) clusters. The difference spectrum = (black) minus (gray) is displayed at a magnified scale (middle). A Raman spectrum of calf thymus DNA was included for comparison (bottom). (b) A photomicrograph and (c) a Raman image of epithelium in a colon tissue section. See text for cluster assignments.

8.2.4 Human Lung Tissue

Raman and FTIR imaging were applied to study the spectral features and derive molecular information of normal lung tissue,[26] congenital cystic adenomatoid malformation (CCAM),[27] and bronchopulmonary sequestration (BPS).[28] Other Raman work in the context of lung have been reported for the detection and grading of neoplasia in cell samples[29] and for the characterization of normal bronchial tissue.[30] CCAMs are benign masses of nonfunctional lung tissue developing during the embryonic gestation from an overgrowth of the terminal bronchioles with subsequent suppressing alveolar growth. BPS involves abnormal pulmonary tissue lacking trancheobronchial connection and having a separate vascular supply that arises from the thoracic and abdominal aorta. Cryosections were mounted on CaF_2 substrates. Raman images were collected at a step size of 66 to 100 μm to assess the whole tissue section. The step size was reduced to 10 μm to resolve details in selected areas. Datasets were segmented by k-means cluster analysis and the averaged spectra of cluster were compared. Photomicrographs and Raman microscopic images of normal lung tissue (Fig. 8.7b) and congenital cystic adenomatous malformation (Fig. 8.7c) are compared. Normal lung tissue shows the typical sponge-like morphology. Morphological changes are visible in solid-type CCAM. Here, the tissue sections have fewer holes which are consistent with nonfunctional, nonaerating, abnormal masses of lung tissue. A cluster analysis of the combined Raman images distinguishes six tissue classes. Normal lung tissue is dominated by the classes represented by red and green clusters. The averaged Raman spectrum is shown in Fig. 8.7a as trace (A). CCAM is dominated by the tissue class represented by the blue cluster. The averaged Raman spectrum is shown in Fig. 8.7a as trace (C). Three additional tissue classes in CCAM were assigned to smooth muscles of vessels (yellow, olive) and microcrystals (black). The Raman spectrum of these microcrystals indicated a high content of the phospholipid phosphatidylcholine. Whereas it was not possible to identify the lipid type using infrared spectroscopy, cluster analysis of FTIR images revealed a further tissue type with spectral contributions of glycogen. This confirms a general tendency that components having high content of polar C—O, O—H and C—H groups such as glycogen show usually stronger infrared than Raman bands. Furthermore, these results clearly demonstrate that Raman and FTIR complement each other.

A typical Raman spectrum of BPS is included in Fig. 8.7a as trace (B). As compared with Fig. 8.4 and 8.5, main differences between the Raman spectra are assigned to spectral contributions of hemoglobin that decreases in the order normal lung tissue > BPS > CCAM. As described in more detail in Sec. 8.3.2 for red blood cells, Raman bands of the heme group are enhanced by a resonance effect. On the one hand, this enables to detect changes in the state or content of hemoglobin more sensitive than using infrared spectroscopy. On the other hand,

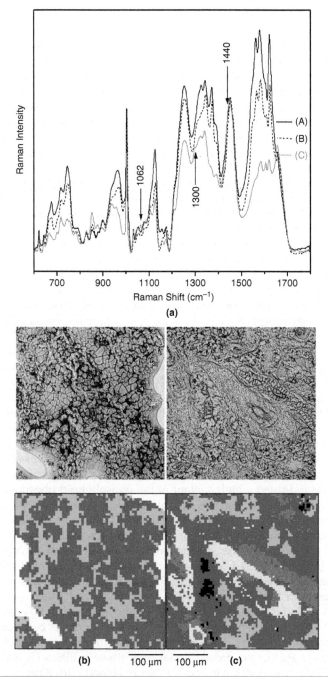

Figure 8.7   (a) Raman spectra from 600 to 1800 cm⁻¹ of normal lung tissue (A), bronchopulmonary sequestration (B), and congenital cystic adenomatous malformation (C). (b) Photomicrographs and Raman images of normal lung tissue and (c) congenital cystic adenomatous malformation.

spectral contributions of other constituents such as proteins and lipids are partly masked by hemoglobin bands. Careful inspection of spectrum (C) in Fig. 8.7a reveals that lipid-associated bands at 1062, 1300, and 1440 cm^{-1} increase in CCAM. Increased spectral contributions of lipids were more pronounced in infrared spectra of CCAM which confirmed the Raman result. We conclude that each pathology is characterized by a distinct biochemical composition which can be probed by Raman and infrared spectroscopy. The power of the vibrational spectroscopic fingerprint was demonstrated in one BPS patient where the coexistence of CCAM was detected and confirmed by histopathology.[28] This diagnosis was overseen during the first histopathological inspection.

8.3 Raman Imaging of Cells

Raman imaging of cells can complement current techniques in cell biology to study cellular and subcellular processes and structures or to identify cell differentiation and cell type. Raman and infrared microspectroscopic studies of individual cells were summarized.[31] Frequently used methods for single cell studies are electron microscopy, fluorescence microscopy, and autoradiography. However, complicated preparations and manipulations have to be performed to fulfill the specific requirements of these analytical methods before they can be applied. Furthermore, the environments are sometimes not physiological, e.g., the vacuum in experiments with electron beams. Autoradiography and fluorescence depend on exogenous fluorophores or other contrast enhancing agents because most biomolecules cannot directly be detected. A decision has to be made in advance: Which property should be probed and which marker should be used? Problems result from the limited stability, bleaching, and restricted accessibility of external markers. In principle, Raman imaging can overcome many restrictions. It combines molecular specificity with diffraction limited spatial resolution in the sub-micrometer range. Due to its nondestructivity using near-infrared wavelengths for excitation, it can be applied under in vivo conditions without staining or other markers because the Raman signals are based on inherent vibrational properties of the cells' biomolecules. All vibrational signatures overlap giving complex spectra. High signal to noise ratios are required to identify subtle spectral differences. Furthermore, Raman spectroscopy is a rapid technique because single Raman spectra can be recorded within seconds. Raman signals can be enhanced using special techniques such as resonance Raman spectroscopy, surface-enhanced Raman spectroscopy (SERS) and coherent anti-Stokes Raman spectroscopy. All these techniques have successfully been applied for single cell studies. Single cells are very suitable objects for Raman spectroscopy because of the high concentrations of biomolecules in their condensed volume. Protein concentration as high as 250 μg/μl, and DNA and RNA concentrations in the range of 100 μg/μl have been reported.[32] These values depend on the cell type, the phase of the cell cycle and the location within the cell.

Data acquisition of small objects with high spatial resolution such as single cells and their subcellular organelles require coupling of Raman spectrometers with microscopes. The coupling of a Raman spectrometer with a microscope offers two main advantages. First, lateral resolutions can be achieved up to the Abbe's limit of diffraction below 1 μm according to the formula $\delta_{lat} = 0.61 \cdot \lambda / NA$. High axial resolutions are obtained by confocal microscopes. The diffraction limit in the axial dimension can be calculated as $\delta_{lat} = 2 \cdot \lambda n / NA^2$ (wavelength λ, refractive index n, numerical aperture NA). Second, maximum sensitivity can be achieved because the photon flux (photon density per area) at the focused laser beam onto the sample is at maximum and the collection efficiency of scattered photons from the sample is at maximum. The lateral resolution, the diameter of the focused laser and the collection efficiency depend on the NA of the microscope objective which is the product of the sine of the aperture angle and the diffraction index. High NA objectives have small working distances below 1 mm in air. The numerical aperture of water immersion objectives can be enlarged due to the larger diffraction index of water. Oil immersion objectives in combination with oil should be avoided because spectral contributions of oil overlap with the Raman spectrum of the sample.

8.3.1   Lung Fibroblast Cells

Lung fibroblast cells were studied because cell cultivation can easily be controlled, cell geometry and morphology seemed comparatively suitable for depicting and discrimination of visible compartments, they grow adherent to quartz slides and they keep their shape over time after fixation by formalin. Although living cells can be studied by Raman spectroscopy, fixation prevents the cell changes morphology or composition during time-consuming acquisition of images by linear Raman spectroscopy. Furthermore, the cell environment should be carefully controlled in long-term experiments, e.g., cell culture media, temperature, or carbon dioxide. Using the Raman microspectrometer HoloSpec with 785 nm excitation (Kaiser Optical System, United States), Raman images were collected with a step size of 1 μm.[33] Acquisition of a 60 × 60 raster with 1 second exposure time per spectrum takes more than 1 hour. Beside exposure time, additional time is needed to move the sample stage. Using the confocal Raman microscope CRM300 (Witec, Germany), a Raman image of a single cell was recorded with 0.2 second exposure time per spectrum and 300 nm step size.[34] Various cell compartments could be identified in this highly resolved Raman image. In spite of the low-exposure time, Raman spectra of reasonable signal to noise ratio were obtained after k-means cluster analysis. Calculating cluster averaged spectra gives reasonable results under the assumption that the spectral signatures within each compartment (e.g., cell nucleus) do not deviate much. This commercial confocal Raman microscope was also applied for single cell studies by other groups, e.g., to image liposomal drug carrier systems.[35]

The cluster membership map of a Raman image (Fig. 8.8c) and the photomicrograph of the cells in PBS buffer (Fig. 8.8b) are compared. The data were recorded using a 60×/NA 1.0 water immersion objective. The clusters were assigned to the nucleus, cytoplasm, lipid vesicles, and cytoplasmic inclusions. The assignments of 34 Raman bands to proteins, lipids, cholesterol, and nucleic acids have previously been summarized.[36] DNA bands are evident at 669 [thymine (T)], 680 [guanine (G)], 727 [adenine (A)], 786 [cytosine (C), T], 1092 (backbone), 1375 (T), and 1577 cm^{-1} (G, A). After normalization to the protein bands, DNA bands were most intense in the nucleus (spectrum in Fig. 8.3a) and lipid bands most intense in vesicles (arrows in spectrum in Fig. 8.3a). DNA bands in the nucleus are distinguished from RNA bands in cytoplasm by marker bands for the phosphate backbone conformation (811 and 1100 cm^{-1}), the geometry of the sugar pucker (shift of the 680 cm^{-1} band), and the nucleotide thymine (no 1375 cm^{-1} band) which is replaced by uridin in RNA. The positions of the changes are marked by arrows in spectrum B in Fig. 8.8a.

Cell stress response is the expression for the reaction of living cells to environmental changes that are potentially harmful such as increase or decrease of temperature, pH value, salt concentration or the presence of toxins. These stress-induced processes cause various modifications within cells that can lead to morphological and biochemical changes or even cell death. In Ref. 31 Raman imaging was applied to study the morphology and chemical composition of normal, stressed, and apoptotic cells. Cell stress was induced by adding 1 mM glyoxal to the medium for 24 hours. Glyoxal is toxic as it inhibits the DNA and protein synthesis. Comparing the cells on quartz slides after fixation with formalin indicated shrinkage of the nucleus in stressed cells and absence of lipid vesicles. Instead, more inclusion particles were present. A stressed cell showed blisters at the surface and it adopted a round shape which pointed to a more advanced stress condition. Another cell showed further shrinkage of the nucleus with fragmentation which is typical for apoptosis. The Raman spectra of the cell compartments enabled to obtain chemical information. All spectra were normalized to the phenylalanine band of proteins at 1003 cm^{-1} because of its high intensity, low overlap with other bands and insensitivity of changes in structure and environment. After normalization a decrease of nucleic acid bands as a function of cell stress was observed whereas protein and lipid bands did not change significantly. These changes are consistent with decomposition or condensation of chromatin. Higher condensed chromatin induced stronger interactions between DNA bases which decreases intensity

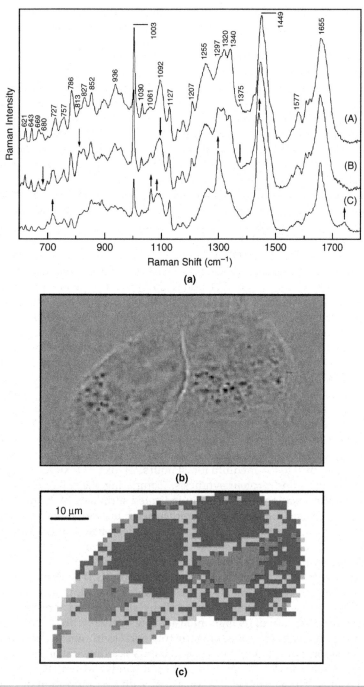

Figure 8.8   Raman spectra from 600 to 1800 cm⁻¹ of nucleus (A), cytoplasm (B), vesicle (C) of the lung fibroblast cells in Fig. 8.8*b*. Photomicrograph and cluster analysis of Raman image separating nucleus (red) cytoplasm (blue/cyan), vesicles (green) and cytoplasmic inclusions (magenta) (*c*).

of Raman bands. This hypochromic effect was already reported in isolated DNA more than 30 years ago.[37] Similarly, a reduced ratio of RNA to protein bands was detected in the cytoplasm as a function of cell stress. Finally, more intense nucleic acid bands were detected in blisters at the cell surface. This result is consistent with apoptotic particles by which cells export material out of the cell. In summary, all observations agree with the key processes of apoptosis, the term for programmed cell death. Among them are shrinkage of the cell, reorganization of cell organelles, condensation of chromatin, decomposition of cell nucleoli and fragmentation of the nucleus.[38] The cell stress process was selected for the first application of Raman imaging because the morphology and molecular composition change in a well-known way. Such experiments are important before new techniques can be applied to unknown processes.

8.3.2   Red Blood Cells

Resonance Raman (RR) spectroscopy is a variant of Raman spectroscopy that can be used to enhance its specificity and sensitivity. Here, the excitation wavelength is matched to an electronic transition of the molecule of interest so that vibrational modes associated with the excited state are enhanced (by as much as a factor of 10^6). Electronic transitions of proteins with prosthetic groups are found in the visible spectral region. As the enhancement is restricted to vibrational modes of the chromophore, the complexity of Raman spectra from cells is reduced. Functional red blood cells (also known as erythrocytes) are an excellent subject for RR microspectroscopy and imaging because of the high concentration of hemoglobin (22 mM) and its unique spectral properties. The resonance-enhanced Raman signals of the chromophore heme are rich in information that enabled a deeper understanding of the structure and function of red blood cells. RR spectroscopy of red blood cells has recently been reviewed.[39] Unlike traditional histopathological methods (e.g., Gimsa staining), this approach allows studying unlabeled red blood cells.

Malaria research is an interesting diagnostic application of RR spectroscopy which has been reviewed by Wood and McNaughton.[40] Malaria is one of the most common infectious diseases and an enormous public health problem. The disease is caused by protozoan parasites of the genus *Plasmodium* that are transmitted by mosquitoes. The parasites multiply within red blood cells, where they digest a major proportion of red blood cell hemoglobin. As free heme, a product of hemoglobin digestion, is toxic for the parasite it detoxifies free heme by conversion into an insoluble crystal called hemozoin or "malaria pigment." The spatial distribution of heme species in red blood cells was probed by RR imaging.[41–43] An example is shown in Fig. 8.9. Typical RR spectra of hemozoin and hemoglobin inside a parasitized red blood cell are compared with RR spectra of β-hematin and hemoglobin as reference compounds. The synthetic compound β-hematin is a structural analogue of hemozoin with a strong marker

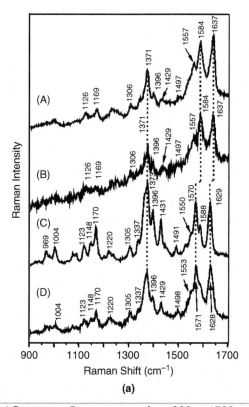

(a)

Figure 8.9   (a) Resonance Raman spectra from 900 to 1700 cm⁻¹ of hemozoin inside a parasitized red blood cell (A), β-hematin powder (B), hemoglobin inside a parasitized red blood cell (C) and hemoglobin powder (D). (b) The photomicrographs of red blood cells are overlaid with the color-coded results of a hierarchical cluster analysis: background (gray), low hemoglobin (green), high hemoglobin (blue), hemozoin (red).

band at 1371 cm⁻¹. Raman images were collected using an argon ion laser emitting 1 mW at 514.5 nm and segmented by hierarchical cluster analysis.[43] Three clusters in images of nonparasitized red blood cells were assigned to background, and strong and weak spectral contributions of hemoglobin, respectively. A forth cluster in the images of the parasitized cells identified the malaria pigment hemozoin. The hierarchical cluster analysis calculates the symmetric distance matrix (size $n \times n$) between all considered spectra (number n) as a measure of their pair-wise similarity. The algorithm then searches for the minimum distance, collects the two most similar spectra into a first cluster and recalculates spectral distances between all remaining spectra and the first cluster. In the next step the algorithm performs a new search for the most similar objects which now can be spectra or clusters. This iterative process is repeated $n-1$ times until all spectra have been merged into one cluster. The result is displayed

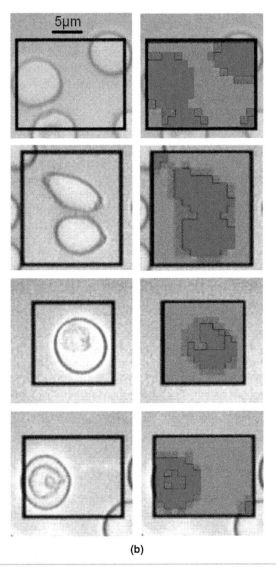

5µm

(b)

Figure 8.9   (Continued)

in a tree-like, two-dimensional dendrogram in which one axis refers to the reduction of clusters with increasing number of iterations and the other axis to the respective spectral distances.

The detoxification of free heme into hemozoin is an important drug target. The function of antimalarials based on quinoline is believed to inhibit hemozoin formation and consequently build up toxic heme. A better understanding of the mechanism of hemozoin formation and the drug interaction would be extremely valuable for the design of new drugs to increase efficacy and overcome resistance which is presently the major drawback of antimalarial therapy. Using resonance Raman spectroscopy the characterization of hemozoin under different physiological conditions,[44] as well as the influences of quinoline additives[45] or chloroquine[46] to β-hematin were investigated. The structural changes of hemozoin during chloroquine exposure were monitored by RR spectroscopy.[47]

8.4   Raman Spectroscopy of Bacteria

Prokaryotic bacteria are generally smaller than single (eukaryotic) cells. While some of them can cause serious infectious diseases, others are vital parts of the gut flora or live on the skin; again other bacteria can spoil food, while other strains are utilized in food, pharmaceutical and chemical industry to produce cheese, vinegar, yoghurt, antibiotics, hormones, lactic acids, and many other products. Fast and reliable identification methods are needed to distinguish the useful and harmless bacteria from the unwanted and toxic ones, and to provide an efficient and appropriate treatment of infections. In times of growing antibiotic resistances it is furthermore desirable to identify new target structures for new and effective drugs.

Besides the classical microbiological identification a variety of alternative methods has evolved, ranging from immunological assays with different labeling techniques to molecular biological methods. An overview of the identification of microorganisms using infrared spectroscopy has recently been given.[48] Simultaneous to the microbial classification by infrared spectroscopy, Raman spectroscopy was used as a nondestructive fingerprint method to identify and classify bacteria.[49] In general, sample preparation is easier than for infrared spectroscopy because bacteria grown in liquid cultures as well as bacteria on agar plates can directly be investigated.[50] Different wavelengths can be employed for resonance and nonresonance excitation and only a minimal sample volume, down to a single cell, is necessary, which means a tremendous reduction in analysis time. An important target structure of many antibiotics is the bacterial cell wall. Tip-enhanced Raman spectroscopy is a relatively new way how to assess information about the outer bacterial layer with lateral resolution down to few nanometers and with high chemical specificity.

8.4.1   Species Classification

In medicine the exact identification of bacteria is necessary in order to provide the appropriate therapy and to prevent antibiotic resistance which may further delay administration of the most appropriate narrow-spectrum antibiotic.[51] Traditional microbiological identification and infection diagnoses are time-consuming and labor-intensive processes: at least 12- to 24 hour incubation is required to obtain an accurate

colony count. In addition 12 ± 24 hours are needed for organism identification and susceptibility testing. As a first differentiation criteria relatively unspecific morphological parameters, such as size,
shape, and color of the bacteria as well as of the colonies, are used
which need to be complemented with expensive and tedious metabolic tests, such as the ability to grow in various media under different conditions, degradation of certain substrates and enzyme
activity.[52]

Raman spectroscopy offers a fast and reliable way to obtain
detailed information about the chemical composition in a noninvasive manner. The Raman spectra provide complex and detailed
"fingerprint-like" information of the overall molecular composition
of living bacterial cells in an extremely brief time span. An early
report described the classification of 42 candida strains comprising
five species by confocal Raman microscopy to demonstrate the feasibility of the technique for the rapid identification.[53] With Raman
microspectroscopy it is furthermore possible to focus the excitation
laser down to about 1 µm in diameter, which is on the order of magnitude of the size of bacteria. Therefore, even single cells can be
probed, which makes the time-consuming cultivation unnecessary
and reduces the amount of potentially hazardous biomaterial.

Excitation in the Visible Wavelength Range

Sample preparation for the Raman measurements is straight forward
and easy. To create a database with known bacterial strains, the bacteria are grown on agar plates at varying environmental conditions
(temperature, nutrition) which mimics the unknown history of individual bacterial cells obtained from patients or found in hospitals,
clean room environments or food production lines. After varying
growth time, the bacteria are harvested and smeared on a fused silica
plate. Raman spectra of single bacterial cells are collected after excitation with 532 nm from a frequency doubled Nd:YAG laser focused
down with a 100× objective and resulting in 10 mW laser power at the
sample. Typical exposure times are 60 seconds per spectrum.

Raman spectra of single bacterial cells from nine different species
are shown in Fig. 8.10a. The most intense Raman band is centered
around 2940 cm^{-1} which is assigned to symmetric and antisymmetric
C—H stretching vibrations of the CH_2 and CH_3 groups from lipids,
proteins and carbohydrates. The scissoring and deformation vibrations of the C—H bond are found around 1450 cm^{-1} and 1337 cm^{-1},
respectively. Vibrations of the peptide linkage of proteins are located
around 1660 cm^{-1} (amide I) and with less intensity around 1242 cm^{-1}
(amide III). The most prominent spectral contribution of the aromatic
amino acid phenylalanine is found at 1003 cm^{-1}. The band at 1575 cm^{-1}
is assigned to the nucleotides guanine and adenine. The band at
1128 cm^{-1} is due to C—N and C—C stretching vibrations. Colored
bacteria, such as *Micrococcus luteus*, exhibit additional sharp signals

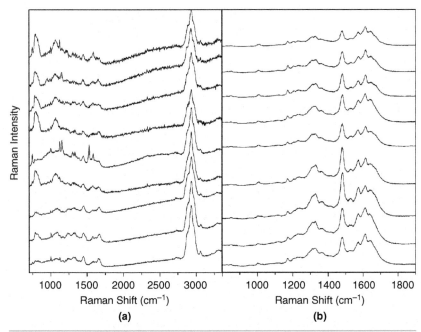

Raman Intensity

1000 1500 2000 2500 3000
Raman Shift (cm⁻¹)
(a)

1000 1200 1400 1600 1800
Raman Shift (cm⁻¹)
(b)

FIGURE 8.10   (*a*) Raman spectra of single bacterial strains from 700 to 3350 cm⁻¹ using λ_{ex} = 532 nm and 60 seconds acquisition time. (*b*) Resonance Raman spectra of the same bacterial strains from 800 to 1900 cm⁻¹ using λ_{ex} = 244 nm and 120 seconds acquisition time. Bacterial strains (from top to bottom): *B. pumilus* DSM 27, *B. sphaericus* DSM 28, *B. subtilis* DSM 10, *E. coli* DSM 423, *M. luteus* DSM 20030, *M. lylae* DSM 20315, *S. warneri* DSM 20316, *S. epidermidis* ATCC 35984, *S. cohnii* DSM 6669.

at 1524, 1159, and 1003 cm⁻¹ which originate from the carotenoid sarcinaxanthin. Uncompensated spectral contributions of the quartz substrate are visible near 800 and 1100 cm⁻¹.

Microbial cells from different species or strains vary in their chemical composition, e.g., in the concentration, structure, and type of proteins, carbohydrates, lipids, and DNA/RNA sequences. These variations can be monitored by Raman spectroscopy and used for spectral-based classification of the bacterial strain. Sophisticated algorithms for data analysis are required to identify the spectral variances which are usually small and distributed over a broad spectral range. Unsupervised statistical methods, such as principal component analysis, and hierarchical cluster analysis utilize the intrinsic variation in the spectra for segmentation. If the strain and species of the bacteria of interest are known supervised statistical methods such as discriminant function analysis, support vector machines (SVMs), *k*-nearest neighbor, near mean centering, or artificial neural networks can train a model which in the next step is used to classify unknown bacteria based on their Raman spectra, provided they are included in the database. Here, results obtained from SVMs will be presented. During the SVM training functions are found which divide the classes (spectra of bacteria from different strains) with a maximum margin. These decision boundaries between the classes are described by so-called support vectors. A robust training of the data set can be achieved with (the time-consuming) leave-one-out testing. That means, all spectra, except one spectrum, are used for calculating the model and the left-out spectrum is used for classification. Then, the same procedure is repeated for another data point to be left out, until every object was removed once from the training set. For the subsequent classification of new data not the whole training dataset is needed anymore, but only the support vectors are used.[54,55]

Bacterial strains which could be found in a clean room environment were chosen to create a database using support vector machines with a nonlinear gaussian radial basis function (rbf) kernel. These spectra include the spectra shown in Fig. 8.10a plus additional bacterial strains within same species. To ensure a stable classification, bacteria grown under varying environmental conditions were included in the training set.[56] The results of the SVM classification of the Raman spectra from the single bacterial cells excited at 532 nm are shown in Table 8.1. Altogether the dataset consists of 2545 spectra from 20 strains of nine species. The bacteria were identified with recognition rates of 96 percent on strain and with 97 percent on species level.[53,56,57]

Once the database is created new single bacterial cells can be investigated within a minimum of time. Table 8.2 shows the classification results for an independent set of different single bacteria from 16 different strains. The spectra were recorded anonymously and classified with the previously created SVM. One hundred and twenty-five out of 130 bacteria were correctly identified. Out of the five incorrectly classified spectra four were assigned to the wrong strain within the right genus, and only one was misclassified on the species level. This proves the high potential of automated Raman-based assignment. Single Raman spectra acquired with 60 seconds exposure time enable to assign 96 percent of the bacteria on the strain level and 99 percent on the species level.[58] A prototype of a fully automated device for the fast, reliable, and nondestructive online-identification of microorganisms from clean room environment has been manufactured.[59] The approach utilizes fluorescence labeling of single bacterial cell to differentiate live/dead bacteria or biotic/abiotic particles prior the Raman identification step.[60] First instruments which could speed up the identification of microbial pathogens are already on the market (SpectraCell RA™ Bacterial Strain Typing Analyzer) or under development (rap.ID Particle Systems for the analysis of microparticles from fluids or air). These classification methods can also be extended to eukaryotic cells.[55]

Bacterial Strain	Total Number of Spectra	Misclassified Strain Spectra	Recognition Rate for Strains, %	Misclassified Species Spectra	Recognition Rate for Species, %
B. pumilus DSM 27	57	9	84.2	4	93.0
B. pumilus DSM 361	43	10	76.7	4	90.7
B. sphaericus DSM 28	53	9	83.0	5	90.6
B. sphaericus DSM 396	42	9	78.6	6	85.7
B. subtilis subsp. subtilis DSM 10	306	8	97.4	6	98.0
B. subtilis subsp.spizizenii DSM 347	42	3	92.9	2	95.2
E. coli DSM 423	51	6	88.2	5	90.2
E. coli DSM 498	21	1	95.2	1	95.2
E. coli DSM 499	20	1	95.0	1	95.0
M. luteus DSM 348	619	3	99.5	2	99.7
M. luteus DSM 20030	48	6	87.5	4	91.7
M. lylae DSM 20315	20	0	100.0	0	100.0
M. lylae DSM 20318	20	1	95.0	1	95.0
S. cohnii subsp. cohnii DSM 6669	67	0	100.0	0	100.0
S. cohnii subsp. cohnii DSM 20260	65	2	96.9	0	100.0
S. cohnii subsp. urealyticum DSM 6718	65	9	86.2	7	89.2

TABLE 8.1   Classification of Raman Spectra of Single Bacteria

Bacterial Strain	Total Number of Spectra	Misclassified Strain Spectra	Recognition Rate for Strains, %	Misclassified Species Spectra	Recognition Rate for Species, %
S. cohnii subsp. urealyticum DSM 6719	63	7	88.9	3	95.2
S. epidermitis ATCC 35984	805	6	99.3	6	99.3
S. warneri DSM 20036	67	3	95.5	2	97.0
S. warneri DSM 20316	71	9	87.3	3	95.8
Average recognition rate	2545	102	95.9	62	97.5

TABLE 8.1   (Continued)

Excitation in the Ultraviolet Wavelength Range

Raman spectra excited at 532 nm represent the overall chemical composition with contributions from all molecules (according to their cross sections) which can be used for (phenotypic) classification. However, biomolecular identification can also be achieved using the structure and abundance of certain macromolecules (DNA, RNA, proteins, cytochromes) inside the bacteria. For example, a widely applied genotypic identification method is based on amplification of DNA by polymerase chain reaction (PCR). Most of the "biomarkers" (DNA, RNA, proteins) absorb in the ultraviolet spectral region between 190 and 280 nm. Excitation of Raman spectra in this wavelength region makes use of the resonance Raman effect. Due to the coupling of the Raman scattering to the molecular absorption the observed Raman bands are increased in intensity by a factor of 10^3 to 10^5. This enhancement allows the detection of molecules that occur only at low concentrations in the presence of other, higher concentrated molecules. When using 244 nm as excitation wavelength, in particular vibrations of aromatic amino acids, as well as of the DNA/RNA bases are enhanced. Furthermore, fluorescence is usually energetically far enough away from the wavelength region where the Raman signal is recorded. Fig. 8.10b shows UV resonance Raman spectra of the same bacterial strains discussed in the section. "Excitation in the Visible Wavelength Range" with excitation at 532 nm. In order to suppress photodamage, the exposure time of the sample to the UV light was minimized by rotating dried bacterial layers on fused silica while accumulating the Raman signal for 120 seconds. That means that the Raman spectra shown in Fig. 8.10b originate not from a single

Bacterial Strain	Number of Spectra	Correctly Identified	Incorrectly Identified as
B. subtilis DSM 347	8	8	
B. sphaericus DSM 28	8	8	
B. sphaericus DSM 396	7	7	
E. coli DSM 423	7	7	
E. coli DSM 498	7	7	
E. coli DSM 1058	20	17	E. coli DSM 499, E. coli DSM 423, E. coli DSM 2769
M. luteus DSM 20030	6	6	
M. lylae DSM 20315	5	5	
M. lylae DSM 20318	5	5	
S. cohnii DSM 6669	8	8	
S. cohnii DSM 6718	5	5	
S. cohnii DSM 6719	5	5	
S. cohnii DSM 20260	7	7	
S. epidermidis ATCC 35984	7	7	
S. epidermidis 195	20	18	S. warneri, E. coli
S. warneri DSM 20036	5	5	
Identification	**130**	**125**	

TABLE 8.2   Classification of Raman Spectra of Single Bacteria: Identification of an Independent Dataset

bacterial cell, but from about 10^5 cells, which is usually obtained from a microcolony after 5 hours of cultivation. The most intense bands at 1475 and 1570 cm^{-1} are assigned to vibrations of purine bases adenine and guanine. A vibration due to uracil is found at 1229 cm^{-1}. Thymine and adenine vibrations are responsible for the Raman band at 1355 cm^{-1}. The band at 1521 cm^{-1} is assigned to cytosine. The aromatic amino acids tyrosine and tryptophan contribute to the band at 1609 cm^{-1}, and tyrosine also contributes to the band at 1324 cm^{-1}.

Analogous to the Raman spectra with 532 nm excitation, the 244-nm excited spectra were classified using a SVM with nonlinear rbf kernel. Recognition rates in Table 8.3 ranged from 96 percent to 100 percent correct classification on the strain level, which gives an average recognition rate of 98.7 percent. Only one out of 1150 spectra was misclassified on the species level, resulting in a recognition rate of 99.6 percent for *Staphylococcus epidermidis* and an overall recognition rate of 99.9 percent.[61] In another study, UV resonance Raman spectroscopy was applied to identify eight different strains of lactic acid bacteria from yoghurt.[62] Classification was accomplished using different chemometric methods. In a first attempt, the unsupervised methods hierarchical cluster analysis and principal component analysis were applied to investigate natural grouping in the data. In a second step the spectra were analyzed using several supervised methods: *k*-nearest neighbor classifier, nearest mean classifier, linear discriminant analysis, and SVMs.

8.4.2 Imaging Single Bacteria

After bacterial identification, the infectious ones require treatment with efficient antibiotics. A suitable target structure of many antibiotics, e.g., of the group of the β-lactams, penicillines and glycopeptides is the bacterial cell wall. An intact cell wall is crucial for the correct function of many vital processes such as, e.g., signal transduction, mass transport, adhesion on surfaces, cell recognition, and enzyme reactions. Many of these processes that take place at and through the cell membrane are not completely understood so far. Furthermore, some extracellular substances were found to form a biofilm around the bacteria which may alter the effectiveness of the antibiotics. Therefore, a detailed knowledge about the chemical composition of the bacterial surface and its spatial arrangement on a molecular level, as well as an in-depth understanding of the dynamics of the cell membrane are important for the development of more effective drugs.

TERS is a recently developed technique which combines SERS with atomic force microscopy (AFM).[63] In SERS a metal surface with roughness in the nanometer scale is brought into close contact to a sample to experience electromagnetic and chemical enhancement factors up to 10^8 to 10^{14} upon illumination with appropriate wavelengths. For TERS, the SERS active metal is reduced to the size of an AFM tip with apex sizes of less than 50 nm in diameter (Fig 8.11*a*). This tip is moved across the sample to record the surface features. At the same time, that part of the sample which is in the close vicinity of the tip apex experiences the enhanced electromagnetic field if the wavelength of the laser excitation is tuned to the plasmon absorption of the nanoparticle. This evanescent electromagnetic field decays very rapidly with increasing distance from the metal surface, so that at only 50 nm away from a metal tip no

Species	Number of Strains	Total Number of Spectra	Misclassified Strain Spectra	Recognition Rate for Strains (%)	Misclassified Species Spectra	Recognition Rate for Species (%)
B. pumilus	2	112	0	100	0	100
B. sphaericus	2	95	1	98.8	0	100
B. subtilis	2	97	0	100	0	100
E. coli	10	271	8	96.4	0	100
M. luteus	2	107	0	100	0	100
M. lylae	2	64	0	100	0	100
S. cohnii	4	111	0	100	0	100
S. epidermidis	8	239	2	99.22	1	99.6
S. warneri	2	54	0	100	0	100
	34	**1150**		**99.3**		**99.9**

TABLE 8.3   Classification of UV-Resonance Raman Spectra of Bacteria

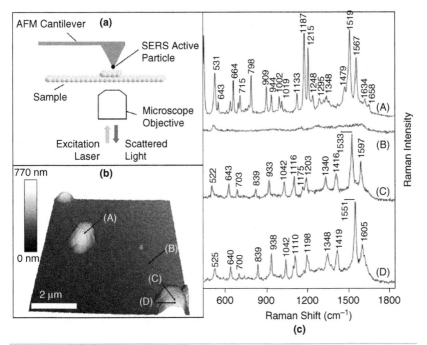

FIGURE 8.11   Schematic setup of tip-enhanced Raman spectroscopy (*a*). Pseudo three-dimensional topographic image (7 × 7 μm²) of single *S. epidermidis* cells on a glass surface (*b*). TERS spectra recorded from positions [(A) to (D)] with 1 second acquisition time. While traces A, C, and D show spectra from the bacterial surface, trace B depicts a typical background spectrum.

Raman scattering enhancement can be observed anymore. Thus, TERS combines chemical information from the Raman spectra with near-field spatial resolution. TERS not only overcomes the low scattering efficiency but also the finite spatial resolution due to the diffraction limit of Raman microspectroscopy. TERS experiments with emphasis on life sciences have recently been reviewed.[64] TERS opens the way for a detailed spatially resolved study of nanometer-scaled structures, such as the bacterial surface and might lead to an understanding of the adhesion of cells to surfaces, biofilm-formation, and the mode of action of antibiotics like β-lactams, penicillins or glycopeptides which attack the cell wall and interfere with its synthesis.

Staphylococcus epidermidis ATCC 35984 was selected as a sample organism for first TERS studies. *S. epidermidis* evolved to a major cause of nosocomial infections, especially associated with the use of implanted medical devices. The pathogenic potential of this strain mainly results from binding to polymer surfaces and biofilm formation. From the biofilms, especially associated with implanted medical devices, the bacteria get into the blood of the patients and cause a septic disease pattern. The biofilm matrix is also responsible for reduced susceptibility of the bacteria for antibiotics. Therefore, information about the bacterial surface might lead to a more profound understanding of biofilm formation and contribute to an improved treatment of *S. epidermidis* infections.

For the TERS investigations single bacterial cells of *S. epidermidis* were drop casted on a glass slide and placed under the AFM. A silver-coated AFM tip was approached from the top and operated in the intermittent contact mode. A 568-nm laser beam from a krypton ion laser was focused from below onto the AFM tip via an oil immersion objective to excite the surface plasmons of the silver at the point of the tip apex. The sample was moved in xy-direction by a piezo-driven stage, while the enhancing tip was maintained in the laser focus, so that changing areas of the sample experienced the high evanescent field and TERS spectra could be recorded.[65,66] An example of a pseudo three-dimensional topographic image of single *S. epidermidis* cells on a glass surface is shown in Fig. 8.11b. The scanned area was $7 \times 7\ \mu m^2$ and the bacterial cells were visible as round features on the glass surface with a diameter of about 1 μm. When the silver-coated tip approached on the cells, high signal-to-noise-ratio TERS spectra could be recorded in only 1 second [Fig. 8.11(c), traces (A), (C), (D)], while a flat baseline was observed when the tip was on the glass substrate without cell [Fig. 8.11(c), trace (B)]. The enhanced Raman signal originates just from a very small area in the vicinity of the tip, corresponding roughly to the tip apex size. Taking this nominal tip radius into account, chemical information from the vibrational spectra of different spots on the cell can be resolved with a spatial resolution down to a few 50 nm. Due to the much higher spatial resolution, the chemical information contained in the TERS spectra results from much less chemical species and is confined to the outer most surface layer. In standard (confocal) Raman microspectroscopy a much larger volume element (circa $0.8 \times 0.8 \times 3\ \mu m^3$) is probed.

Most of the TERS bands of the staphylococcus cells (Fig. 8.11(c), traces (A), (C), (D)) can tentatively be assigned to contributions from peptides and polysaccharides. Protein contributions are evident for the amide I band around 1660 cm^{-1}, and N-acetyl-related bands (amide II) around 1533 to 1567 cm^{-1} and 1519 cm^{-1}. For the amide III band a wide spectral range is given in the literature, so that the bands between 1348 and 1187 cm^{-1} might have contributions of the amide III band from different amino acids. Bands at 1206 and 1198 are furthermore reported to contain major contributions from tyrosine and phenylalanine. Further peptide Raman bands are present with the C—C mode of phenylalanine at 1002 cm^{-1}, the C—C skeletal modes in proteins between 930 and 938 cm^{-1}, and the NCO deformation of tyrosine at 643 cm^{-1}. Raman bands of different carbohydrate moieties are found with the CH bending of the CH_2OH group around 1200 cm^{-1}, and the OH deformation vibration around 1340 cm^{-1} up to 1360 cm^{-1}

in oligo- and polysaccharides. The observed dominance of vibrational bands due to protein and sugar moieties is consistent with the known chemical composition of the staphylococcus surface. As for all Gram-positive bacteria the cytoplasmic membrane is surrounded by peptidoglycan, also known as murein, which is made up by linear chains of the two alternating amino sugars N-acetyl glucosamine (NAG) and N-acetyl muramic acid (NAM). The NAM chains are cross-linked by short (4 to 5 residues) amino acid chains. The peptidoglycan layer which is assumed to be around 50 nm thick is pervaded by other polysaccharides like teichoic acid and a variety of surface proteins. Furthermore, on the cell membrane there are also different catalysis centers of enzymes and anchoring and binding sites for adhesion on surfaces and cell recognition. Therefore, at the surface the cell exposes a mixture of sugar derivatives from the peptidoglycan layer and the teichoic acids and polysaccharide intercellular adhesin (PIA), and several different proteins.[67]

Tip-enhanced Raman spectroscopy bears a high potential for the noninvasive investigation of surfaces because it allows the recording of the topography of the investigated surface with highest spatial resolution below the diffraction limit, while at the same time rich chemical information from those surface structures can be obtained via vibrational spectra. Measuring times per spectrum can be kept very short (1 second or less), because the silver particle at the tip apex enhances the Raman signals. From the signal-to-noise ratio and the probed sample area the enhancement factor can be estimated to be around 10^4 to 10^5. If the reduced contact time between tip and sample due to the intermittent contact mode oscillations is taken into account enhancement factors of 10^6 to 10^8 can be calculated.[68] Very recently, the application of TERS was extended to the investigation of a single tobacco mosaic virus.[69] The development of fast identification techniques is another important research topic which takes advantage of the unique prospects of TERS.

8.5 Conclusions

This contribution described recent biomedical applications of Raman spectroscopy that have been reported since 2005. Raman images were collected from dried sections and nondried specimens of murine and human brain tumors, colon tissue and lung malformations. Raman microscopic imaging was also applied to study single cells at the subcellular level. Classification models were developed to determine the strain and species of bacteria based on Raman spectra. The topology and chemical composition of bacterial surfaces were probed by tip-enhanced Raman spectroscopy. Further progress of Raman spectroscopy in life sciences is expected within the next years. Dedicated miniaturized, biomedical, fiber-optic probes will enable to collect Raman spectra and images of tissue in vivo under minimal invasive conditions. Nonlinear enhancement effects such as CARS and stimulated Raman scattering microscopy[70] will improve the sensitivity which is the key to collect Raman images at real-time video-time frame rates. These techniques will not only provide label-free chemical contrast in biomedical imaging of tissue, but also of single cells. The diffraction limit of light can be overcome by coupling Raman spectroscopy with near-field microscopy. The most promising technique is TERS which combines signal enhancements in the close vicinity of metal nanoparticles and AFM tips for excitation and far-field microscopy for effective detection.

A major obstacle for the broader dissemination of Raman-based methods in the medical field is their technical complexity. Therefore, user-friendly Raman instruments are another important requirement. Besides the Raman systems for the detection and classification of microorganisms (see section "Excitation in the Visible Wavelength Range"), other Raman systems have been introduced for skin studies, e.g., to detect carotenoid levels (Pharmanex, United States) or to record depth concentration profiles (River Diagnostics, The Netherlands). Although in its early stages, these developments clearly demonstrate that Raman spectroscopy has the potential to be fully accepted as a complementary diagnostic tool for rapid and nondestructive in vitro, ex vivo, and in vivo analyses of tissues, cells and bacteria.

Acknowledgments

Financial support of the European Union via the Europäischer Fonds für Regionale Entwicklung (EFRE) and the "Thüringer Kultusministerium (TKM)" (project: B714-07037) is highly acknowledged.

References

1. S. W. Hell, "Microscopy and Its Focal Switch," *Nature Methods*, **6**:24–32, 2009.
2. C. Krafft and V. Sergo, "Biomedical Applications of Raman and Infrared Spectroscopy to Diagnose Tissues," *Spectroscopy*, **20**:195–218, 2006.
3. D. I. Ellis and R. Goodacre, "Metabolic Fingerprinting in Disease Diagnosis: Biomedical Applications of Infrared and Raman Spectroscopy," *Analyst*, **131**:875–885, 2006.
4. C. Krafft, G. Steiner, C. Beleites, and R. Salzer, "Disease Recognition by Infrared and Raman Spectroscopy," *Journal of Biophotonics*, **2**:13–28, 2008.
5. I. Notingher and L. L. Hench, "Raman Microspectroscopy: A Noninvasive Tool for Studies of Individual Living Cells In Vitro," *Expert Review of Medical Devices*, **3**:215–234, 2006.
6. R. J. Swain and M. M. Stevens, "Raman Microspectroscopy for Non-Invasive Biochemical Analysis of Single Cells," *Biochemical Society Transactions*, **35**:544–549, 2007.
7. C. Krafft, B. Dietzek, and J. Popp, "Raman and CARS Spectroscopy of Cells and Tissues," *Analyst*, **134**:1046–1052, 2009.
8. C. Krafft, M. Kirsch, C. Beleites, G. Schackert, and R. Salzer, "Methodology for Fiber-Optic Raman Mapping and FT-IR Imaging of Metastases in Mouse Brains," *Analytical and Bioanalytical Chemistry*, **389**:1133–1142, 2007.

9. C. Krafft, L. Neudert, T. Simat, and R. Salzer, "Near Infrared Raman Spectra of Human Brain Lipids," *Spectrochimica Acta A*, **61**:1529–1535, 2005.
10. C. Krafft, L. Shapoval, S. B. Sobottka, K. D. Geiger, G. Schackert, and R. Salzer, "Identification of Primary Tumors of Brain Metastases by SIMCA Classification of IR Spectroscopic Images," *Biochimica et Biophysica Acta*, **1758**:883–891, 2006.
11. C. Krafft, L. Shapoval, S. B. Sobottka, G. Schackert, and R. Salzer, "Identification of Primary Tumors of Brain Metastases by Infrared Spectroscopic Imaging and Linear Discriminant Analysis," *Technology in Cancer Research and Treatment*, **5**:291–298, 2006.
12. S. Koljenovic, T. C. Bakker Schut, R. Wolthuis, A. J. P. E. Vincent, G. Hendriks-Hagevi, L. F. Santos, J. M. Kros, and G. J. Puppels, "Raman Spectroscopic Characterization of Porcine Brain Tissue Using a Single Fiber-Optic Probe," *Analytical Chemistry*, **79**:557–564, 2007.
13. D. N. Louis, H. Ohgaki, O. D. Wiestler, W. K. Cavenee, P. C. Burger, A. Jouvet, B. W. Scheithauer, and P. Kleihues, "The 2007 WHO Classification of Tumours of the Central Nervous System," *Acta Neuropathologica*, **114**:97–109, 2007.
14. S. Koljenovic, L. P. Choo-Smith, T. C. Bakker Schut, J. M. Kros, H. van den Bergh, and G. J. Puppels, "Discriminating Vital Tumor from Necrotic Tissue in Human Glioblastoma Tissue Samples by Raman Spectroscopy," *Laboratory Investigation*, **82**:1265–1277, 2002.
15. S. Koljenovic, T. C. Bakker Schut, A. Vincent, J. M. Kros, and G. J. Puppels, "Detection of Meningeoma in Dura Mater by Raman Spectroscopy," *Analytical Chemistry*, **77**:7958–7965, 2005.
16. C. Krafft, S. B. Sobottka, G. Schackert, and R. Salzer, "Raman and Infrared Spectroscopic Mapping of Human Primary Intracranial Tumors: A Comparative Study," *Journal of Raman Spectroscopy*, **37**:367–375, 2006.
17. C. Krafft, S. B. Sobottka, G. Schackert, and R. Salzer, "Near Infrared Raman Spectroscopic Mapping of Native Brain Tissue and Intracranial Tumors," *Analyst*, **130**:1070–1077, 2005.
18. M. Köhler, S. Machill, R. Salzer, and C. Krafft, "Characterization of Lipid Extracts from Brain Tissue and Tumors Using Raman Spectroscopy and Mass Spectrometry," *Analytical and Bioanalytical Chemistry*, **393**:1513–1520, 2009.
19. S. B. Sobottka, K. D. Geiger, R. Salzer, G. Schackert, and C. Krafft, "Suitability of Infrared Spectroscopic Imaging as an Intraoperative Tool in Cerebral Glioma Surgery," *Analytical and Bioanalytical Chemistry*, **393**:187–195, 2009.
20. J. C. Taylor, C. A. Kendall, N. Stone and T. A. Cook. "Optical adjuncts for enhanced colonscopic diagnosis" *Br. J. Surg.* **94**: 6–16, 2007.
21. C. Krafft, D. Codrich, G. Pelizzo, and V. Sergo, "Raman and FTIR Microscopic Imaging of Colon Tissue: A Comparative Study," *Journal of Biophotonics*, **1**:154–169, 2008.
22. J. X. Cheng and X. S. Xie, "Coherent Anti-Stokes Raman Scattering Microscopy: Instrumentation, Theory, and Applications," *Journal of Physical Chemistry B*, **108**:827–840, 2004.
23. A. Volkmer, "Vibrational Imaging and Microspectroscopies Based on Coherent Anti-Stokes Raman Scattering Microscopy," *Journal of Physics D: Applied Physics*, **38**:R59–R81, 2005.
24. M. Müller and A. Zumbusch, "Coherent Anti-Stokes Raman Scattering Microscopy," *ChemPhysChem*, **8**:2156–2170, 2007.
25. C. Krafft, A. Ramoji, C. Bielecki, N. Vogler, T. Meyer, D. Akimov, P. Rösch, et al., "A Comparative Raman and CARS Imaging Study of Colon Tissue," *Journal of Biophotonics*, **2**:303–312, 2009.
26. C. Krafft, D. Codrich, G. Pelizzo, and V. Sergo, "Raman and FTIR Imaging of Lung Tissue: Methodology for Control Samples," *Vibrational Spectroscopy*, **46**:141–149, 2008.
27. C. Krafft, D. Codrich, G. Pelizzo, and V. Sergo, "Raman Mapping and FTIR Imaging of Lung Tissue: Congenital Cystic Adenomatoid Malformation," *Analyst*, **133**:361–371, 2008.
28. C. Krafft, D. Codrich, G. Pelizzo, and V. Sergo, "Raman and FTIR Imaging of Lung Tissue: Bronchopulmonary Sequestration," *Journal of Raman Spectroscopy*, **40**:595–603, 2009.

29. P. R. Jess, M. Mazilu, K. Dholakia, A. C. Riches, and C. S. Herrington, "Optical Detection and Grading of Lung Neoplasia by Raman Microspectroscopy," *International Journal of Cancer*, **124**:376–380, 2009.
30. S. Koljenovic, T. C. Bakker Schut, J. P. van Meerbeek, A. P. Maat, S. A. Burgers, P. E. Zondervan, J. M. Kros, and G. J. Puppels, "Raman Microspectroscopic Mapping Studies of Human Bronchial Tissue," *Journal of Biomedical Optics*, **9**:1187–1197, 2004.
31. M. Romeo, S. Boydston-White, C. Matthäus, M. Miljkovic, B. Bird, T. Chernenko, P. Lasch, and M. Diem, "Infrared and Raman Microspectroscopic Studies of Individual Human Cells" in: M. Diem, P. R. Griffiths, and J. M. Chalmers (eds.), *Vibrational Spectroscopy for Medical Diagnosis*, John Wiley & Sons, Chichester, 2008, pp. 27–70.
32. E. R. Hildebrandt, N. R. Cozzarelli, "Comparison of Recombination in vitro and in E. coli cells: measure of the effective concentration of DNA in vivo", *Cell*, **81**:331–340, 1995.
33. C. Krafft, T. Knetschke, R. H. Funk, and R. Salzer, "Studies on Stress-Induced Changes at the Subcellular Level by Raman Microspectroscopic Mapping," *Analytical Chemistry*, **78**:4424–4429, 2006.
34. C. Krafft, T. Knetschke, R. H. Funk, and R. Salzer, "Identification of Organelles and Vesicles in Single Cells by Raman Microspectroscopic Mapping," *Vibrational Spectroscopy*, **38**:85–93, 2005.
35. C. Matthäus, A. Kale, T. Chernenko, V. Torchilin, and M. Diem, "New Ways of Imaging Uptake and Intracellular Fate of Liposomal Drug Carrier Systems Inside Individual Cells, Based on Raman Microscopy," *Molecular Pharmacology*, **5**:287–293, 2008.
36. C. Krafft, T. Knetschke, A. Siegner, R. H. Funk, and R. Salzer, "Mapping of Single Cells by Near Infrared Raman Microspectroscopy," *Vibrational Spectroscopy*, **40**:240–243, 2009.
37. S. C. Erfurth and W. L. Peticolas, "Melting and Premelting Phenomenon in DNA by Laser Raman Scattering," *Biopolymers*, **14**:247–264, 1975.
38. J. D. Robertson and S. Orrenius, and B. Zhivotovsky, "Review: Nuclear Events in Apoptosis," *Journal of Structural Biology*, **129**:346–358, 2000.
39. B.R. Wood and D. McNaughton, "Resonance Raman Spectroscopy of Red Blood Cells Using Near-Infrared Laser Excitation," *Analytical and Bioanalytical Chemistry* **387**:1691–1703, 2007.
40. B. R. Wood and D. McNaughton, "Resonance Raman Spectroscopy in Malaria Research," *Expert Review of Proteomics*, **3**:525–544, 2006.
41. B. R. Wood, S. J. Langford, B. M. Cooke, F. K. Glenister, J. Lim, and D. McNaughton, "Raman Imaging of Hemozoin within the Food Vacuole of Plasmodium Falciparum Trophozoites," *FEBS Letters*, **554**:247–252, 2003.
42. T. Frosch, S. Koncarevic, L. Zedler, M. Schmitt, K. Schenzel, K. Becker, and J. Popp, "In Situ Localization and Structural Analysis of the Malaria Pigment Hemozoin," *Journal of Physical Chemistry B*, **111**:11047–11056, 2007.
43. A. Bonifacio, S. Finaurini, C. Krafft, S. Parapini, D. Taramelli, and V. Sergo, "Spatial Distribution of Heme Species in Erythrocytes Infected with Plasmodium Falciparum by Use of Resonance Raman Imaging and Multivariate Analysis," *Analytical and Bioanalytical Chemistry*, **392**:1277–1282, 2008.
44. T. Frosch, M. Schmitt, G. Bringmann, W. Kiefer, and J. Popp, "Structural Analysis of the Anti-Malaria Active Agent Chloroquine under Physiological Conditions," *Journal of Physical Chemistry B*, **111**:1815–1822, 2007.
45. I. Solomonov, M. Osipova, Y. Feldman, C. Baehtz, K. Kjaer, I. K. Robinson, G. T. Webster, et al., "Crystal Nucleation, Growth, and Morphology of the Synthetic Malaria Pigment β^2-Hematin and the Effect Thereon by Quinoline Additives: The Malaria Pigment as a Target of Various Antimalarial Drugs," *Journal of the American Chemical Society*, **129**:2615–2627, 2007.
46. T. Frosch, B. Kuestner, S. Schlücker, A. Szeghalmi, M. Schmitt, W. Kiefer, and J. Popp, "In Vitro Polarization-Resolved Resonance Raman Studies of the Interaction of Hematin with the Antimalarial Drug Chloroquine," *Journal of Raman Spectroscopy*, **35**:819–821, 2004.
47. G. T. Webster, L. Tilley, S. Deed, D. McNaughton, and B. R. Wood, "Resonance Raman Spectroscopy Can Detect Structural Changes In Haemozoin (Malaria

Pigment) Following Incubation with Chloroquine in Infected Erythrocytes," *FEBS Letters*, **582**:1087–1092, 2008.

48. M. Wenning, S. Scherer, and D. Naumann, "Infrared Spectroscopy in the Identification of Microorganisms," in: M. Diem, P. R. Griffiths, and J. M. Chalmers (eds.), *Vibrational Spectroscopy for Medical Diagnosis*, John Wiley & Sons, Chichester, 2008, pp. 71–96.

49. K. Maquelin, C. Kirschner, L. P. Choo-Smith, N. van den Braak, H. P. Endtz, D. Naumann, and G. J. Puppels, "Identification of Medically Relevant Microorganisms by Vibrational Spectroscopy," *Journal of Microbiological Methods*, **51**:255–271, 2002.

50. K. Maquelin, L.-P. I. Choo-Smith, T. Van Vreeswijk, H. P. Endtz, B. Smith, R. Bennett, H. A. Bruining, and G. J. Puppels, "Raman Spectroscopic Method for Identification of Clinically Relevant Microorganisms Growing on Solid Culture Medium," *Analytical Chemistry*, **72**:12–19, 2000.

51. R. Goodacre, E. M. Timmins, R. Burton, N. Kaderbhai, A. M. Woodward, D. B. Kell, and P. J. Rooney, "Rapid Identification of Urinary Tract Infection Bacteria Using Hyperspectral Whole-Organism Fingerprinting and Artificial Neural Networks," *Microbiology*, **144**:1157–1170, 1998.

52. D. Ivnitski, I. Abdel-Hamid, P. Atanasov, and E. Wilkins, "Biosensors for Detection of Pathogenic Bacteria," *Biosensors and Bioelectronics*, **14**:599–624, 1999.

53. K. Maquelin, L. P. Choo-Smith, H. P. Endtz, H. A. Bruining, and G. J. Puppels, "Rapid Identification of Candida Species by Confocal Raman Microspectroscopy," *Journal of Clinical Microbiology*, **40**:594–600, 2002.

54. P. Rösch, M. Harz, M. Krause, R. Petry, K.-D. Peschke, O. Ronneberger, H. Burkhardt, et al., "Online Monitoring and Identification of Bioaerosols (OMIB)," in: J. Popp, and M. Strehle (eds.), *Biophotonics: Vision for a Better Health Care*, Wiley-VCH, Weinheim, 2006, pp. 89–165.

55. P. Rösch, M. Harz, K.-D. Peschke, O. Ronneberger, H. Burkhardt, and J. Popp, "Identification of Single Eukaryotic Cells with Micro-Raman Spectroscopy," *Biopolymers*, **82**:312–316, 2006.

56. M. Harz, P. Rösch, K.-D. Peschke, O. Ronneberger, H. Burkhardt, and J. Popp, "Micro-Raman Spectroscopical Identification of Bacterial Cells of the Genus Staphylococcus in Dependence on Their Cultivation Conditions," *Analyst*, **130**:1543–1550, 2005.

57. P. Rösch, M. Harz, M. Schmitt, K.-D. Peschke, O. Ronneberger, H. Burkhardt, H.-W. Motzkus, et al., "Chemotaxonomic Identification of Single Bacteria by Micro-Raman Spectroscopy: Application to Clean Room Relevant Biological Contaminations," *Applied Environmental Microbiology*, **71**:1626–1637, 2005.

58. P. Rösch, M. Harz, K.-D. Peschke, O. Ronneberger, H. Burkhardt, A. Schüle, G. Schmautz, et al., "Online Monitoring and Identification of Bio Aerosols," *Analytical Chemistry*, **78**:2163–2170, 2006.

59. J. Popp and M. Strehle, *Biophotonics—Visions for Better Health Care*, Wiley-VCH Verlag GmbH & Co. KGaA, Weinheim, 2006.

60. M. Krause, B. Radt, P. Rösch, and J. Popp, "The Identification of Single Living Bacteria by a Combination of Fluorescence Staining Techniques and Raman Spectroscopy," *Journal of Raman Spectroscopy*, **38**:369–372, 2007.

61. N. Tarcea, M. Harz, P. Rösch, T. Frosch, M. Schmitt, H. Thiele, and J. Popp, "UV Raman Spectroscopy—a Technique for Biological and Mineralogical In Situ Planetary Studies," *Spectrochimica Acta A*, **68**:1029–1035, 2007.

62. K. Gaus, P. Rösch, R. Petry, K.-D. Peschke, O. Ronneberger, H. Burkhardt, K. Baumann, and J. Popp, "Classification of Lactic Acid Bacteria with UV-Resonance Raman Spectroscopy," *Biopolymers*, **82**:286–290, 2006.

63. R. M. Stöckle, Y. D. Suh, V. Deckert, and R. Zenobi, "Nanoscale Chemical Analysis by Tip-Enhanced Raman Spectroscopy," *Chemical Physics Letters*, **318**:131–136, 2000.

64. T. Deckert-Gaudig, E. Bailo, and V. Deckert, "Perspectives for Spatially Resolved Molecular Spectroscopy—Raman on the Nanometer Scale," *Journal of Biophotonics*, **1**:377–389, 2008.

65. A. Rasmussen and V. Deckert, "Surface- and Tip-Enhanced Raman Scattering of DNA Components," *Journal of Raman Spectroscopy*, **37**:311–317, 2006.

66. U. Neugebauer, P. Rösch, M. Schmitt, J. Popp, C. Julien, A. Rasmussen, C. Budich, V. Deckert, "On the Way to Nanometer-Sized Information of the Bacterial Surface by Tip-Enhanced Raman Spectroscopy," *ChemPhysChem*, 7:1428–1430, 2006.

67. U. Neugebauer, U. Schmid, K. Baumann, W. Ziebuhr, S. Kozitskaya, V. Deckert, M. Schmitt, and J. Popp, "Towards a Detailed Understanding of Bacterial Metabolism: Spectroscopic Characterization of Staphylococcus Epidermidis," *ChemPhysChem*, 8:124–137, 2007.

68. C. Budich, U. Neugebauer, J. Popp, and V. Deckert, "Cell Wall Investigations Utilizing Tip-Enhanced Raman Scattering," *Journal of Microscopy*, 229:533–539, 2008.

69. D. Cialla, T. Deckert-Gaudig, C. Budich, M. Laue, R. Möller, D. Naumann, V. Deckert, and J. Popp, "Raman to the Limit: Tip-Enhanced Raman Spectroscopic Investigations of a Single Tobacco Mosaic Virus," *Journal of Raman Spectroscopy*, 40:240–243, 2009.

70. C. W. Freudiger, W. Min, B. G. Saar, S. Lu, G. R. Holtom, C. He, J. C. Tsai, J. X. Kang, and X. S. Xie, "Label-Free Biomedical Imaging with High Sensitivity by Stimulated Raman Scattering Microscopy," *Science*, 322:1857–1861, 2008.

CHAPTER 9

The Current State of Raman Imaging in Clinical Application

Mariya Sholkina, Gerwin J. Puppels, and Tom C. Bakker Schut

Center for Optical Diagnostics and Therapy
Department of Dermatology
Erasmus Medical Center Rotterdam
The Netherlands

9.1 Introduction

Raman spectroscopy is a noninvasive, nondestructive optical technique that provides information about the composition, configuration, and interactions of the molecules in the measurement volume. Raman spectroscopy is a vibrational spectroscopic technique, like infrared absorption, which does not need labeling or other preparation of the sample. This makes Raman spectroscopy suitable for in vitro and in vivo measurements of living cells and tissues.[1]

The investigation of the behavior and metabolism of biomolecules in cells and tissues has become center to scientific fields such as medical, pharmaceutical, and microbiological diagnostics. Thorough understanding of (intra)cellular processes is necessary for the development of medical diagnostics, smart designed drugs, and in food and environmental technology. Standard techniques such as optical microscopy and fluorescence spectroscopy used to dominate the field of bioanalysis because of their ease of use and high sensitivity. However, all these methods suffer from a lack of specificity and reveal only little or no molecular information. Vibrational spectroscopic techniques like infrared (IR) absorption and Raman spectroscopy provide quantitative molecular specific information of the sample.

IR absorption has been very popular in the early periods of biomedical vibrational imaging due to its relatively high-signal strength, but Raman spectroscopy is currently becoming the method of choice because a higher-spatial resolution can be achieved, smaller sensitivity to artifacts and presence of water, and possibility of application in vivo. Technological improvements in the last two decades have helped to overcome the sensitivity problem of Raman spectroscopy and have made the technique broadly applicable in biomedical research.[2]

Because of the heterogeneous nature of cells and tissues, single point Raman microspectroscopy cannot adequately describe the chemical microstructure of cells and tissues. Also spatial information about molecular concentrations is needed. By measuring Raman spectra as a function of the position within the sample, a three-dimensional (x, y, and λ), or even four-dimensional dataset (x, y, z, and λ) can be obtained. Because Raman spectroscopy is an optical technique, the spatial resolution is determined by the diffraction limit.

From such a multidimensional spectral dataset, one can generate many Raman images, depending on which part of the spectral information is used for generating the contrast in the image. Studying the detailed molecular composition and dynamics of complex organized systems such as cells and tissues using Raman spectroscopic imaging is becoming increasingly popular. There is no need for dyes or labels, and one can look at different biochemical aspects of the sample simultaneously.[3,4]

Raman imaging has high potential as a clinical diagnostic technique that offers hematologists and pathologists spatially resolved molecular information on their patient material.[5–9] It can also be used by surgeons to determine whether the resection margins of excised tumors are free of cancerous cells. However, such applications of the technology are hampered by the time it currently takes to obtain a Raman image with a sufficient resolution and spectral quality.[10,11]

In this chapter an overview is given of the current state of the art in instrumentation and methodologies for biomedical Raman imaging. We have limited the technical overview to dispersive Raman instrumentation (leaving out FT Raman spectroscopy and techniques like CARS and SERS) as this measurement technique is most effective in collecting Raman photons and therefore in our view most suited for Raman imaging of biomedical samples, which is mostly signal intensity limited. Potential areas of application of Raman imaging in (bio) medical research and diagnosis are discussed and illustrated with examples from the recent literature. Lastly, current limitations in (clinical) application of the technique, and perspectives to overcome these limitations are reviewed.

9.1.1   History

Observing the wonderful blue opalescence of the Mediterranean Sea during a voyage to Europe in the summer of 1921, gave Sir

Chandrasekhara V. Raman the idea that this phenomenon owed its origin to the scattering of sunlight by the molecules of the water. After his return to Calcutta, he started investigations with his student Krishnan. In 1922, Raman published his first observations, ideas, and physical concepts as *Molecular Diffraction of Light*. In April 1923, he experimentally showed that "associated with the Rayleigh-Einstein type of molecular scattering, there was another and still feebler type of secondary radiation, the intensity of which was of the order of magnitude of a few hundredths of the classical scattering, and differed from it in not having the same wavelength as the primary or incident radiation." Many experiments followed, and later, in 1928 Raman developed the theory, that they observed the optical analogue of the Compton effect.[12] At the same time several other laboratories around the world (Rayleigh, Robert Wood, Landsberg, and Mandelstam) were investigating the same subject. The Russian scientist Mandelstam reported the effect in crystalline quartz and calcite, and called that inelastic light-scattering phenomenon "combinatorial scattering."[13] In 1930, Raman won the physics Nobel Prize for his work on the scattering of light and for the discovery of the Raman effect.

In 1975, Delhaye and Dhamelincourt introduced the first "Raman microprobe" or "Raman microscope" and outlined several approaches to Raman imaging.[14,15] In the last three decades,[16] with further development of lasers and detectors, the technology became much more sensitive and interest in using Raman spectroscopy in studies of complex biological systems and in biomedical applications strongly increased, leading to the first applications in cell and tissue studies.[17] The first Raman-based intracellular results were obtained in 1990–1991 when Puppels et al. developed a highly sensitive confocal Raman spectrometer enabling high-resolution single-cell studies.[18] Nowadays numerous applications of Raman spectroscopy on cells and tissues have been developed, both in vitro and in vivo.[19–24]

9.1.2 Principles

Light can interact with atoms and molecules in different ways. Photons can be absorbed (in some cases followed by emission of another photon, like in fluorescence), or they can be elastically or inelastically scattered, as can be depicted in the Jablonski diagram of Fig. 9.1. In the elastic scattering process there is no energy transfer between the light and the scattering molecules. The wavelength of the scattered light has the same frequency as the incoming light. This scattering process is known as Rayleigh scattering.

If the photon is inelastically scattered by the molecule, some energy is transferred from the photon to the molecule or vice versa. This energy is used to increase or decrease the vibrational energy of the molecule and the wavelength of the scattered light is different from the incident light. This scattering process is known as Raman scattering. If the vibrational energy of a molecule is increased, the scattered photon

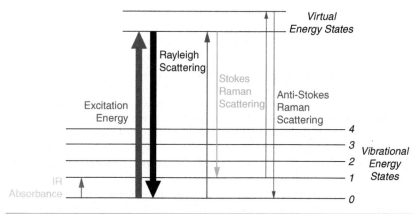

FIGURE 9.1   Jablonsky diagram of Rayleigh and Raman scattering.

loses some energy and will have a longer wavelength than the incident photon (Stokes shift). If the vibrational energy of a molecule is decreased, the scattered photon gains some energy and will have a shorter wavelength than the incident photon (anti-Stokes shift). The vibrations of a molecule in its surrounding can be described as harmonic oscillators: Each molecule has $3N$-6 independent vibrational modes, where N is the number of atoms of the molecule. As these vibrations are quantized, the molecules can acquire or lose discrete amounts of energy, dependent on the energy of the vibration. A Raman spectrum of a molecule is a representation of the emitted intensity as a function of vibrational modes energy and therefore is highly characteristic for a specific molecule in a specific surrounding.

The intensity of a Raman spectrum is linearly dependent on the concentration of molecules in the measurement volume. Raman spectrum of a sample with different volume concentrations of different molecules will be a linear combination of the Raman spectra of the different molecules times their volume concentration (apart from any molecular interaction effects).

Although Raman spectroscopy is a technique characterized by a low-signal intensity, as the probability of a Raman scattering event is about 1 to 10 million times lower than that of an elastic (Rayleigh) scattering event, and Raman spectra can be obscured by fluorescence effects, it can provide quantitative molecular information without destruction of the sample that makes it a powerful technique.

9.2   Instrumentation

Raman spectroscopic imaging combines spatial (structure) and spectral (chemical) information. For each point of the sample (x, y or x, y, z) a spectrum (λ) is measured. This is achieved by either scanning in spatial

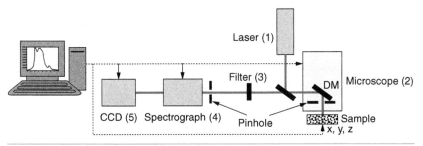

FIGURE 9.2   Confocal dispersive Raman microscopic setup.

direction(s) or by scanning the spectral range of interest. Figure 9.2 shows the basic setup for dispersive Raman spectroscopic imaging. The main elements are the laser, the microscope, the filter, the spectrograph and the detector.

9.2.1   Laser

The choice of the excitation wavelength is very important for Raman spectroscopic applications. The scattering efficiency of Raman scattering is dependent on the excitation wavelength with a power of $1/\lambda^4$. So UV and visible excitation will provide a more intense Raman signal than excitation with near-infrared (NIR) and IR wavelengths and are often used for imaging experiments of inorganic matters. Moreover, UV excitation will also lead to resonance enhancement if the wavelength of the exciting laser coincides with an electronic absorption of a molecule the intensity of Raman-active vibrations can be enhanced by a factor of 10^2 to 10^4.

In most biomedical applications, however, excitation in the UV or visible region results in a high-fluorescence background which can drown the Raman signal. In the deep UV (<250 nm) there is a fluorescence-free region extending over 4000 wavenumbers above the excitation wavelength providing very high-detection sensitivities and low-background noise. However, due to absorption UV and visible excitation can photodamage the biological sample, so the best choice is NIR (650 to 1400 nm) or IR excitation (1400 to 3000 nm). Both are much used for biomedical samples. But toward the IR the water absorption band in the spectrum starts to interfere strongly, resulting in lower scattering efficiency and heating of the sample.

Between 650 and 850 nm there is a window where there is little absorption by biochemical molecules (see Fig. 9.3). Therefore, the best choice for cell and tissue Raman investigation is NIR excitation. Here the absorption and background fluorescence are relatively low, and the detection efficiency of charge-coupled-device (CCD) detectors is still high. Many gas lasers like He-Ne (633 nm), Ar-Kr (647 nm), etc., have been used, but in recent years the NIR diode and solid-state lasers (671, 785, 830 nm, etc.) have become powerful and stable

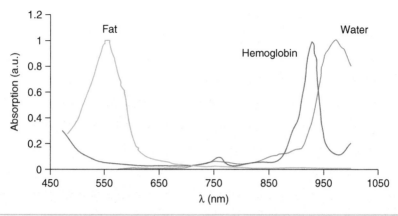

FIGURE 9.3   Absorption spectra of fat, water, and hemoglobin.

enough for Raman experiments, and they are small, cheap, and easy to operate compared to most gas lasers. In order to avoid photodamage of the sample, the excitation power is mostly limited to 10 to 50 mW when focusing the laser to a diffraction limited spot.

9.2.2   Microscope

In most Raman imaging instrumentation a confocal microscope is used to enable high resolution in the axial direction and to reduce the influence of fluorescence and other uninformative background signals. For scanning in spatial directions a motorized microscope stage and motorized objectives are used. Most commercially available microscopes can be easily adapted for Raman measurements because a laser can be coupled in through one of the auxiliary parts. As only a small part of the available spectrum is used for Raman, it can easily be combined with transmission or reflection imaging—by choosing appropriate dichroic mirrors and filters, they can even be used simultaneously, facilitating precise targeting in the sample.

The choice of objective is dependent on the application and excitation wavelength used. Mostly objective magnifications of 40 and higher are used. The objective should be well corrected for chromatic aberrations to prevent different depth dependent collection efficiency over the spectral range.

9.2.3   Filters

Rayleigh scattering of the sample and other back reflected laser light cause the light intensity at the laser wavelength to be stronger than the intensity of the Raman scattered light by many orders of magnitude. As this light can generate Raman signals in the measurement setup, the Rayleigh light needs to be filtered out as close to the sample as possible. The laser rejection filter should combine a strong suppression of light at the laser wavelength (preferably by 8 or more orders of magnitude) with a very high transmission of Raman signal, starting at wavelengths close to the laser line.

The first filters that were used were edge filters, which transmit low-wavelength side from excitation, Stokes Raman lines. Later holographic notch filters were made to suppress stray light and very effectively filter the Rayleigh line (10^6). They have a very high laser line rejection which enables measurements of bands close to the laser line, but they are sensitive to temperature and aging. The same high laser line rejection ratio can be achieved with a chevron-type filter, which is a combination of two dielectric high-pass filters, between which the light is reflected multiple times—in every pass the most of the signal reflects and the laser line is absorbed.[17,25] For global illumination, when the whole field of view is illuminated and the whole image is collected in a narrow spectral interval, commonly used filter types include dielectric, acousto-optic, and liquid-crystal tunable filters (LCTFs). Using a high-quality refractive microscope objective with LCTF-based Raman imaging systems provides sufficient magnification, so the resolution becomes diffraction limited (around 250 nm for 514 nm laser). Nowadays narrowband LCTFs (5 to 8 cm^{-1}) are available that allow high-spectral-resolution imaging.[26] However, for recording the whole spectrum, this type of imaging is very inefficient, since most generated Raman photons will be rejected by the filter. Therefore this technique seems only useful for band or band-ratio imaging.

9.2.4 Spectrometer

The standard configuration for a dispersive spectrometer is either a triple monochromator or a single monochromator in combination with high-quality laser line suppression filters. In a triple monochromator the first two stages are used for the elimination of the Rayleigh scattered light, whereas the third monochromator, the so-called spectrograph, disperses the collected Raman radiation onto a multichannel detector or a CCD camera. Then the Raman spectra can be recorded very close to the Rayleigh line (within a few wavenumbers), but the transmittance of such systems is low. Much higher collection efficiency can be obtained with a single grating monochromator in combination with a notch or chevron-type filter.

9.2.5 CCD

The detectors that are now used are multichannel detectors, either intensified diode arrays but in most cases CCDs. A standard silicon CCD is sensitive from 400 to 1000 nm and has a quantum efficiency (QE) in the NIR of around 40 percent. To make it more sensitive for NIR signal, deep depletion silicon is used. Nowadays back-thinned (or back-illuminated) deep depletion CCDs are used with a QE up to 80 percent. Cooling of the CCD (down to −70°C) helps to reduce the influence of heat generated noise (dark current). To reduce the noise generated in reading out the camera, slow readout speeds are used.

A next generation of CCD detectors that can be useful in Raman imaging is now emerging: the electron multiplied CCD (EMCCD). It has been available for some time for UV and VIS excitation, but now EMCCDs that have a better sensitivity in the NIR have become available. This type of detector is especially useful for high-speed, low-signal applications that are limited by the readout noise of the CCD. By amplification of the signal, the readout noise that increases with higher readout speed becomes much smaller compared to the signal noise (shot noise). However, the amplification process itself also generates noise and there is an additional (spurious) noise induced when shifting the charge toward the readout register, so for applications that are limited by signal noise this is not an improvement, although they may benefit from enabling a higher read out speed.

9.3   Imaging Techniques

Several approaches to Raman imaging have been invented to simultaneously record spatial and spectral information, as depicted in Figure 9.4. Raman imaging data consists of 2 spatial (x, y) and one spectral dimension—the Raman intensity as a function of Raman frequency. Most modern Raman imaging methods employ a multi-channel CCD that can record two dimensions of the three-dimensional information. Raman imaging systems can be differentiated by the way they collect the third dimension. Most Raman imaging systems use a spectrograph coupled to a CCD and a scanning stage for either two-dimensional point scanning, one-dimensional line scanning, or spatial multiplexing.[27]

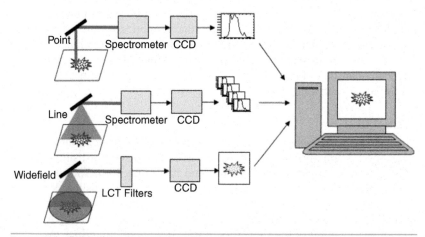

FIGURE 9.4   Point, line, and widefield Raman spectroscopic imaging.

Point and line Raman mapping, involve a laser spot or a laser line sample illumination. In point mapping, a laser spot is scanned in two spatial dimensions (x and y) with small step increments and a spectrum is collected at each pixel. The laser can be focused to a diffraction-limited spot size, and at such a high-spatial-resolution, spectra of subcellular organelles can be obtained with features dominated by only a single biochemical component like DNA or lipids. These point-to-point measurements may exhibit large spectral variation, and collection of hundreds or thousands of spectra is required to completely characterize a cellular system. Since a complete Raman spectrum is associated with each pixel, such maps can be used to generate detailed chemical images revealing the distribution of different molecular components in a cell. For many tissue-imaging applications such a high-spatial resolution is not required. If a lower-spatial resolution in the image plane is desired while maintaining the confocality, the laser spot or the stage can be moved during measurement, thus averaging out over the illuminated area.

Line mapping is an extension of point mapping that makes full use of a two-dimensional image sensor such as a CCD. By illuminating the sample with a line using a cylindrical lens or scanning mirror(s), the whole line can be imaged on the CCD- the spatial data are registered on the detector parallel to the entrance slit of the spectrometer, while the spectral information is dispersed perpendicularly.

In the direct (also called global or widefield) imaging approach, all spatial points of the Raman image at a specific wavenumber (range) are imaged simultaneously on the detector. In contrast to point mapping approaches, direct Raman imaging methods employ global or widefield sample illumination. The Raman scattered light from the sample is collected, filtered, and magnified onto a two-dimensional CCD detector. Commonly used filter types include dielectric, acousto-optic, and liquid-crystal tunable filters (LCTFs). Using high-quality refractive microscope objective with LCTF-based Raman imaging systems provides sufficient magnification, the resolution can be diffraction limited (around 250 nm for 514 nm laser, that is $d = 1.22\lambda/\text{NA}$).[27]

An interesting comparison between the different Raman imaging techniques was made by Schlucker et al.[28]; they implemented the three different Raman imaging modalities (point mapping, line mapping, and direct global imaging) within a single instrumental configuration that shares a similar optical path and the same microscope objective and CCD detector. As a well-defined test sample for use with all three techniques, they employed a scanning electron microscopy (SEM) standard consisting of a grid of 5-μm squares of silicon. All imaging was done with 514 nm excitation. A detailed experimental comparison was made of the various Raman imaging methodologies with a sample-limited excitation power, in which the advantages and limitations of each method based on their spectral SNR, spatial resolution, and data acquisition times were considered. In Table 9.1 the parameters and results are summarized.

	Point Mapping		Line Mapping		Global Mapping
	Sample limited	Laser limited	Sample limited	Laser limited	Sample limited
Power density (mW/cm^2)	0.008	13.5	0.008	0.096	0.008
Spectral resolution	4 cm^{-1}		4 cm^{-1}		7 cm^{-1}
Spatial resolution	1.1 µm		1.1 µm (h) 1.06 µm (h)		0.313 µm
Signal-to-noise ratio	29	9	77	40	14
Total acquisition time	11 h 24 min	1 h 16 min	12 min 34 s	3 min 47 s	46 min 22 s
Exposure time only	(61 × 61)(10) s = 10 h 20 min	(61 × 61)(0.1) s = 6 min 12 s	61(10) s = 10 min 10 s	61(1) s = 1 min 1 s	194(10) s = 32 min 20 s

TABLE 9.1 Parameters of Point, Line, and Global Raman Imaging

For point and line Raman mapping approaches, the duration of the experiment depends critically on the laser excitation power density and the number of points to be scanned. The experimental comparison demonstrates that line mapping represents the fastest method for acquiring spectral information at a reasonable spatial resolution, typically 1 µm, while yielding reconstructed Raman images of good quality.

In contrast to the two mapping approaches, the time needed for global imaging depends primarily on the number of spectral channels at which an entire image is recorded. Global Raman imaging employing LCTF technology is the method of choice for obtaining sub-micrometer, essentially diffraction-limited, spatially resolved high-quality images from flat samples. To characterize a sample's chemical heterogeneity, only a relatively few global Raman images need to be recorded at well-defined wavenumber positions, which are known either a priori or from a spectral analysis of data obtained in point or line scanning determinations.[28]

Further development of LCTF will lead to better transmission so smaller accumulation times for increased S/N ratio that will make that method more used in wider applications. However, this technique is not optimal for full spectral imaging as most generated Raman photons are rejected by the filter and only a small part is imaged on the detector.

There are other methods that try to enhance existing techniques. Hadamard transform imaging also employs global illumination, as in direct imaging. In this method, a series of images encoded with Hadamard transform mask (binary spatial light modulation) are recorded, and from these images high-definition and high-spatial-resolution Raman images can be reconstructed.[27]

Another approach is the fiber-bundle image compression method that uses a two-dimensional array of fibers to collect a Raman spectra from a grid of sample points. The fibers can then be arranged in a linear array and imaged on the CCD so that all three dimensions (two spatial and one spectral) are being measured simultaneously.[29] Although this is an attractive method to obtain immediate Raman images, the imaging on fiber bundles results in a rather large signal loss compared to normal line imaging.

Therefore, if one wants to obtain high-resolution Raman spectra with a high resolution, line imaging seems the most efficient method. As mentioned,[28] there are three parameters to consider, the signal acquisition time, the exposure time and the signal-to-noise ratio of the spectra. For low-intensity signals, the total signal exposure time in combination with the readout speed determines the amount of noise in the spectra. Therefore, longer integration times per pixel are used, in combination with a slow readout speed, resulting in long collection times per image. For high-intensity signals, as is the case in Hadamard transform imaging, the integration time can be much shorter and the readout speed higher. Although this seems attractive, it should be noted that in Hadamard transform imaging, as is the case in global illumination with a scanning filter, the generated Raman photons are not all detected and many are rejected by the mask or the filter, so these techniques can never be as efficient as the point or line imaging techniques and are only attractive if a high spatial resolution is required.

9.4   Data Analysis: Spectra to Image(s)

From a multidimensional spectral imaging dataset like a Raman map, one can generate many images, depending on which part of the spectral information is used for generating the contrast in the image. The simplest images use only the (scaled) intensities of a single wavenumber channel or a wavenumber band as a gray value (band imaging). One can extend this using the ratio of two Raman bands as gray value, or combine three Raman bands (or band ratios) to generate a RGB (false or pseudo) color image (band ratio imaging). In order to make use of all the information in the spectra at once, more powerful spectral data analysis methods, based on multivariate statistical techniques, can be applied.

For large images, multivariate analyses can become challenging in terms of computer power and analysis time because of the amounts of data that need to be processed. Because the information within a spectrum is correlated (intensities at different wavenumbers are not independent of each other), and because the spectra within a Raman map often have a high correlation with each other, it is in many cases useful to apply a data reduction and orthogonalization technique like principal components analysis (PCA).[30] PCA finds orthogonal directions (principal components or PCs) in the spectral data space that correspond with the major sources of variation within that dataset. The amount of variation captured in each PC diminishes with increasing PC number. Often the last PCs only represent noise components and can be omitted, which can give a significant data reduction and noise filtering. Each individual spectrum can be expressed as a set of scores on these PCs. The scores of all spectra on a certain PC can again be used as a gray value to generate an image, now using information from the whole spectrum to generate image contrast. With the scores of the first three PCs, being the PCs that represent the major sources of variation, one can generate an RGB image that has the highest spectral contrast information density possible, but this need not always be the most informative contrast possible.

Multivariate spectral data analysis techniques can be separated in two categories: *classification* techniques to group spectra on basis of spectral similarities and *quantification* techniques for (bio)chemical composition analyses. For each, two subcategories are distinguished: *supervised* and *unsupervised*. Supervised analysis makes use of models built on previously measured data, where the unsupervised models only use an algorithm working on the current dataset. Below, the basic principles of the currently used methods are described. For more extensive reviews on spectral data analysis methods the reader is referred to existing literature, since that is beyond the scope of this chapter.[31]

9.4.1   Classification Techniques

The currently most used technique to generate Raman images is cluster analysis. Cluster analysis is an unsupervised classification technique that identifies groups of spectra, based on their spectral similarity. Group membership can be color coded, giving similar spectra the same color. Different tissue structures and tissue pathologies will have a different biochemical makeup, which is reflected in their Raman spectra, and therefore the similarity of spectra within a certain structure will be greater than the similarity between spectra of two different structures. The resulting Raman image therefore shows tissue structures and pathologies as a pseudocolor image. The spectral information can further be used to analyze the difference between the mean spectra of different structures and pathologies and thus provide chemical information about these differences. Note that no spatial information is used in this type of analysis.

Cluster analysis results are dependent on both the clustering algorithm used, and on the similarity (or distance) measure used. There are two types of cluster algorithms: hierarchical and nonhierarchical, partitioning algorithms. Hierarchical algorithms like single linkage, complete linkage and Ward's method, calculate the similarity or distance between any two of the spectra in the Raman map (full similarity or distance matrix). For large maps this becomes a computational intensive procedure. Partitioning methods like k-means and ISODATA are less computationally demanding, because they do not require a full distance matrix. These algorithms assign all spectra of a map to the closest of a number of predefined spectral cluster centers (which, for instance, can be spectra from a number of different locations in the map). Then they iterate through all data points, updating the cluster centers as spectra are assigned to the clusters, until a stable solution is reached. Partitioning methods are generally faster for big Raman maps, but they are heuristic (result is dependent on the choice for the initial cluster centers) and not complete (the number of clusters has to be predefined, whereas hierarchical cluster results in a complete membership matrix, and the desired amount of clusters can be determined afterward).

If one wants to classify the spectra of an image into predefined groups, techniques like linear discriminant analysis[32] and artificial neural networks[33] can be used. These are supervised techniques that make use of a multivariate model based on a gold standard to classify the spectra that are used for building that model. The model then extracts the spectral features that are most useful to discriminate between these different groups, and this information can then be used to classify new spectra into one of the groups. Using these techniques automated analysis applications can be built which for instance can discriminate between tumor and non tumor cells.

9.4.2   Quantification Techniques

Quantification techniques are mostly supervised techniques; although there are also unsupervised techniques, like independent component analysis (ICA) and multivariate curve resolution (MCR), they are difficult to apply for generating Raman images because of the complexity of the spectra and the generally low signal-to-noise ratio of the spectra. Supervised quantification methods like least squares (LS) or partial least squares (PLS) can extract direct concentration information from the spectra. LS fitting is often used to determine and map the biochemical composition of cells.[34] The use of PLS to generate Raman images is not widespread but has been applied to predict the concentration of necrotic tissue in a section of brain tumor tissue.[35]

9.5   Raman Mapping and Imaging in Bioscience

In this paragraph a number of applications are reviewed that show the current state of art of (nonresonant Stokes shift) Raman imaging. This overview is in no way complete—the applications were chosen to illustrate the broad range of possible applications and to illustrate different methods to extract useful data from the images. The applications are divided into two sections: Secs. 9.5.1 and 9.5.2.

9.5.1   Single Cells

Raman spectroscopy provides a way of studying cellular processes on a molecular level and in vivo, without having to change the cellular interior and environment. Raman imaging adds spatial information to the chemical information, so that the molecular concentrations and changes therein can be localized, providing unique insight in the biochemical functioning of cells.[5,6,9,36–43]

For intracellular Raman imaging a confocal detection setup is used with a high-spatial resolution of 1 μm or less and a laser power in the order of 10 to 50 mW (although higher powers are used in the NIR without measurable damage). Because of the high-spatial resolution, spectra obtained from within a cell resemble pure compound spectra more than spectra obtained from whole tissues. Often the spectra are a simple mixture of a limited number of compounds.

To determine the biochemical composition of the cell, a "basic set" of single molecule Raman spectra is often used, consisting of the spectra of DNA, RNA, and some sugars, proteins, and lipids. An example of such a spectral basis set is given in next figures. Figure 9.5 shows part of the so called "fingerprint region" (from 400 to 1800 cm⁻¹, a very molecule specific spectral region) and Fig. 9.6 shows the high

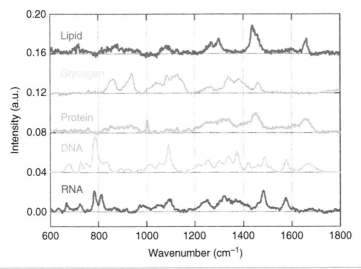

FIGURE 9.5 Spectra of the most prevalent biochemical components of cells (lipid, glycogen, protein, DNA, and RNA) in the low wavenumber fingerprint region. (*Reproduced from Ref. 44, with permission of the Elsevier Limited publisher.*)

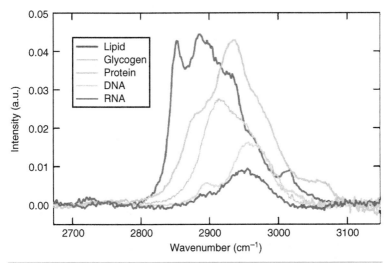

FIGURE 9.6 Spectra of the most prevalent biochemical components of cells (lipid, glycogen, protein, DNA, and RNA) in the high-wavenumber region. (*Reproduced from Ref. 44, with permission of the Elsevier Limited publisher.*)

wavenumber region (from 2600 to 3200 cm^{-1}, containing more general bands like CH and CH$_2$ stretching modes).[44]

With a simple fitting routine, using such a kind of "basis set," it is possible to estimate the relative amounts of biochemical components in cells and nuclei.

Many cell biological processes, such as division, differentiation, apoptosis, necrosis, and phagocytosis, are accompanied by large spatial reorganizations of the molecular components that constitute the cell. The chemical composition of the cell too can strongly change over the time. In work of Uzunbajakava et al.,[42] nonresonant confocal Raman point imaging has been used to map the DNA and the protein distributions in individual apoptotic cells. Raman imaging of a single cell took time about 15 to 20 minutes, with 0.5- to 2-second accumulation per pixel, and a resolution of 250 nm. They investigated the distribution of cytoplasmic and nuclear materials of the cells in a late stage of apoptosis using spectral band intensity images after PC filtering of the spectra. In these experiments a high laser power (647 nm) was used of 60 to 120 mW in a confocal spot (0.5 μm), but no damage to the fixed cells was observed.

In work by Yu et al.[45] confocal Raman microspectroscopic mapping (with an excitation wavelength of 785 nm, and 10 mW laser power on the sample) was used to investigate various HBE (human breast epithelia) cell lines (one normal or untransformed and three tumorigenic or transformed). Tumorigenesis is a genetic process and it is expected that the early cancer effects will be found in the nuclei. They extracted DNA, RNA, and proteins from the nuclei of HBE cells and developed spectra-fitting models in order to identify the compositional changes associated with tumorigenesis in cell nuclei via in vivo investigation of single cells. Their results showed that transformed and untransformed cells indeed show distinguishable differences in the Raman spectra of their nuclei.

High-spatial-resolution Raman mapping can reveal the structure and arrangement of subcellular organelles in detail and can provide insight into cellular biochemical dynamics.[46] For example, in a study by van Manen et al.[47] single-cell confocal Raman microscopy (with an excitation wavelength of 647 nm, and 100 mW laser power on the sample) was used to analyze the chemical composition of lipid bodies, cytoplasmic organelles rich in triglycerides and phospholipids that are associated with phagosomes in professional phagocytes. They found that incubation of phagocytes with AA and latex beads leads to the accumulation of lipid bodies, rich in esterified arachidonate, near the phagocytosed microspheres (Fig. 9.7).

Krafft et al.[39] compared two different commercially available Raman imaging instruments to investigate the chemical composition organelles and vesicles in single human lung fibroblast cells. The setups were both dispersive point mapping instruments, but had a different laser excitation wavelength and power, a different spectral and spatial resolution, and a different signal integration time per pixel. System 1 had 785 nm/100 mW excitation and 1 μm spatial resolution, system 2 had 532 nm/20 mW excitation and 0.3 μm resolution.

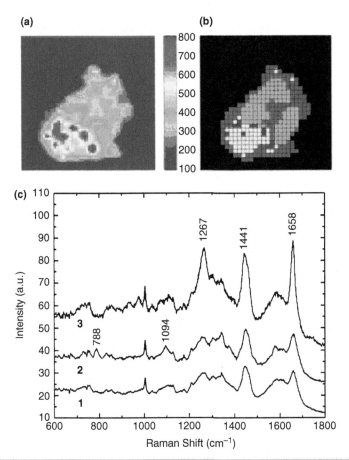

FIGURE 9.7 Formation of lipid bodies in neutrophils after incubation with 10 MAA for 1 hour. (*a*) Raman image (14.1 × 14.1 μm^2) in the 1635- to 1680-cm^{-1} region; (*b*) corresponding HCA image; (*c*) average Raman spectra extracted from the magenta (spectrum 1), green (spectrum 2), and blue (spectrum 3) clusters displayed in (*b*.) Strong bands at 1267, 1441, and 1658 cm^{-1} in spectrum 3 are assigned to lipids. (*From Ref. 47, with permission of the National Academy of Sciences, United State*s.)

Figure 9.8 shows the typical differences between the (cluster analysis) images generated by the two instruments and Fig. 9.9 shows typical spectra of single points in the map for the two instruments. The differences are clear: Where one instrument produces low-resolution images with high-quality spectra, and the other produces high-resolution images, but low-quality spectra. For each map the average spectra for each cluster were calculated. As shown in Fig. 9.9, the quality of these cluster-averaged spectra is very comparable for both instruments. In the same article two different multivariate statistical techniques were used to generate the images. An unsupervised cluster

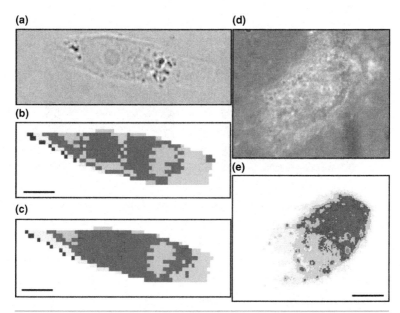

FIGURE 9.8   Photomicrographs of two lung fibroblast cells in buffer (*a*, *d*). Color-coded results of cluster analyses in the spectral range 3000 to 2700 cm^{-1} (*b*) and in the spectral range 1800 to 980 cm^{-1} (*c*) of Raman map (69 × 24, 1-μm step size). Color-coded result of cluster analysis in the spectral range 2700 to 3600 cm^{-1} (*e*) of Raman map (150 × 120, 0.3-μm step size). Bar = 10 μm. (*Figure provided by C. Krafft and adapted from Ref. 39.*)

analysis was performed to classify spectra according to their similarity, and cluster membership was used for color coding the points in the map (as shown in Fig. 9.8). The fitting procedure for determination of the subcellular composition also was used. The small fitting set of reference compounds contained calf thymus DNA, cholesterol, a lipid extract from brain and the protein bovine serum albumin. The resulting fit coefficients can be directly plotted as gray values to give compound distribution plots, as shown in Fig. 9.10. The fit results were only shown for the system with high-quality spectra.

The possibility of in vivo confocal Raman microspectroscopy and imaging is illustrated by a study of Naito et al.[34] They used a setup for imaging single cells (5 to 10 μm) that was equipped with a 633-nm laser, using 4 mW on the sample. The cells were imaged by point mapping with a 0.3 μm-spatial resolution with 1-second exposure time per point. It took around 9 minutes to image one cell (19 × 21 pixels).

Using this setup they studied the process of spontaneous death of a single budding yeast (*Saccharomyces cerevisiae*) cell.[34,50] A strong and sharp Raman band at 1602 cm^{-1} was occasionally observed from positions within the cytoplasm. It always appeared concomitant with the bands of phospholipids. As it sharply reflected the metabolic activity

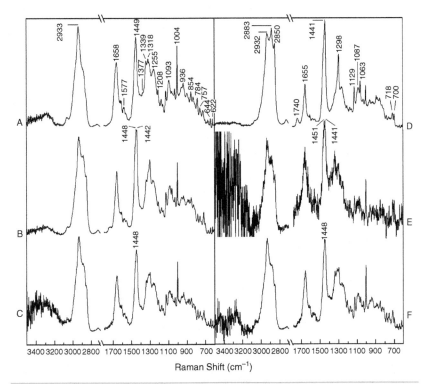

FIGURE 9.9   Cluster-averaged spectra from Raman map nucleus (A), cytoplasm (B), cytoplasm (C), vesicles (D), peripheral membrane (E), peripheral membrane (F). The intensity scale of region 1800 to 600 cm^{-1} is two times amplified compared to region 3500 to 2700 cm^{-1}. (*Figure provided by C. Krafft and adapted from Ref. 39.*)

of mitochondria in a living cell, Huang et al. called it "the Raman spectroscopic signature of life".[50] By using this signature, this group was able to quantitatively trace the process of cell growth and death at the molecular level. Time-resolved Raman images indicate that the mitochondrial metabolic activity of the cell is first markedly lowered and that the subsequent disappearance of the vacuole is followed by an inevitable cell death within a few hours (see Fig. 9.11).[34]

9.5.2   Tissues

Raman imaging has proven to be a powerful tool studying the molecular architecture of tissue. It has been extensively tested for its capability to detect and characterize diseases, tumors and other pathologies. Tissue imaging studies can also serve as a first step in developing in vivo applications by investigating what specific changes occur during disease, and where they are located, so that this information can be used to optimize in vivo probes and data analysis models. Clinically, Raman tissue imaging may be used by pathologists to help characterizing

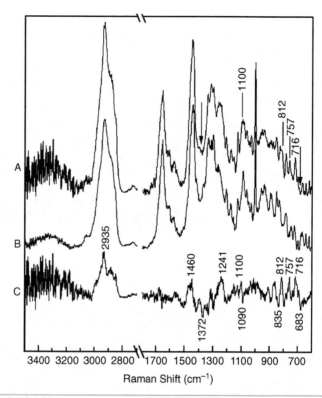

FIGURE 9.10   Cluster-averaged spectrum of inclusion body from Raman map in Fig. 9.8C (A), fit spectrum as linear composition of DNA, cholesterol, protein, and lipid (B) and difference spectrum (C = A – B). The intensity scale of region 1800 to 600 cm^{-1} is two times amplified compared to region 3500 to 2700 cm^{-1}. (*Figure provided by C. Krafft and adapted from Ref. 39.*)

tissues, grading tumors, identifying tissue inclusions, screening for rare cells etc. Clinical application of Raman spectroscopy is still in its infancy and hampered by the cost and lack of speed of most instruments.

For Raman tissue imaging a spatial resolution in the order of a (few) cell(s) in combination with a confocal detection setup is mostly used. Most instruments are equipped with a motorized stage in order to image large tissue sections. Point and line mapping methods are mostly used for tissue imaging experiments. A laser power of 50 to 100 mW is used to illuminate the sample. As most tissues exhibit high-fluorescent backgrounds when excited in the blue green part of the spectrum, most used excitation wavelengths are 633, 785, and 847 nm. For bone and teeth tissue shorter wavelengths are used typically 532 nm.

The data collection time for an image with reasonable spatial resolution and spectral quality is in the order of several hours to days. High-spatial-resolution Raman mapping is therefore only suitable for

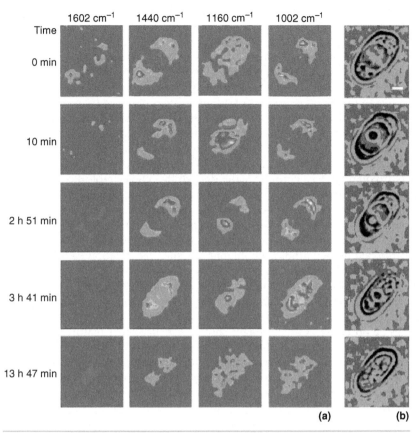

FIGURE 9.11 (a) Time-resolved Raman imaging and (b) the corresponding optical microscope images of a dying S. cerevisiae cell. Images of four Raman bands at 1602, 1440, 1160, and 1002 cm^{-1} are aligned in the same column. The length of the inserted bar is 1 μm (*From Ref. 34, reproduced with permission of John Wiley & Sons Ltd.*)

the analysis of fixed and/or dried cells and tissue. Fixing biological systems with alcohols or aldehydes is invasive, and can lead to distorted spectral features and artifacts. That pretreatment also eliminates two fundamental and attractive features of Raman microspectroscopy: noninvasive analysis and minimal sample preparation.[46]

To date Raman tissue imaging has been used for investigation of breast,[17,45,50–53] skin,[54–59] bone,[60–63] lung and bronchial,[48,64–66] brain,[35,67–69] bladder,[70] prostatic,[71] cervix,[72] oesophagus,[73,74] arteries,[75,76] and parathyroid.[77] Most of these applications are related to detection and characterization of (pre)cancers with the brain as primary research area, but also other diseases as obstruction of arteries or urinary tract have been studied with Raman imaging. In the field of cancer diagnosis and therapy, there are several problems that Raman imaging can help to resolve, e.g., the detection and characterization of neoplastic changes, discrimination of the cancer type and staging of the cancer, and the effects of drug an/or radiation therapy on cancer development.

Understanding the chemical basis of the tissue spectra can be very useful to extract diagnostic parameters from these spectra. Shafer-Peltier et al.[52] demonstrated that they could discriminate between macroscopic spectra of different types of breast tissue on by fitting them with a set of reference spectra derived from Raman imaging. The reference set of nine independent basis spectra consisted of various breast tissue morphological structures, such as the epithelial cell cytoplasm, cell nucleus, fat, β-carotene, collagen, calcium hydroxyapatite, calcium oxalate dehydrate, cholesterol-like lipid deposits, and water. The resulting fitting coefficients yielded the contribution of each basis spectrum to the macroscopic tissue spectrum, thereby elucidating the chemical/morphological makeup of the tissue.[58] The fitting coefficients were shown to be different for the different breast tissue pathologies, enabling tissue identification based on a thorough understanding of the chemical composition.

Raman images of tissue are often compared with hematoxylin and eosin (H&E)-stained images of the same or adjacent section that are routinely used in pathology. Although similar tissue structures can be observed, Raman images show more also other features that are not present in the H&E-stained sections. This is illustrated by the work of Koljenovic et al. on healthy human bronchial tissue.[66] Using big maps, up to 1.5×0.5 mm, with a high spectral (8 cm^{-1}) and spatial resolution (1 to 12 µm), she identified the chemical composition of different morphologic structures and provided the basis for further in vitro and in vivo investigations of the biochemical changes that accompany pathologic transformation of tissue Fig. 9.12. All maps were made with a point mapping system using 847 nm, 100 mW for excitation and 10 seconds signal collection time.

As mentioned above, much work has focused on the different aspects of cancer research. Much of this work is fundamental research on the differences between normal and (pre)cancerous tissue in order to develop clinical applications. Biochemical chances that occur can be spatially identified which is important for targeting the most discriminative areas using fiber-optic Raman probes.

An example of the kind of information that Raman imaging can provide is given in Fig. 9.13, published in a study by Amareff et al.[67] Using 785 nm, 160 mW laser light, focused to a spot of 3 to 4 µm, the sampled large areas (20 to 30 mm^2) of mouse glioma tissue with a relatively low resolution of 50 µm and 5 seconds signal integration time per spectrum. She showed that the cholesterol and phospholipid contents were highest in the corpus callosum and decreased gradually toward the cortex surface as well as in the tumor. Using a combination of subsequent k-means and hierarchical cluster analysis techniques, she demonstrated clear distinction between normal, tumor, two types of necrotic

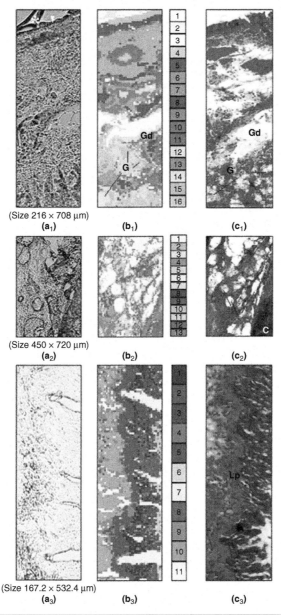

(Size 216 × 708 µm)
(a₁)   (b₁)   (c₁)

(Size 450 × 720 µm)
(a₂)   (b₂)   (c₂)

(Size 167.2 × 532.4 µm)
(a₃)   (b₃)   (c₃)

FIGURE 9.12   Raman microspectroscopic mapping experiments. (a_1) Microphotograph of an unstained section of bronchial wall. (b_1) A gross overview Raman map based on KCA using 16 clusters. (b_2) A pseudocolor map (480 spectra, 8-mm resolution) based on KCA using 13 clusters. (c_2) HE staining of the same tissue section showing secretory glands (G) and slightly blue stained cartilage (C). The gland secretions were washed out after HE procedure. (a_3) Unstained section of an area consisting of a part of the bronchial epithelium and a part of the submucosa. (b_3) Raman spectra (4598), obtained at a 4.4-mm resolution, were analyzed by KCA using 11 clusters resulting in a pseudocolor map. (c_3) HE-stained section with the clearly visible lamina propria (Lp) that separates bronchial epithelium from underlying submucosal tissue. (*From Ref. 66. Reproduced with the permission of SPIE publication.*)

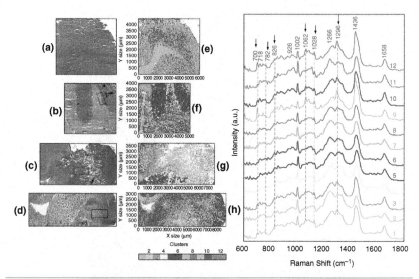

Figure 9.13   Photomicrographs (H&E staining) of healthy (*a*) and glioma (*b* to *d*) brain tissue sections. Pseudocolor Raman maps (*e* to *h*) are based on 12-means cluster analysis on sections *a* to *d*, respectively. As per contra, representative cluster-averaged Raman spectra collected from healthy and glioma brain tissue sections. Spectra are shown with the same color than in the pseudocolor maps (*e* to *h*). (*From Ref. 67, reproduced with permission of Elsevier.*)

tissue, and she demonstrated the presence of edematous tissues around the tumor.

Krafft et al.[68] showed that using fiber optic probes Raman images can be obtained in which tumor cells can be observed (Fig. 9.14). Using a commercially available Raman probe with a 90-mW 785-nm laser with a spot size of about 60 μm they imaged cross sections of mouse brains containing metastases of malignant melanomas with a resolution of about 120 μm and compared the results with Fourier transform infrared absorption maps of adjacent tissue. They found that spectral contributions of melanin in tumor cells were resonance enhanced in Raman spectra on excitation at 785 nm which enabled their sensitive detection in Raman maps. These metastatic cells of malignant melanomas could not be identified in FTIR images.

A very well developed field of Raman imaging application is the investigation of high-structured hard tissues, like bone and teeth. The mechanical properties of bone are influenced by a variety of material and structural properties, such as the tissue organization, the amount of mineral, and the orientation and cross-linking of the collagen component. All these structural aspects contribute to the quality of bone tissue and to the resulting biomechanical properties of the bone organ. Raman spectroscopy may be applied in a microspectroscopic fashion, enabling the determination of bone material properties at the

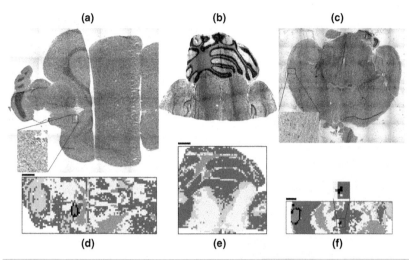

(a)   (b)   (c)

(d)   (e)   (f)

FIGURE 9.14   Raman mapping of pristine mouse brain tissue with metastases of malignant melanomas: (*d*) 37 × 93 map of front side and (*e*) 61 × 59 map of reverse side of the first specimen; (*f*) 88 × 26 map of the second specimen. H&E-stained tissue sections (*a*), (*b*), and (*c*) correspond to Raman maps (*d*), (*e*), and (*f*), respectively. Insets in (*a*) and (*c*) at higher magnification indicate the presence of pigmented tumor cells. Colors represent class assignments by cluster analysis-tissue of high (orange), medium high (yellow), medium low (cyan), and low (blue) lipid-to-protein ratios, and high (gray) and medium (black) concentrations of tumor cells. Bar = 1 mm. (*Figure provided by C. Krafft and adapted from Ref. 68.*)

microscopic level as a function of anatomical location and bone surface metabolic activity.[62,63]

In studies of bone diseases, Raman imaging has proven to be a very informative research tool. Much Raman imaging technique and analysis development was realized in the group of Michael Morris. A perfect example of their work is a study of Tarnowski et al.[63] on the earliest mineral and matrix changes in force-induced craniosynostosis (premature fusion of the skull bones at the sutures). Using a Raman line imaging system with a 785-nm diode laser (with 50 mW on the sample) they took Raman images of (mechanically) loaded and unloaded sutures and studied the changes over time. All Raman images were acquired with 2.6 × 2.6 μm spatial resolution and ranged in a rectangular field of view from 180 × 210 to 180 × 325 μm. A factor analysis technique, using also the spatial information in the images, helped identify the major components in the spectra such as the bone matrix and the mineral component. The results are shown in Fig. 9.15. It was the first work on studying musculoskeletal disease by Raman imaging, and they found that osteogenic fronts subjected to uniaxial compression had a decreased relative mineral content compared with unloaded osteogenic fronts, presumably because of new and incomplete mineral deposition at the osteogenic front across the suture as the suture was undergoing fusion.

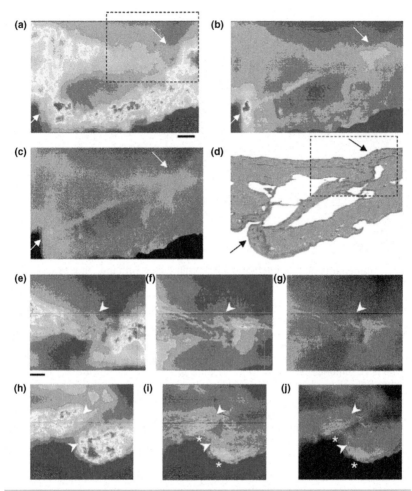

FIGURE 9.15   Raman score images and RGB images for 4-day loaded and unloaded specimens. The endocranial surfaces of the calvaria are at the bottom of the images. (a to g) Loaded specimen. (h to j) Unloaded specimen. (a, e, and h) Mineral score image. (b, f, and i) Matrix score image. Scale for Raman score images: red, high intensity; blue, low intensity. (c, g, and j) RGB contrast-enhanced images (blue areas indicate mineral presence; green areas indicate matrix presence; light blue-green areas indicate the presence of mineral and matrix). (d) Reference hematoxylin and eosin stained section from the same specimen imaged in a to g. The boxed area of the specimen shown in (a to d) was reimaged to show the osteogenic fronts (OF) on a smaller intensity scale and show more detail; the Raman score images are shown in (e to g). The loaded mineral score image shows a decreased amount of mineral at the OF (a and e), whereas the unloaded mineral score image (h) shows that the mineral is evenly dispersed throughout the unloaded calvaria. (b and f) Increased matrix is present beyond the mineral at the OFs in the loaded specimens. Arrows in (b to d) indicate areas of increased collagen production and cell proliferation in the loaded specimens [observed as increased intensity in the Raman matrix score image (b) and heavy purple staining in the stained section (d)]. (i and j) Increased protein presence is seen along the surfaces of the unloaded calvaria and especially along the endocranial surface. (From Ref. 62, with permission of the American Society for Bone and Mineral Research.)

(a)

Polarization
Direction

Interstitial Bone

Osteonal Bone

(b)

Polarization
Direction

Polarization
Direction

60 µm

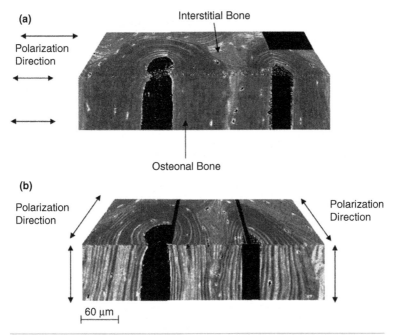

FIGURE 9.16   (a) Three-dimensional view of v1 PO_4/amide I ratio for different polarizations of the incident laser beam as indicated in the figure by double arrows. Same lamellae show different contrast depending on polarization direction of the beam in panel (b). (*From Ref. 63, copyright © 2007 Elsevier Inc., reproduced with permission of Elsevier.*)

Polarization information is an useful parameter when studying mineralized tissues like teeth and bone. This is nicely illustrated in work of Kazanci et al.[62] where bone sections were scanned in different ways to investigate the influence of polarization on Raman spectroscopic bone analysis. Using 532 nm excitation and 2 seconds signal integration time per point they collected images of up to 200×200 µm with a resolution of 1 µm and from these images they constructed three-dimensional images of the section by using band ratios as gray values. Figure 9.13 shows the (PO_4/amide I)-ratio image for the two different polarizations used for excitation and detection. Although the construction of this three-dimensional representation must have taken some time, it nicely illustrates the sort of extra information that can be obtained from polarization measurements.

9.6   Limitations and Perspectives

Raman spectroscopy has proven to be a powerful analytical technique for biomedical applications. Its noninvasive character and information-rich features make it an attractive modality to monitor biochemical dynamics and processes within living cells and tissues.

Raman spectroscopic imaging perfectly combines spatial and spectroscopic (biochemical) information and enables locating the origin of macroscopic Raman tissue signals and changes therein caused by disease. Biomedical application of Raman imaging is currently for a large part focused on more fundamental cancer research, where Raman spectroscopic imaging has shown to be a powerful research tool that can provide molecular information at subcellular resolution that is difficult to obtain by other methods. In this role its results can be used to develop optimal probes[10,43,73,78,79] and data-analysis models that can be applied clinically in vivo. The fact that acquisition of these maps may take hours to days is of minor importance.

For clinical in vitro application, Raman imaging should overcome its main limiting factor, the time needed to generate images that is useful in clinical diagnosis. For tissue section screening applications, one would like to have diffraction-limited spatial resolution (0.5 µm, or even higher using the accuracy of a motorized sample stage) in a total field of view in the order of a cm^2. This would result in a total image of 20.000×20.000 pixels, i.e., 400 million pixels. With a point mapping technique and 1 second signal collection time, this would take about 12.7 years. Most of this image is uninformative and will never be inspected, at least not in that resolution. When a pathologist inspects a stained tissue section, he starts with a low magnification, i.e., a low-spatial-resolution image (in the order of 50 µm). He/she then selects areas for further inspection at a higher magnification (order of 5 µm). Only if he wants to examine the way a cell is build up, he would use the highest resolution, i.e., 0.5 µm. This zooming ability is typically what an useful Raman imaging instrument also would need to provide. This means the ability to generate images in the order of 200×200 pixels, i.e., 40 thousand pixels with a pixel size of 0.5, 5, or 50 µm. With a point mapping technique and 1 second signal collection time this would take about 11.1 hours per image which is way long for practical applications. By using a line imaging technique to measure 128 pixels at a time, and by bringing the signal acquisition time back to 0.2 second (and make full use of noise filtering aspects of multivariate techniques), this time can be reduced to somewhat more than a minute, which comes close to something practical.

There are several ways to bridge this gap from research tool to clinical application. The instrumentation can still be improved and extended.[80–82] The amount of light that can be focused on a point is limited because of photodamage to the tissue. However, if spread, higher-laser powers can be used. New EMCCDs can help to lower the needed CCD readout time which is especially useful when short-acquisition times are used. Multiplexing techniques like Hadamard transform imaging can also help to reduce the influence of detector noise, but at the cost of not detecting all generated Raman photons.

With respect to data analysis of biomedical Raman images there is still much room for progress. Robust data-analysis techniques that make use of the full spectral information, and that are less sensitive to noise, can also help lower the needed spectral quality, and thus acquisition time, to extract the clinical information from the spectra. Although simple techniques like calculating band ratios can offer informative and easily interpretable images, the spectral quality that is needed to get a detailed image is much higher than for multivariate techniques that use the whole spectrum. Of the multivariate techniques, cluster analysis has become popular because of its easy of use and power to yield meaningful images without the need for any subjective input. However, techniques strain the large variations that are present in the data, which need not be the interesting ones. More sophisticated techniques exist that also use the spatial distribution of the spectra to obtain image segmentations, but these have not yet been extensively used in biomedical imaging.[83]

In conclusion, we think that Raman imaging can develop from a research tool into a widely used clinical diagnostic tool as many applications have shown (part of) its diagnostic potential. However, more technical progress is needed to reduce the acquisition time of tissue Raman images and better extract all the clinically relevant information from the multidimensional images. Considering the long way from the discovery of the Raman effect to its application in biomedical research, and considering the increasing interest in the technique, we expect that this can be realized in a relatively short period of time.

References

1. C. A. Owen, I. Notingher, R. Hill, M. Stevens, and L. L. Hench, "Progress in Raman Spectroscopy in the Fields of Tissue Engineering, Diagnostics and Toxicological Testing," *Journal of Material Science Materials in Medicine*, 17:1019–1023, 2006.
2. R. J. Meier, "Vibrational Spectroscopy: A Vanishing Discipline?" *Chemical Society Reviews*, 34:743–752, 2005.
3. R. Petry, M. Schmitt, and J. Popp, "Raman Spectroscopy—A Prospective Tool in the Life Sciences," *ChemPhysChem*, 4:14–30, 2003.
4. Y.-L. Wang, K. M. Hahn, R. F. Murphy, and A. F. Horwitz, "From Imaging to Understanding: Frontiers in Live Cell Imaging," *The Journal of Cell Biology*, 174(4):481–484, 2006.
5. J. W. Chan, D. S. Taylor, T. Zwerdling, S. M. Lane, K. Ihara, and T. Huser, "Micro-Raman Spectroscopy Detects Individual Neoplastic and Normal Hematopoietic Cells," *Biophysical Journal*, 90:648–656, 2006.
6. C. M. Krishna, "Combined Fourier Transform Infrared and Raman Spectroscopic Approach for Identification of Multidrug Resistance Phenotype in Cancer Cell Lines," *Biopolymers*, 82:462–470, 2006.
7. C. M. Krishna, G. D. Sockalingum, G. Kegelaer, S. Rubin, V. B. Kartha, and M. Manfait, "Micro-Raman Spectroscopy of Mixed Cancer Cell Populations," *Vibrational Spectroscopy*, 38:95–100, 2005.
8. A. Kudelski, "Analytical Applications of Raman Spectroscopy," *Talanta* (2008), doi:10.1016/j.talanta.2008.02.042.
9. K. Ramser, W. Wenseleers, S. Dewilde, S. Van Doorslaer, L. Moens, and D. Hanstorp, "A Combined Micro-Resonance Raman and Absorption Set-Up Enabling in Vivo Studies under Varying Physiological Conditions: The Nerve Globin in the Nerve Cord of Aphrodite Aculeate," *Journal of Biochemical and Biophysical Methods*, 70:627–633, 2007.

10. T. Vo-Dinh, P. Kasili, and M. Wabuyele, "Nanoprobes and Nanobiosensors for Monitoring and Imaging Individual Living Cells," *Nanomedicine: Nanotechnology, Biology, and Medicine*, **2:**22–30, 2006.

11. R. Wolthuis, T. C. Bakker Schut, P. J. Caspers, G. J. Puppels, H. P. J. Buschman, T. J. Römer, and H. A. Bruining, "Raman Spectroscopic Methods for In Vitro and in Vivo Tissue Characterization," *Fluorescent and Luminescent Probes*, 2d ed., Chap. 32, pp. 433–455, 1999.

12. C. V. Raman, "The Molecular Scattering of Light," *Nobel Lecture*, December 11, (1930).

13. F. Adar, "Evolution and Revolution of Raman Instruments," *Handbook of Raman Spectroscopy*, I. R. Lewis (ed.), 2001, Vol. 2, pp. 11–40.

14. M. M. Delhaye, "Optical Raman Microprobe with Laser." U.S. Patent 4195930, 1997.

15. M. Delhaye and P. Dhamelincourt, "Raman Microprobe and Microscope with Laser Excitation," *Journal of Raman Spectroscopy*, **3**(1):33–43, 1975.

16. M. Diem, M. Romeo, S. Boydston-White, M. Miljkovic, and C. Matthaus, "A Decade of Vibrational Micro-Spectroscopy of Human Cells and Tissue," *Analist*, **129:**880–885, 2004.

17. G. J. Puppels, J. Greve, R. J. H. Clark (ed.), "Whole Cell Studies and Tissue Characterization by Raman Spectroscopy," *Biomedical Application of Spectroscopy*, 1:1–47, 1996.

18. G. J. Puppels, F. F. M. De Mul, C. Otto, J. Greve, M. Robert-Nicoud, D. J. Arndt-Jovin, and T. M. Jovin, "Studying Single Living Cells and Chromosomes by Confocal Raman Microspectroscopy," *Nature (Lond.)*, **347:**301–303, 1990.

19. T. C. Bakker Schut, R. Wolthuis, P. J. Caspers, and G. J. Puppels, "Real-Time Tissue Characterization on the Basis of In Vivo Raman Spectra," *Journal of Raman Spectroscopy*, **33:**580–585, 2002.

20. B. F. Brehm-Stecher and E. A. Johnson, "Single-Cell Microbiology: Tools, Technologies, and Applications," *Microbiology and Molecular Biology Reviews*, 68(3):538–559, 2004.

21. D. Clark and S. Sasic, "Chemical Images: Technical Approaches and Issues," *Cytometry Part A*, **69A:**815–824, 2006.

22. E. V. Efremov, F. Ariese, and C. Gooijer, "Achievements in Resonance Raman spectroscopy Review of a Technique with a Distinct Analytical Chemistry Potential," *Analytica Chimica Acta*, **606:**119–134, 2008.

23. M. F. Escoriza, J. M. VanBriesen, S. Stewart, J. Maier, and P. J. Treado, "Raman Spectroscopy and Chemical Imaging for Quantification of Filtered Waterborne Bacteria," *Journal of Microbiological Methods*, **66:**63–72, 2006.

24. P. J. Lambert, A. G. Whitman, O. F. Dyson, and S. M. Akula, "Raman Spectroscopy: The Gateway into Tomorrow's Virology," *Virology Journal*, 3:51, 2006.

25. G. J. Puppels, "Confocal Raman Microspectroscopy," *Fluorescence and Luminescence Probes*, 2d ed., 22, 1999.

26. D. Tuschel, P. J. Treado, and J. E. Demuth, "Method for Raman Imaging of Semiconductor Materials." U.S. Patent 7123358, October 17, 2006.

27. P. J. Treado, and M. P. Nelson, "Raman Imaging," *Handbook of Raman Spectroscopy*, I. R. Lewis, (ed.), 2001, pp. 191–249.

28. S. Schlucker, M. D. Schaeberle, S. W. Huffman, and I. W. Levin, "Raman Microspectroscopy: A Comparison of Point, Line, and Wide-Field Imaging Methodologies," *Analytical Chemistry*, **75:**4312–4318, 2003.

29. A. D. Gift, J. Ma, K. S. Haber, B. L. McClain, and D. Ben-Amotz, "Near-Infrared Raman Imaging Microscope Based on Fiber-Bundle Image Compression," *Journal of Raman Spectroscopy*, 30:757–765, 1999.

30. I. T. Jolliffe, "Principal Component Analysis," *Series: Springer Series in Statistics*, 2d ed., Vol. XXIX, 2002, p. 487.

31. I. Notingher, G. Jell, P. L. Notingher, I. Bisson, O. Tsigkou, J. M. Polak, M. M. Stevens, and L. L. Hench, "Multivariate Analysis of Raman Spectra for In Vitro Non-Invasive Studies of Living Cells," *Journal of Molecular Structure*, **744–747:**179–185, 2005.

32. B. G. Tabachnik, and L. S. Fidell, *Using Multivariate Statistics*, 5th ed., Harper & Row, New York, 2006.

33. T. Udelhoven, M. Novozhilov, and J. Schmitt, "The NeuroDeveloper®: A Tool for Modular Neural Classification of Spectroscopic Data," *Chemometrics and Intelligent Laboratory System*, 66(2):219–226, 2003.
34. Y. Naito, A. Toh-e, and H.-o. Hamaguchi, "In Vivo Time-Resolved Raman Imaging of a Spontaneous Death Process of a Single Budding Yeast Cell," *Journal of Raman Spectroscopy*, 36:837–839, 2005.
35. S. Koljenovic, L. P. Choo-Smith, T. C. Bakker Schut, J. M. Kros, H. J. van den Berge, and G. J. Puppels, "Discriminating Vital Tumor from Necrotic Tissue in Human Glioblastoma Tissue Samples by Raman Spectroscopy," *Laboratory Investigation*, 82(10):1265–1277, 2002.
36. C. M. Creely, G. Volpe, G. P. Singh, M. Soler, and D. V. Petrov, "Raman Imaging of Floating Cells," *Optics Express*, 13(16):6105–6110, 2005.
37. P. R. T. Jess, V. Garce s-Chavez, A. C. Riches, C. S. Herrington, and K. Dholakia, "Simultaneous Raman Micro-Spectroscopy of Optically Trapped and Stacked Cells," *Journal of Raman Spectroscopy*, 38:1082–1088, 2007.
38. C. Krafft, T. Knetschke, A. Siegner, R. H. W. Funk, and R. Salzer, "Mapping of Single Cells by Near Infrared Raman Microspectroscopy," *Vibrational Spectroscopy*, 32:75–83, 2003.
39. C. Krafft, T. Knetschke, R. H. W. Funk, and R. Salzer, "Identification of Organelles and Vesicles in Single Cells by Raman Microspectroscopic Mapping," *Vibrational Spectroscopy*, 38:85–93, 2005.
40. G. J. Puppels, H. S. P. Garritsen, G. M. J. Segers-Nolten, F. F. M. de Mul, and J. Greve, "Raman Microspectroscopic Approach to the Study of Human Granulocytes," *Biophysical Journal*, 60:1046–1056, 1991.
41. K. W. Short, S. Carpenter, J. P. Freyer, and J. R. Mourant, "Raman Spectroscopy Detects Biochemical Changes due to Proliferation in Mammalian Cell Cultures," *Biophysical Journal*, 88:4274–4288, 2005.
42. N. Uzunbajakava, A. Lenferink, Y. Kraan, E. Volokhina, G. Vrensen,y J. Greve, and C. Otto, "Nonresonant Confocal Raman Imaging of DNA and Protein Distribution in Apoptotic Cells," *Biophysical Journal*, 84:3968–3981, 2003.
43. Y. S. Yamamoto, Y. Oshima, H. Shinzawa, T. Katagiri, Y. Matsuura, Y. Ozaki, and H. Sato, "Subsurface Sensing of Biomedical Tissues Using a Miniaturized Raman Probe: Study of Thin-Layered Model Samples," *Analytica Chimica Acta* 619(1):8–13, 2008.
44. J. Mourant, P. Kunapareddy, S. Carpenter, and J. P. Freyer, "Vibrational Spectroscopy for Identification of Biochemical Changes Accompanying Carcinogenesis and the Formation of Necrosis," *Gynecologic Oncology*, 99:S58–S60, 2005.
45. C. Yu, E. Gestl, K. Eckert, D. Allara, and J. Irudayaraj, "Characterization of Human Breast Epithelial Cells by Confocal Raman Microspectroscopy," *Cancer Detection and Prevention*, 30:515–522, 2006.
46. S. Swain, "Raman Microspectroscopy for Non-Invasive Biochemical Analysis of Single Cells," *Biochemical Society Transaction*, 35:544–549(3), 2007.
47. H.-J. van Manen, Y. M. Kraan, D. Roos, and C. Otto, "Single-Cell Raman and Fluorescence Microscopy Reveal the Association of Lipid Bodies with Phagosomes in Leukocytes," *PNAS*, 102(29):10159–10164, 2005.
48. C. Krafft, D. Codrich, G. Pelizzo, and V. Sergo, "Raman and FTIR Imaging of Lung Tissue: Methodology for Control Samples," *Vibrational Spectroscopy*, 46:141–149, 2008.
49. A. S. Haka, K. E. Shafer-Peltier, M. Fitzmaurice, J. Crowe, R. R. Dasari, and M. S. Feld, "Diagnosing Breast Cancer by Using Raman Spectroscopy," *PNAS*, 102(35):12371–12376, 2005.
50. Y.-S. Huang, T. Karashima, M. Yamamoto, T. Ogura, and Hiro-o. Hamaguchi, "Raman Spectroscopic Signature of Life in a Living Yeast Cell," *Journal of Raman Spectroscopy*, 35:525–526, 2004.
51. R. E. Kast, G. K. Serhatkulu, A. Cao, A. K. Pandya, H. Dai, J. S. Thakur, V. M. Naik, et al., "Raman Spectroscopy Can Differentiate Malignant Tumor from Normal Breast Tissue and Detect Early Neoplastic Changes in a Mouse Model," *Biopolymers*, 89(3):235–241, 2008.

52. J. Kneipp, T. B. Schut, and G. Puppels, "Characterization of Breast Duct Epitelia," *Vibrational Spectroscopy*, 32:67–74, 2003.
53. K. E. Shafer-Peltier, A. S. Haka, M. Fitzmaurice, J. Crowe, J. Myles, R. R. Dasari, and M. S. Feld, "Raman Microspectroscopic Model of Human Breast Tissue: Implications for Breast Cancer Diagnosis In Vivo," *Journal of Raman Spectroscopy*, 33:552–563, 2002.
54. P. J. Caspers, G. W. Lucassen, and G. J. Puppels, "Combined In Vivo Confocal Raman Spectroscopy and Confocal Microscopy of Human Skin," *Biophysical Journal*, 85:572–580, 2003.
55. M. Gniadecka, P. A. Philipsen, S. Sigurdsson, S. Wessel, O. F. Nielsen, D. H. Christensen, J. Hercogova, et al., "Melanoma Diagnosis by Raman Spectroscopy and Neural Networks: Structure Alterations in Proteins and Lipids in Intact Cancer Tissue," *Journal of Investigative Dermatology*, 122:443–449, 2004.
56. A. Nijssen, T. C. Bakker Schut, F. Heule, P. J. Caspers, D. P. Hayes, M. H. A. Neumann, and G. J. Puppels, "Discriminating Basal Cell Carcinoma from its Surrounding Tissue by Raman Spectroscopy," *Journal of Investigative Dermatology*, 11:964–969, 2002.
57. G. N. Stamatas, J. de Sterke, M. Hauser, O. von Stetten, and A. van der Pol, "Lipid Uptake and Skin Occlusion Following Topical Application of Oils on Adult and Infant Skin," *Journal of Dermatological Science*, 50(2):135–142, 2008.
58. G. Zhang, D. J. Moore, C. R. Flach, and R. Mendelsohn, "Vibrational Microscopy and Imaging of Skin: From Single Cells to Intact Tissue," *Analytical and Bioanalytical Chemistry*, 387:1591–1599, 2007.
59. G. Zhang, C. R. Flach, and R. Mendelsohn, "Tracking the Dephosphorylation of Resveratrol Triphosphate in Skin by Confocal Raman Microscopy," *Journal of Controlled Release*, 123(2):141–147, 2007.
60. A. Carden, R. M. Rajachar, M. D. Morris, and D. H. Kohn, "Ultrastructural Changes Accompanying the Mechanical Deformation of Bone Tissue: Raman Imaging Study," *Calcified Tissue International*, 72:166–175, 2003.
61. K. Golcuk, G. S. Mandair, A. F. Callender, N. Sahar, D. H. Kohn, and M. D. Morris, "Is Photobleaching Necessary for Raman Imaging of Bone Tissue Using a Green Laser?" *Biochimica et Biophysica Acta*, 1758:868–873, 2006.
62. M. Kazanci, H. D. Wagner, N. I. Manjubala, H. S. Gupta, E. Paschalis, P. Roschger, and P. Fratzl, "Raman Imaging of Two Orthogonal Planes within Cortical Bone," *Bone*, 41:456–461, 2007.
63. C. P. Tarnowski, M. A. Ignelzi Jr., W. Wang, J. M. M. Taboas, S. A. Goldstein, and M. D. Morris, "Earliest Mineral and Matrix Changes in Force-Induced Musculoskeletal Disease as Revealed by Raman Microspectroscopic Imaging," *Journal of Bone and Mineral Research*, 19(1):64–71, 2004.
64. T. C. Bakker Schut, G. J. Puppels, Y. M. Kraan, J. Greve, L. L. J. van der Maas, and C. G. Figdor, "Intracellular Cartenoid Levels Measured by Raman Microspectroscopy: Comparison of Lymphocytes from Lung Cancer Patients and Healthy Individuals," *International Journal of Cancer (Pred. Oncol.)*, 74:20–25, 1997.
65. Z. Huang, A. Mcwilliams, H. Lui, D. I. Mclean, S. Lam, and H. Zeng, "Near-Infrared Raman Spectroscopy for Optical Diagnosis of Lung Cancer," *International Journal of Cancer*, 107:1047–1052, 2003.
66. S. Koljenovic and T. C. Bakker Schut, "Raman Microspectroscopic Mapping Studies of Human Bronchial Tissue," *Journal of Biomedical Optics*, 9(6):1187–1197, 2004.
67. N. Amharref, A. Beljebbar, S. Dukic, L. Venteo, L. Schneider, M. Pluot, and M. Manfait, "Discriminating Healthy from Tumor and Necrosis Tissue in Rat Brain Tissue Samples by Raman Spectral Imaging," *Biochimica et Biophysica Acta*, 1768:2605–2615, 2007.
68. C. Krafft, M. Kirsch, C. Beleites, G. Schackert, and R. Salzer, "Methodology for Fiber-Optic Raman Mapping and FTIR Imaging of Metastases in Mouse Brains," *Analytical and Bioanalytical Chemistry*, 389:1133–1142, 2007.
69. R. Rabah, R. Weber, G. K. Serhatkulu, A. Cao, H. Dai, A. Pandya, R. Naik, G. Auner, J. Poulik, and M. Klein, "Diagnosis of Neuroblastoma and Ganglioneuroma Using Raman Spectroscopy," *Journal of Pediatric Surgery*, 43:171–176, 2008.

70. B. W. D. de Jong, T. C. Bakker Schut, J. Coppens, K. P. Wolffenbuttel, D. J. Kok, and G. J. Puppels, "Raman Spectroscopic Detection of Changes in Molecular Composition of Bladder Muscle Tissue Caused by Outlet Obstruction," *Vibrational Spectroscopy*, **32**(1):57–65, 2003.

71. P. Crow, B. Barrass, C. Kendall, M. Hart-Prieto, M. Wright, R. Persad, and N. Stone, "The Use of Raman Spectroscopy to Differentiate between Different Prostatic Adenocarcinoma Cell Lines," *British Journal of Cancer*, **92**:2166–2170, 2005.

72. C. M. Krishna, G. D. Sockalingum, B. M. Vadhiraja, K. Maheedhar, A. C. K. Rao, L. Rao, L. Venteo, et al., "Vibrational Spectroscopy Studies of Formalin-Fixed Cervix Tissues," *Biopolymers*, **85**(3):214–221, 2006.

73. I. A. Boere, T. C. Bakker Schut, J. van den Boogert, R. W. F. de Bruin, and G. J. Puppels, "Use of Fibre Optic Probes for Detection of Barrett's Epithelium in the Rat Oesophagus by Raman Spectroscopy," *Vibrational Spectroscopy*, **32**:47–55, 2003.

74. C. Kendall, N. Stone, N. Shepherd, K. Geboes, B. Warren, R. Bennett, and H. Barr, "Raman Spectroscopy, a Potential Tool for the Objective Identification and Classification of Neoplasia in Barrett's Oesophagus," *Journal of Pathology*, **200**:602–609, 2003.

75. S. W. E. van de Poll, K. Kastelijn, T. C. Bakker Schut, C. Strijder, G. Pasterkamp, G. J. Puppels, and A. van der Laarse, "On-Line Detection of Cholesterol and Calcification by Catheter Based Raman Spectroscopy in Human Atherosclerotic Plaque Ex Vivo," *Heart*, **89**:1078–1082, 2003.

76. S. W. E. van de Poll, T. C. Bakker Schut, A. van der Laarse, and G. J. Puppels, "In Situ Investigation of the Chemical Composition of Ceroid in Human Atherosclerosis by Raman Spectroscopy," *Journal of Raman Spectroscopy*, **33**:544–551, 2002.

77. K. Das, N. Stone, C. Kendall, C. Fowler, and J. Christie-Brown, "Raman Spectroscopy of Parathyroid Tissue Pathology," *Lasers in Medical Science*, **21**:192–197, 2006.

78. L. Fu and M. Gu, "Fibre-Optic Nonlinear Optical Microscopy and Endoscopy," *Journal of Microscopy*, **226**(3):195–206, 2007.

79. T. D. Wang and J. van Dam, "Optical Biopsy: A New Frontier in Endoscopic Detection and Diagnosis," *Clinical Gastroenterology and Hepatology*, **2**(9):744–753, 2004.

80. T. Dieing and O. Hollricher, "High Resolution, High Speed Confocal Raman Imaging," *Vibrational Spectroscopy*, **48**(1):22–27, 2008.

81. L. Duponchel, P. Milanfar, C. Ruckebusch, and J.-P. Huvenne, "Super-Resolution and Raman Chemical Imaging: From Multiple Low Resolution Images to a High Resolution Image," *Analytica Chimica Acta*, **607**:168–175, 2008.

82. G. Evans, "High Quality Imaging and Spectrographic Data Available in Minutes, Not Hours," *Renishaw Biophotonics International*, **2**:28–32, 2008.

83. H. Grahn and P. Geladi (eds), *Techniques and Applications of Hyperspectral Image Analysis*, Wiley, Chichester, 2007, ISBN: 978-0-470-01086-0.

CHAPTER 10

Vibrational Spectroscopic Imaging of Microscopic Stress Patterns in Biomedical Materials

Giuseppe Pezzotti

Ceramic Physics Laboratory and Research Institute for
  Nanoscience
Kyoto Institute of Technology
Kyoto, Japan

The Center for Advanced Medical Engineering and Informatics
Osaka University
Osaka, Japan

In the technological practice, the evaluation of the quality of biomaterials comprises a number of standardized parametric components that contribute to material structural reliability but are only partially captured by measurements of chemical and structural characteristics. Advances in Raman spectroscopy in recent years have provided new insights into the micromechanical behavior of biomaterials, including the origin of improved fracture toughness in natural and synthetic inorganic biomaterials and the visualization of residual stress patterns stored on load bearing surfaces. The translation of these notions into biomedical practice is an important priority for future research and it may eventually lead to better prediction of fracture risks and to an improved understanding of the mechanisms by which pharmacological interventions affect strength in natural biomaterials.

This chapter describes some microscopic aspects of internal stress states in which biomaterials quality is directly associated with increased fracture risk, sometimes despite improved chemical purity and structural design. Raman techniques for the measurement of biomaterials quality will be described and discussed together with how these have advanced our understanding of the microscopic mechanisms by which biomaterials strength may be improved.

10.1   Introduction

Understanding modern biomedical materials requires a multidisciplinary approach involving characterizations by different analytical approaches and advanced spectroscopic techniques. Studies of new processing procedures directed at the preparation of biocompatible materials caused an increased interest in natural biomaterials and in new investigation methods that allow analyzing their behavior in order to rationally explain their unique properties. Traditionally, the applicability of implant materials in medicine is determined by testing procedures that thoroughly examine structural reliability, biocompatibility, cellular response, and lifetime in the human body. In case of inorganic materials for bone or joint replacement, it is necessary to complement biological methods with mechanical studies of structural resistance and durability in the presence of a biological environment. These studies should also give information about phase composition and its interaction with structural properties, as well as about the kinetics of fracture propagation. Vibrational spectroscopy allows studying the processes occurring in biomaterials under strain and analyzing the interaction between implant and biological systems.[1,2] Vibrational spectroscopy methods have been extensively used for in vitro and in vivo investigations of degradation mechanisms and kinetics of several biomedical devices;[3-5] they have also been used for characterizing the crystalline and amorphous domains in biomineralization processes.[6,7] In particular, Raman spectroscopy has proved to be a valuable tool in the field of biomaterials engineering, allowing the study of the processes occurring during their preparation and use.[8] When coupled with an optical microscope, Raman spectroscopy has been widely applied to determine phase compositions,[9] to analyze the degree of crystallinity in bone tissue,[7] and the rate of tissue regeneration.[10] Raman mapping on the microscopic scale is very helpful in the study of biomaterial surfaces and events at biomaterial surfaces such as wear and elastic/plastic strain intensification.[1,2,11] A Raman spectrum typically contains a large amount of information in its sharp, well-resolved spectral bands. The band spectral positions, intensities, and shapes provide an interpretable and fairly unique fingerprint for qualitative analysis. The Raman microscopy provides us with the essential information needed to verify the reliability of biomaterials. Provided that standardized analytical references and protocols are established for confocal microscopy techniques, the Raman microspectroscopy also allows visualization as to how stresses and strains intensify or relax at the very surface or in the bulk of structures and devices.[11] From a purely spectroscopic point of view, stress assessments deal with the wavenumber shift of selected Raman bands upon stress. This phenomenon is known as the piezo-spectroscopic effect (henceforth referred to as PS effect after Grabner).[12] For this effect to be quantitatively exploited, the probed biomaterial should be Raman sensitive. Luckily, many biomaterials (e.g., hydroxyapatite, alumina, zirconia, etc.) and biomedical devices (e.g., artificial joints, artificial bones, etc.) of interest possess intense and sharp-featured Raman spectra. Therefore, they can be characterized with respect to their internal residual stress state without requiring any alteration or manipulation of the material.[1,2,13,14]

Throughout our Raman microspectroscopy studies of biomaterials, we have analyzed microscopic fracture mechanisms in natural biomaterials (e.g., cortical bone[15]) and synthetic biomaterials (e.g., alumina, zirconia and related composites[13,14,16]). In both natural and synthetic biomaterials, a conspicuous amount of toughening was found to arise from microscopic mechanisms operating both at the crack tip and along the crack wake. A crack-tip mechanism conspicuously enhanced the resistance of cortical bone to fracture initiation. The microscopic mechanism responsible for stress relaxation at the crack tip in the natural bone material inspired the development of new synthetic biomaterials toughened by extrinsic mechanisms. Furthermore, an understanding of how cracks grow in bone is important for locating the main cause of clinical stress fractures and for developing new methods of crack healing.

Nowadays, synthetically prepared ceramic materials, particularly alumina and zirconia, are widely used in joint replacement resulting in reduction in wear particles as compared to metallic and polymeric biomaterials.[17,18] Obviously, surgeons are concerned about the risk of using a *brittle* ceramic material for heavily loaded artificial joints (e.g., hip and knee joints). In answer to such concerns, Raman spectroscopy may provide definitive solutions to several of the problems relating to the chemical and structural reliability of biomaterials commonly employed in arthroplasty. Quality control regimens based on Raman techniques can be systematically applied to identify surface and subsurface residual stress fields, as well as phase transformations and chemical alteration of biomaterials prior to implantation in the patient. Therefore, the potential for manufacturing errors and other material reliability-related problems can be greatly reduced.

In this chapter, we demonstrate that Raman microspectroscopy possesses considerable potential in biomaterials science because it gives a chance to raster samples with high-spatial resolution and to characterize nondestructively their micromechanical characteristics in addition to chemical and physical properties. Although Raman spectroscopy is a known technique in material physics and chemistry, there is conspicuous margin left for expanding this technique to a fully nondestructive three-dimensional analysis of the micromechanical characteristics of biomaterials and related devices. In this, we believe our studies hold some novelty. In addition, the usefulness of Raman spectroscopy is particularly relevant in the field of biomedical devices and we sincerely hope that this chapter will help improving biomaterials and their design in such an important field for human health and welfare. In this context, we feel that if a rationally conceived biomaterials design as well as an improved quality control employing a powerful nondestructive tool like Raman spectroscopy is applied, many of the problems presently affecting the biomaterials field can be solved. With the aim of seeking the assistance of the international scientific and orthopaedic communities for more advanced studies employing Raman spectroscopy in inorganic biomaterials, we show in this chapter some examples of how Raman techniques may provide definitive answers to so far unanswered questions in the field of chemical and structural reliability of this important class of materials.

10.2 Principles of Raman Spectroscopy

Raman spectroscopy is based on the Raman effect, which is the inelastic scattering of light from a molecule or a crystal, due to the interaction of the incident radiation with the vibrational energies of the atoms involved. A Raman spectrum is obtained, which is shifted in frequency with respect to the incident radiation, and is a characteristic fingerprint of the substance under examination. The Raman spectrum conveys information about the chemical and crystallographic structure, mechanical properties, and crystallographic orientation of the involved compounds. Raman spectroscopic experiments are useful for substance characterization, orientation measurements, phase and crystalline fractions in crystals, and residual stress assessment on a wide range of biomaterials. Moreover, the coupling of a Raman device with a confocal microscope allows obtaining in-depth-resolved analyses and information also from thin layers of material, which is particularly advantageous when bulk samples are not supplied. In principle, it is possible to separate the energy of an atom (or a group of atoms) in a crystal into three additive components: (1) rotational (for molecules), (2) vibrational (referred to vibrations between the constituent atoms), and (3) electronic (due to the electronic motion inside atoms). The basis for this separation lies in the fact that the velocity of electrons is much greater than the vibrational velocity of nuclei, which is in turn much greater than the velocity of molecular rotation. If a crystal is analyzed under incident light, a transfer of energy from the electromagnetic field of light to the atom will occur, which equals the energy gap between two quantized states.

Stress assessments in the Raman spectroscope are possible because of the PS effect.[12,19,20] The PS effect may be simply defined as the shift in the wavelength of a spectroscopic transition in a solid in response to an applied strain or stress. This definition has been given as a general one, independent of the spectral signal arising from luminescence or Raman scattering. In linear elastic materials, the PS effect can be represented with a linear (tensorial) relationship between the second-rank stress tensor and the observed Raman wavenumber shift with respect to an unstressed (reference) wavenumber value characteristic of a selected Raman band. Upon applying preliminary calibrations, it is possible to determine the proportionality constant (i.e., a constant usually referred to as the PS coefficient) between various kinds of stress tensors (e.g., uni-axial, equi-biaxial, etc.) and wavenumber shifts of each selected Raman band. Based on this knowledge, it is then possible to quantitatively ana-lyze unknown stress states. A rigorous treatment of the tensorial PS for-malism applied to Raman bands, a description of the related calibration methods and a verification of its validity in various natural and synthetic biomaterials have been given elsewhere.[21]

Throughout our studies, Raman spectra were collected with a triple monochromator spectrometer (T-64000, ISA Jovin-Ivon/Horiba Group, Tokyo, Japan) equipped with a charge-coupled detector (high-resolution CCD camera). All the spectra were recorded at room temperature. The laser excitation source was a monochromatic blue line emitted by an Ar laser at a wavelength 488 nm or by a Kr laser at a wavelength of 614 nm. The optical microscope was connected to a video monitor that allowed scanning of the sample surface to locate the selected location for the spectroscopic measurement. The wavenumbers of Raman band-maxima were obtained by fitting the CCD raw data to mixed gaussian/lorentzian curves with commercially available software. Samples were placed on a mapping device (lateral resolution of 1 μm), which was con-nected to a personal computer to drive highly precise in-plane displace-ments (along both x and y axes) on the sample. Given the curved nature of the investigated surfaces, an autofocus device was adopted through-out automatic mapping experiments. Confocal experiments were con-ducted with focusing the waist of the laser beam either on the surface or on subsurface planes of the investigated biomaterial through an optical lens (100×). Then, the Raman signal was re-focused onto a small confo-cal (pinhole) aperture that acted as a spatial filter (pinhole aperture-diameter of 100 μm), passing the signal excited at the beam waist, but substantially eliminating fluorescence signals produced at other points above and below the beam waist. The filtered signal then returned to the spectrometer (via the probe head) where it was dispersed onto the CCD camera to produce a spectrum. In practice, a pinhole aperture was placed in the optical train of the spectrometer and used to regulate the rejection of out-of-focus light. The penetration depth of the laser depended on the transparency of the investigated material as well as on the selected pinhole aperture. Confocal microscopy was used to

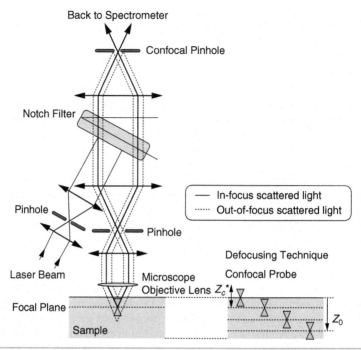

FIGURE 10.1   Schematic draft of the confocal probe configuration and of the probe/sample interaction.

probe selected xyz locations in the samples, with micrometer (lateral) spatial resolution. Two-dimensional data maps were built up by xy scanning the sample, and three-dimensional data sets were produced by sequentially acquiring a set of xy slices at different z depths. A schematic

Materials	Confocal Probe Depth z_{c}^{*}, μm	Normal Probe Depth z_{c}^{*}, μm	PS Coefficient, cm^{-1}/GPa
Bony apatite	4.5	28.3	2.95 ± 0.12 (band at 960 cm^{-1})
Alumina	4.1	30.2	−0.71 ± 0.06 (band at 416 cm^{-1})
Tetragonal zirconia	4.2	23.5	−6.00 ± 0.15 (band at 142 cm^{-1})
Monoclinic zirconia	3.3	10.0	−1.10 ± 0.16 (band at 384 cm^{-1})

TABLE 10.1   Raman Probe and PS Coefficients for Various Biomaterials

draft of the confocal probe configuration and the laser penetration depths for 90 percent Raman emission in normal and confocal probe configurations are given in Fig. 10.1 and Table 10.1, respectively. In Table 10.1, penetration depth for Raman confocal and normal probe (z_c^* and z_n^*, respectively) refres to the threshold depth value for 90 percent Raman intensity emission. PS coefficients and data scatter recorded during calibration refer to the Raman band indicated in brackets.

10.3 Raman Effect in Biological and Synthetic Biomaterials

10.3.1 Spectral Features

According to group theory, the phonon modes of solids decompose as irreducible representations, which split into selected vibrational modes. Such modes can be classified depending on the propagation direction of the selected mode with respect to the selected polarization direction [(e.g., longitudinal optic (LO) or transverse optic (TO)]. Selected Raman bands can be then used for stress assessments, according to the knowledge of their respective PS coefficients.

Hydroxyapatite crystals possess a point group of C_6. The space group $C_6^6(P6_3)$ specifies the unit cell symmetry for hydroxyapatite more correctly than $C_{6h}^2(P6_3/m)$, since the OH ions are not located on the mirror plane σ_h.[22] The phosphate and OH ions are located on sites of C_1 and C_3 symmetries for C_6^6 space groups, respectively. The factor group (or correlation) analysis based on these sites predicts types and numbers of symmetry species due to the crystal field splitting. It has been reported that the space group of C_6^6 symmetry is consistent with the results from the Raman experiments on synthetic hydroxyapatite at low temperature. Factor group analyses have been based on the following irreducible representation, with a total of five OH-related bands expected for space group of P6.[22]

$$\Gamma = A + E_1 + E_2 \tag{10.1}$$

Figure 10.2 shows a typical spectrum of cortical bone (bovine femur), in which the pertinent vibrational modes has been labeled, although not all the theoretically predicted vibrational bands could be observed in the collected spectra.

Alumina materials are of high-technological interest as synthetic biomaterials. They possess the corundum structure, which belongs to the D_{3d}^6 space group with two molecular Al_2O_3 groups per unit cell.[23] The aluminum atoms are octahedrally coordinated with two layers of oxygen atoms, the octahedron being severely distorted. The irreducible representations for the optical modes in the crystal can be expressed as

$$\Gamma = 2A_{1g} + 2A_{1u} + 3A_{2g} + 2A_{2u} + 5E_g + 4E_u \tag{10.2}$$

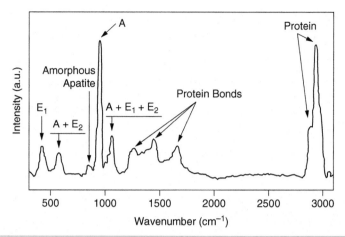

FIGURE 10.2   Raman spectrum of cortical bone (bovine femur), in which the Raman bands belonging to organic (proteins) and inorganic species have been labeled. In the spectrum, vibrational modes have been indicated according to Ref. 22.

while the irreducible representations for the acoustical modes are $1A_{2u} + 1E_u$. A peculiarity of the corundum structure, owing to the existence of a center of symmetry in the unit cell, is that all the allowed Raman vibrational modes are infrared forbidden and vice versa (i.e., vibrations with symmetry A_{1g} and E_g are Raman active and infrared inactive, while the A_{2u} and E_u vibrations are infrared active and Raman inactive). On the other hand, the A_{1u} and A_{2g} vibrational modes are predicted by theory to be both infrared and Raman inactive. Typical Raman spectra of alumina are shown in Fig. 10.3 and the corresponding Raman modes labeled according to the formalism shown in Eq. (10.2).

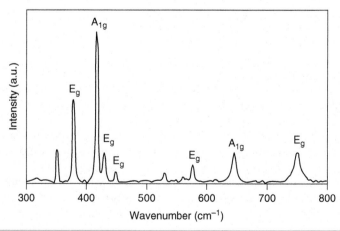

FIGURE 10.3   Raman spectrum of alumina; vibrational modes have been indicated according to Ref. 23.

Zirconia—The monoclinic phase of ZrO_2, with space group $P2_1/2$, can be represented with the point group C_{2h} at Γ, consisting of operations E, I, C_2^y, and M_y. The character table of this point group indicates that there are four symmetry classes and thus four irreducible representations, each of which is one dimensional. Standard group-theoretical analysis indicates that the modes at the Γ point can be decomposed as[24,25]

$$\Gamma^m = 9A_g + 9A_u + 9B_g + 9B_u \qquad (10.3)$$

where the superscript m refers to the monoclinic polymorph. Among the total 36 modes, 18 modes $9A_g + 9B_g$ are Raman active and 15 modes $8A_u + 7B_u$ are infrared active, the remaining three modes being the zero-frequency translational modes. Only the 15 infrared-active modes contribute to the lattice dielectric tensor. In the case of the tetragonal polymorph, the irreducible representations for the optical modes in the crystal can be expressed as [24–26]

$$\Gamma^t = 1A_{1g} + 2A_{2u} + 3E_g + 3E_u + 2B_u + 2B_{1g} \qquad (10.4)$$

where the superscript t refers to the tetragonal polymorph. The E_u and E_g representations are two-dimensional while all other modes are one-dimensional. One A_{2u} mode and one E_u pair are acoustic, leaving one IR-active A_{2u} and two IR-active E_u pairs; A_{1g}, B_{1g}, and E_g are Raman active, while B_{2u} is silent. Raman spectra of tetragonal, monoclinic, and mixed tetragonal/monoclinic polymorphs are shown in Figs. 10.4a, b, and c, respectively. Raman active modes are labeled on the spectra in Figs. 10.4a and b for tetragonal and monoclinic phases, respectively.

10.3.2 PS Behavior

In single-crystalline materials, the PS effect can be represented with a linear (tensorial) relationship between the second-rank stress tensor σ_{ij} and the observed Raman wavenumber shift Δv with respect to the unstressed (reference) wavenumber characteristic of a selected Raman band:[12,21]

$$\Delta v = \Pi_{ij}\sigma_{ij} = \Pi_{ij}^*\sigma_{ij}^* \qquad (10.5)$$

where $i, j = 1, 2, 3$ and Π_{ij} is the tensor of PS coefficients (given in units of cm^{-1}/GPa). Data for the matrix of PS coefficients of single-crystals are typically reported with reference to the main crystallographic axes of the crystal. In order to make this explicitly known a suffix "*" is usually added to all symbols referring to this special choice of coordinate system (e.g., Π_{ij}^* and σ_{ij}^*). The number of independent components of the PS tensor might eventually reduce according to crystal symmetry; for example, in the particular case of a single-crystal with full symmetry the PS tensor reduces to a scalar, Π^*, while in crystals with planar symmetry Π_{ij}^* is purely diagonal.

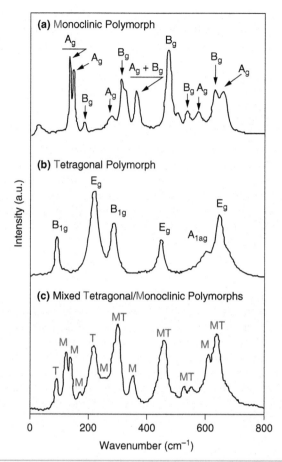

FIGURE 10.4   Raman spectra of tetragonal, monoclinic and mixed tetragonal/ monoclinic polymorphs are shown in (a), (b), and (c), respectively. Vibrational modes have been labeled according to Refs. 24 and 25.

In polycrystalline materials with a random orientation of the grain structure, the Raman probe might cover a volume in which a statistically large number of grains are comprised. In such a case, the value of PS coefficient becomes independent of crystallographic direction and represents an average value along all the possible crystallographic directions. Note that this is the case of many advanced polycrystalline ceramics, which are designed with a finely grained microstructure to improve their structural response. It follows that for a general three-axial stress state, $\sigma_{ij} = \sigma_{ij}^*$, applied to a polycrystalline sample, Eq. (10.5) reduces to a scalar proportionality equation, as follows:[21]

$$\Delta v = \Pi_{av} < \sigma_{ij}^* >$$ (10.6)

where $<\sigma_{ij}^* > = \sigma_{11}^* + \sigma_{22}^* + \sigma_{33}^*$ is the trace of the principal stress tensor, and Π_{av} is the PS coefficient averaged among all the crystallographic directions. In other words, in a randomly oriented polycrystal for which the Raman probe is significantly larger than individual grains, a measurement of Raman spectral shift gives direct access to the trace of the stress tensor. In addition, the tensile or compressive nature of the stress tensor trace can also be deduced from the positive or negative sign of the observed frequency shift Δv with respect to an unstressed state.

Table 10.1 summarizes the stress dependence of vibrational modes of different biomaterials in terms of their average PS coefficients Π_a; the shown results are from calibration tests conducted under uniaxial load in four point bending configuration on finely grained polycrystalline hydroxyapatite, alumina, and zirconia materials. In the above formalism, the numerical values of the PS coefficients end up being altered with respect to those measured in the respective single crystals. In addition, the PS coefficients measured in polycrystals are also affected by the presence of secondary phases. For this reason, in order to obtain precise estimations of the stress tensor trace, calibrations should be preliminary repeated for each individual material.

10.4 Visualization of Microscopic Stress Patterns in Biomaterials

10.4.1 Micromechanics of Fracture and Crack-Tip Stress Relaxation Mechanisms

Unlike the majority of synthetic ceramic materials, bone is a toughened solid whose fracture resistance is cumulatively enhanced by extrinsic mechanisms operating both in the crack wake and ahead of the crack tip (i.e., crack-face bridging and microcracking, respectively).[27,28] Fracture mechanics studies have provided quantitative characterizations of the resistance of bone to fracture in terms of critical stress intensity factor and critical strain energy release rate as measured at the onset of crack initiation. This approach has been coupled with characterizations of toughness as a function of crack length (rising R-curve behavior), which is useful to quantify the toughening contribution arising from microscopic mechanisms occurring behind the advancing crack front.[15,29] This latter approach has also allowed scientists to explicitly demonstrate the differences in crack propagation resistance between cortical bones of different kinds and from different species.[30] The fracture mechanics results obtained in those studies provide an improved understanding of the mechanisms associated with the failure of cortical bone, and as such represent a step forward from the perspective of developing a realistic framework for fracture risk assessment, for determining how the increasing propensity for fracture with age can be prevented, and for developing new synthetic materials whose toughening behavior mimics that of natural cortical bone. From a more general perspective, mechanical factors are known to affect bone remodeling as well as an increased mechanical demand results in bone formation (i.e., while decreased demand results in net bone resorption). Current theories clearly suggest that mechanical stress plays an important role in bone modeling and remodeling.[31] Therefore, microscopic information of the stress patterns developed in bone may lead to a better understanding of how bone functionality and crack healing can be affected by mechanical loading.

The objective of our Raman studies has been that of investigating how macroscopically applied external loading similar in magnitude to that occurring in vivo during bone fracture manifest at the microscopic level in the bone matrix.[2,15] Using the PS approach applied to the 980 cm^{-1} Raman band of hydroxyapatite, a direct evaluation of the stress tensor trace $< \sigma_{ij}^* >$ has been obtained and spatially resolved maps of microscopic stress experimentally determined around the tip of a propagating crack. It was found that stress patterns were highly heterogeneous and strongly related to the locations of the observed microdamages (cf. scanning electron micrograph and stress map in Fig. 10.5a and b, respectively), indicating that the resulting stress field is significantly altered by the microdamages. In a recent study, Thompson et al.[32] reported that a recoverable bond in the collagen molecules might contribute to conspicuous energy dissipation in the postyield deformation of bone. To elucidate the role of collagen in the postyield deformation of bone, microdamage accumulation has been proposed to lead to surface energy dissipation during bone deformation, whereas the degradation and plastic deformation in the collagen network are the major mechanisms in the inelastic and viscoelastic energy consumption.[33–35] We have also confirmed the occurrence of collagen stretching mechanism in a previous study.[2] In Fig. 10.5a, it can be seen that a cloud of microcracks is formed along a constant direction. In addition, in correspondence of locally microcracked areas, the crack-tip stress was released (cf. Fig. 10.5b). As a consequence, the stress field around the crack tip assumed a peculiar stripe-like morphology. In other words, bone is capable to partly release the stress intensification at the tip of a propagating crack through the occurrence of a self-damaging mechanism. It should be noted that the use of a confocal probe enabled us to minimize in depth averaging of the Raman signal, so that a relatively sharp map could be obtained. In the crack-tip experiments shown in this study, the confocal probe was shifted below the sample-free surface by about 10 μm in order to minimize surface effects. It is interesting to compare the crack-tip stress field developed in cortical bone with that recorded in synthetic polycrystalline alumina (scanning electron micrograph and stress map in Fig. 10.5c and d, respectively). In this latter experiment, the stress field was visualized by monitoring the PS behavior of the 418 cm^{-1} band of alumina. In finely grained alumina, no microcracks were found around the crack tip and no stress relaxation mechanism could be observed in the recorded stress pattern. Under such micromechanical conditions, the crack-tip stress field preserves the symmetry characteristics of linear elastic materials, no crack-tip toughening effect is operative in delaying crack propagation, and the material is typically brittle.

Partly stabilized tetragonal zirconia polycrystals are among the toughest synthetic biomaterials and are extensively used in biomedical applications due to their excellent structural reliability, high-wear resistance and compatibility with the human body environment.[17,18] The improvement in mechanical properties has been reported to arise from stress-induced phase transformation from the tetragonal to the monoclinic polymorph, which takes place in the neighborhood of a propagating crack.[36–38] In the context of this study, it could be interesting to visualize the transformation toughening mechanism by means of Raman microspectroscopy, in order to provide microscopic information on the effect of polymorphic transformation on the crack-tip stress pattern ahead of an advancing crack. Figure 10.5e and f shows the monoclinic transformation field in the neighborhood of the crack tip and the corresponding equilibrium stress field pattern, respectively. The depicted stress field, compressive in nature, represents the equilibrium stress computed as the average (weighted by the respective volume fractions) of the stress fields stored in the constituent tetragonal and monoclinic phases. Stress fields were evaluated by exploiting the PS

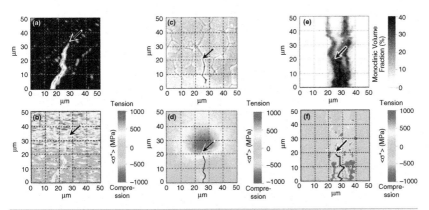

FIGURE 10.5 Scanning electron micrographs of a crack tip in cortical bone (a) and in polycrystalline alumina (c), with the respective stress maps in (b) and (d), respectively. The stress field was visualized by monitoring the PS behavior of the 418 cm^{-1} band in alumina and of the 980 cm^{-1} in the hydroxyapatite. Transformation zone at the tip of a propagating crack in zirconia and the respective equilibrium stress distribution are shown in (e) and (f), respectively. (*With kind permission of Springer Science+Business Media.*)

behavior of the 180 and 260 cm^{-1} Raman bands of monoclinic and tetragonal zirconia polymorphs, respectively. Upon stress-induced tetragonal-to-monoclinic transformation, significant volume expansion occurs, which in turn induces a compressive stress field that shields the crack mouth. In other words, partly stabilized zirconia materials are capable to release the tensile crack-tip stress through a self-induced highly compressive stress field arising from polymorphic transformation.

Besides the importance of such visualization from a basic materials science point of view, the knowledge of the amount of polymorph transformation and of the residual stress magnitude is fundamental for improving the microstructural design of zirconia ceramics and for correctly understanding the difference of their mechanical properties. From this perspective, the Raman spectroscopic evaluation can be considered as complementary to macroscopic fracture mechanics characterizations of critical stress intensity factor at the onset of crack initiation and of rising R-curve behavior.

10.4.2   Residual Stress Patterns on Ceramic-Bearing Surfaces of Artificial Hip Joints

The finite element method has been for many years the main tool used by materials technologists to solve engineering problems related to stress and strain analysis of static or dynamic loaded contacts. In the field of arthroplasty, the finite element method has been used to analyze mechanical and thermal responses of acetabular cup and femoral head components for total hip replacement.[39–41] Numerical evaluations have been based on the concept that the wear volume is proportional to the product of contact load and sliding distance, the proportionality constant being referred to as the specific wear rate between the head and the cup-bearing surfaces. According to such numerical studies, hip joint simulators have been designed to reproduce as much reliably as possible the sliding kinetics taking place between the head and the cup. However, although the results obtained from the model are useful in understanding the causes of wear in hip prostheses and they may contribute to predict the overall joint performance, they have shown insufficient to fully represent the complex kinetics of in vivo loaded hip-bearing surfaces and thus to predict their actual wear rate.[42] In particular, it is very difficult to predict residual stresses by the finite element method, although such type of stresses are very closely related to wear phenomena. It should also be noted that the environment of orthopaedic implants sometimes induces additional effects as head/cup microseparation[43] and systematic multiaxial microdisplacements[44,45] at the contact of the modular prosthesis components. These additional effects may contribute to the total lifetime of the implant in a nonnegligible way, especially when artificial joints are made of brittle materials such as ceramics. The necessary optimization of orthopaedic device lifetime thus requires a better knowledge of the damages induced by wear contact. In this context, confocal Raman spectroscopy might provide scientists and technologists with new insight into contact mechanics, being capable to reveal microscopic patterns of residual stress stored on the bearing surfaces. In other words, the kinetics of surface sliding within the joint (i.e., including microdisplacements) remains stored onto its bearing surfaces and Raman maps of residual stress reveal it with microscopic precision. Figure 10.6 shows maps of residual stress as collected on the entire surface of five different femoral heads, which were retrieved after exposures in human body elapsing from 1 month to 19 years. All the femoral heads were made of monolithic alumina and operated against monolithic alumina acetabular cups. Stress maps were collected with micrometric resolution by placing the focal plane of the probe at the sample surface. The Raman characterization reveals the overall stress patterns with a statistically meaningful sampling, residual stress analysis being performed in toto over the entire surface of the femoral heads. In contrast to the residual stress level after only 1-month exposure (Fig. 10.6a), which was almost zero, a main residual stress areas could be recognized in which residual tensile stresses remain stored onto the load-bearing surface of the head exposed in vivo for 2 years and 6 months (Fig. 10.6b). This area is supposed to undergo a severe impact regime due to microseparation during the initial period of in vivo implantation. However, a mixed trend of tensile and compressive stresses was found by screening two

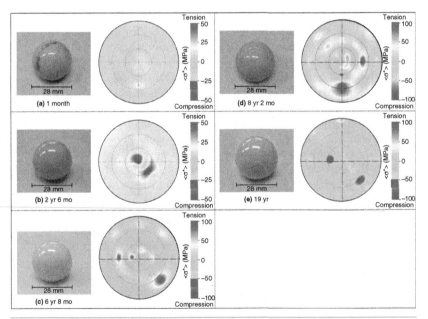

FIGURE 10.6   Photographs and Raman maps of residual stresses stored on the entire surface of five different femoral heads exposed in vivo for different periods of time. All the investigated alumina femoral heads belonged to hip implants in which the acetabular cup (i.e., the bearing counterpart) was also made of alumina.

alumina femoral heads retrieved after medium-term implantation in human body (i.e., 6 years and 8 months and 8 years and 2 months, in Fig. 10.6c and d, respectively). Such a complex pattern of residual stress is clearly affected by microdisplacements occurring between the ceramic bearing surfaces. On the other hand, two areas of strong compressive stress were found in the femoral head exposed in vivo for 19 years (Fig. 10.6e). The general trend in surface residual stress in the wear-zone surface of retrieved balls was increasingly tensile up to several years exposure in vivo, then for implants subjected to longer exposure times in human body residual stress fields in the wear zone first of mixed tensile/compressive nature, and then progressed toward fully compressive trends. The topographic location of areas of stress intensification was not the same for all the retrievals and this was considered to be the consequence of different designs of the artificial joints, different attitudes of the patients, and different angular inclinations selected by the surgeon in positioning the ceramic acetabular cup. Nevertheless, surface residual stress fields showed a trend whose origin should reside in the mechanical interaction between the bearing surfaces. Based on the experimental visualization of stress patterns by Raman PS, we propose that both shock and impingement of the acetabular cup on the femoral head introduce on the ceramic surface a residual stress field whose nature changes from tensile in the short term to compressive in the long-term exposure in vivo and whose highest magnitude is reached after significant long-term exposures. The pattern of residual stress can be referred to as the *loading history* of the implant. A simplified model for explaining the time dependence of surface residual stress in femoral heads can be then given as follows. In the early period of implantation time, the surface of the femoral head is subjected to significant local shocks and impingement arising from severe microseparation phenomena taking place in the hip joint. As a consequence of such a micromechanical situation, intergranular microcracking will take place and will later develop into debris formation in main-wear zones. Cracking, which is a consequence of local shocks and point forces, can introduce in the surface a residual stress field of tensile nature. Cracks selectively develop at the alumina grain boundaries where the stress intensification is higher. Microcracking will successively develop into grain detachment with subsequent development of a significant amount of ceramic debris. This stage is likely accompanied by a release of the tensile residual stress field, while the compressive residual stresses stored on the ceramic joint surface are the consequence of long-term impingement assisted by the presence of third bodies (i.e., the ceramic debris). This latter compressive stress field continuously increases with increasing exposure time in vivo up to a saturation value, above which more extensive grain detachment occurs and abraded areas (i.e., stripe-wear zones[46,47]) may develop.

Standard tribological studies of ceramic-on-ceramic load-bearing surfaces usually mainly rely on phenomenological parameters for describing the wear effects, like as surface topography, friction coefficient, and loss rates.[48] However, we have shown here that advanced spectroscopic analyses can be also employed in order to obtain a deeper understanding of wear mechanisms and load-bearing kinetics. PS Raman analyses can be useful to clarify the actual mechanisms lying behind the wear behavior of ceramic bearings, which typically occur in a complex way, hardly predictable by theoretical simulations and reproducible in artificial joint simulators. Raman spectroscopy advancements are particularly relevant to the future of ceramic-on-ceramic bearings, which are considered as the main protagonists in the new generation of materials for arthroplastic applications.

10.5   Conclusions

A PS Raman technique, based on a microscopy measurement enabled us to quantitatively assess in situ the microscopic stress fields developed during fracture at the crack tip of natural and synthetic biomaterials. Using the Raman PS technique, crack-tip toughening mechanisms could be clearly visualized and assessed quantitatively. This chapter also presents results on microscopic stress analysis of ceramic biomaterials as collected by Raman microspectroscopy on the bearing surfaces of artificial hip joints. Based on the current and previous results, improved designs of more realistic hip joint simulators can be attempted, which take into consideration phenomena like microseparation and surface microdisplacements. Mechanistic Raman spectroscopic analyses may help rationalizing the structural behavior of biomaterials, thus building up new theories that attribute such behavior to clear factors. The possibility of nondestructively and quantitatively measuring stress fields, in addition to phase fractions from Raman spectra of biomaterials definitely offers a chance to materials scientists to rationally address the best microstructural design approach to better synthetic biomaterials.

References

1. G. Pezzotti, "Stress Microscopy and Confocal Raman Imaging of Load-Bearing Surfaces in Artificial Hip Joints," *Expert Review of Medical Devices*, 4:165–189, 2007.
2. G. Pezzotti, "Raman Piezo-Spectroscopic Analysis of Natural and Synthetic Biomaterials," *Analytical and Bioanalytical Chemistry*, 381:577–590, 2005.
3. E. E. Lawson, B. W. Barry, A. C. Williams and H. G. M. Edwards, "Biomedical Applications of Raman Spectroscopy," *Journal of Raman Spectroscopy*, 28:111–117, 1998.
4. G. Pezzotti and A. A. Porporati, "Raman Spectroscopic Analysis of Phase-Transformation and Stress Patterns in Zirconia Hip Joints," *Journal of Biomedical Optics*, 9:372–384, 2004.

5. L. P. Choo-Smith, H. G. M. Edwards, H. P. Endtz, J. M. Kros, F. Heule, H. Barr, J. S. Robinson Jr., H. A. Bruining, and G. J. Puppels, "Medical Applications of Raman Spectroscopy: From Proof of Principle to Clinical Implementation," *Biopolymers*, **67**:1–9, 2002.

6. K. C. Blakeslee and R. A. Condrate, Sr., "Vibrational Spectra of Hydrothermally Prepared Hydroxyapatites," *Journal of the American Ceramic Society*, **54**:559–563, 1971.

7. D. C. O'Shea, M. L. Bartlett, and R. A. Young, "Compositional Analysis of Apatites with Laser-Raman Spectroscopy: (OH, F, Cl) Apatites," *Archives Oral Biology*, **19**:995–1006, 1974.

8. G. Penel, C. Delfosse, M. Descamps, and G. Leroy, "Composition of Bone and Apatitic Biomaterials as Revealed by Intravital Raman Microspectroscopy," *Bone*, **36**:893–901, 2005.

9. G. Katagiri, H. Ishida, A. Ishitani and T. Masaki, "Direct Determination by Raman Microprobe of the Transformation Zone Size in Y_2O_3 Containing Tetragonal ZrO_2 Polycrystals," in: *Advanced in Ceramics, Vol. 24: Science and Technology of Zirconia III*, S. Somiya, N. Yamamoto, and H. Yanagida (eds.) Westerville OH, The American Ceramic Society, 1988, pp. 537–544.

10. P. Matousek, "Deep Non-Invasive Raman Spectroscopy of Living Tissue and Powders," *Chemical Society Reviews*, **36**:1292–1304, 2007.

11. G. Pezzotti, T. Kumakura, K. Yamada, T. Tateiwa, L. Puppulin, W. Zhu, and K. Yamamoto, "Confocal Raman Spectroscopic Analysis of Cross-Linked Ultra-High Molecular Weight Polyethylene for Application in Artificial Hip Joints," *Journal of Biomedical Optics*, **12**:014011-1-14, 2007.

12. L. Grabner, "Spectroscopic Technique for the Measurement of Residual Stress in Sintered Al_2O_3," *Journal of Applied Physics*, **49**:580–583, 1978.

13. G. Pezzotti, T. Tateiwa, W. Zhu, T. Kumakura, and K. Yamamoto, "Fluorescence Spectroscopic Analysis of Surface and Sub-Surface Residual Stress Fields in Alumina Hip Joints," *Journal of Biomedical Optics*, **11**:24009–24018, 2006.

14. W. Zhu and G. Pezzotti, "Spatially Resolved Stress Analysis in Al_2O_3/3Y-TZP Multilayered Composite Using Confocal Fluorescence Spectroscopy," *Applied Spectroscopy*, **59**:1042–1048, 2005.

15. G. Pezzotti and S. Sakakura, "Study of the Toughening Mechanisms in Bone and Biomimetic Hydroxyapatite Materials Using Raman Microprobe Spectroscopy," *Journal of Biomedial Materials Research Part A*, **65**:229–236, 2003.

16. G. Pezzotti, K. Yamada, S. Sakakura, and R. P. Pitto, "Raman Spectroscopic Analysis of Advanced Ceramic Composite for Hip Prosthesis," *Journal of the American Ceramic Society*, **91**:1199–1206, 2008.

17. C. Piconi and G. Maccauro, "Zirconia as a Ceramic Biomaterial," *Biomaterials*, **20**:1–25, 1999.

18. J. Chevalier, "What Future for Zirconia as a Biomaterial?" *Biomaterials*, **27**: 535–543, 2006.

19. E. Anastassakis and E. Burstein, "Morphic Effects: 1. Effects of External Forces on Photon-Optical-Phonon Interactions," *Journal of Physics Chemistry of Solids*, **32**:313–324, 1971.

20. E. Anastassakis, A. Pinczuk, E. Burstein, F. H. Pollak, and M. Cardona, "Effect of Static Uniaxial Stress on the Raman Spectrum of Silicon," *Solid State Communications*, **8**:133–136, 1970.

21. G. Pezzotti and W. H. Mueller, "Micromechanics of Fracture in a Ceramic/Metal Composite Studied by In Situ Fluorescence Spectroscopy I: Foundations and Stress Analysis," *Continuum Mechanics and Thermodynamics*, **14**:113–126, 2002.

22. H. Tsuda and J. Arends, "Orientational Micro-Raman Spectroscopy on Hydroxyapatite Single Crystals and Human Enamel Crystallites," *Journal of Dental Research*, **73**:1703–1710, 1994.

23. G. H. Watson Jr. and W. B. Daniels, "Measurements of Raman Intensities and Pressure Dependence of Phonon Frequencies in Sapphire," *Journal of Applied Physics*, **52**:956–961, 1981.

24. X. Zhao and D. Vanderbilt, "First-Principles Study of Structural, Vibrational, and Lattice Dielectric Properties of Hafnium Oxide," *Physical Review B*, **65**:075105-1-4, 2002.

25. A. Mirgorodsky, M. B. Smirnov, and P. E. Quintard, "Phonon Spectra Evolution and Soft-mode Instabilities of Zirconia During the c–t–m Transformation," *Journal of Physics and Chemistry of Solids*, **60**:985–992, 1997.

26. T. Merle, R. Guinebretiere, A. Mirgorodsky, and P. E. Quintard, "Polarized Raman Spectra of Tetragonal Pure ZrO_2 Measured on Epitaxial Films," *Physical Review B*, **65**:144302-1, 2002.

27. J. Kruzic, J. Scott, R. Nalla and R. O. Ritchie, "Propagation of Surface Fatigue Cracks in Human Cortical Bone," *Journal of Biomechanics*, **39**:968–972, 2006.

28. D. Vashishth, J. C. Behiri, and W. Bonfield, "Crack Growth Resistance in Cortical Bone: Concept of Microcrack Toughening," *Journal of Biomechanics*, **30**:763–769, 1997.

29. R. K. Nalla, J. J. Kruzic, J. H. Kinney, and R. O. Ritchie, "Effect of Aging on the Toughness of Human Cortical Bone: Evaluation By R-Curves," *Bone*, **35**: 1240–1246, 2004.

30. P. Fratzl, H. S. Gupta, E. P. Paschalis, and P. Roschger, "Structure and Mechanical Quality of the Collagen-Mineral Nano-Composite in Bone," *Journal of Materials Chemistry*, **14**:2115–2123, 2004.

31. S. Nomura and T. Takano-Yamamoto, "Molecular Events Caused by Mechanical Stress in Bone," *Matrix Biology*, **19**:91–96, 2000.

32. J. B. Thompson, J. H. Kindt, B. Drake, H. G. Hansma, D. E. Morse, and P. K. Hansma, "Bone Indentation Recovery Time Correlates with Bond Reforming Time," *Nature*, **6865**:773–776, 2001.

33. P. Zioupos, "Accumulation of In-Vivo Fatigue Microdamage and Its Relation to Biomechanical Properties in Ageing Human Cortical Bone," *Journal of Microscopy*, **201**:270–278, 2001.

34. D. B. Burr, M. R. Forwood, D. P. Fyhrie, R. B. Martin, M. B. Schaffler, and C. H. Turner, "Bone Microdamage and Skeletal Fragility in Osteoporotic and Stress Fractures," *Journal of Bone and Mineral Research*, **1**:6–15, 1997.

35. A. C. Courtney, W. C. Hayes, and L. J. Gibson, "Age-Related Differences in Post-Yield Damage in Human Cortical Bone: Experiment and Model," *Journal of Biomechanics*, **11**:1463–1471, 1996.

36. R. H. Hannink and M. V. Swain, "Progress in Transformation Toughening of Ceramics," *Annual Reviews of Materials Science*, **24**:359–408, 1994.

37. L. R. F. Rose, "The mechanics of Transformation Toughening," *Proceedings of the Royal Society London A*, **412**:169–197, 1987.

38. M. G. Cain, S. M. Bennington, M. H. Lewis, and S. Hull, "Study of the Ferroelastic Transformation in Zirconia by Neutron Diffraction," *Philosophical Magazine Part B*, **69**:499–507, 1994.

39. S. J. Hampton, T. P. Andriacchi, and J. O. Galante, "Three Dimensional Stress Analysis of The Femoral Stem of a Total Hip Prostheses," *Journal of Biomechanics*, **13**:443–448, 1980.

40. D. L. Bartel, V. L. Bicknell, and T. M. Wright, "The Effect of Conformity, Thickness, and Material on Stresses in Ultra-High Molecular Weight Components for Total Joint Replacement," *Journal of Bone and Joint Surgery America*, **68**:1041–1051, 1986.

41. F. Bachtar, X. Chen, and T. Hisada, "Finite Element Contact Analysis of the Hip Joint," *Medical and Biological Engineering and Computing*, **44**:643–651, 2006.

42. M. Sundfeldt, L. V. Carlsson, C. B. Johansson, P. Thomsen, and C. Gretzer, "Aseptic Loosening, Not Only a Question of Wear: A Review of Different Theories," *Acta Orthopaedia*, **77**:177–197, 2006.

43. S. Williams, T. D. Stewart, E. Ingham, M. H. Stone, and J. Fisher, "Influence of Microseparation and Joint Laxity on Wear of Ceramic on Polyethylene, Ceramic on Ceramic and Metal on Metal Total Hip Replacements," *Journal of Bone and Joint Surgery British*, **85-B**:57–63, 2003.

44. L. Duisabeau, P. Combrade, and B. Forest, "Environmental Effect of Fretting of Metallic Materials for Orthopaedic Implants," *Wear*, **256**:805–816, 2004.

45. E. Ebramzadeh, F. Billi, S. N. Sangiorgio, S. Mattes, W. Schmoelz, and L. Dorr, "Simulation of Fretting Wear at Orthopaedic Implant Interfaces," *Journal of Biomechanical Engineering*, **127**:357–364, 2005.

46. H. A. McKellop and D. D'Lima, "How Have Wear Testing and Joint Simulator Studies Helped to Discriminate Among Materials and Designs?" *Journal of American Academy of Orthopaedic Surgeons*, **16**(Suppl. 1) S111–S119, 2008.
47. T. D. Steward, J. L. Tipper, G. Insley, R. M. Streicher, E. Ingham, and J. Fisher, "Long-Term Wear of Ceramic Matrix Composite Materials for Hip Prostheses under Severe Swing Phase Microseparation," *Journal of Biomedical Materials Research*, **66B:**567–573, 2003.
48. S. Affatato, M. Spinelli, M. Zavalloni, C. Mazzega-Fabbro and M. Viceconti, "Tribology and Total Hip Joint Replacement: Current Concepts in Mechanical Simulation," *Medical Engineering Physics*, **30:**1305–1317, 2008.

Tissue Imaging with Coherent Anti-Stokes Raman Scattering Microscopy

Eric Olaf Potma

Department of Chemistry & Beckman Laser Institute
University of California
Irvine, California

> As spontaneous Raman spectroscopy has blossomed and grown during one-half century, it may be predicted with some confidence that coherent nonlinear Raman spectroscopy will yield many new results in the next half century.
>
> Nicolas Bloembergen, 1978

11.1 From Spontaneous to Coherent Raman Spectroscopy

Sir Chandrasekhara Venkata Raman was justifiably excited when he noticed that if a monochromatic light beam is passed through a simple transparent liquid, different colors can be detected in the scattered light. Raman called this frequency shifted scattered light "a new type of secondary radiation," because he realized that the phenomenon he discovered was caused by a molecular property different from the property of fluorescence emission.[1] Since Raman's early work in 1928, it has been well established that molecular vibrations are responsible for the observed frequency shifts in the scattered light. The discovery of the Raman effect was only the beginning of a broad and growing branch of molecular spectroscopy, with new applications that continue to emerge to this day, some of which are presented in this book.

Raman's excitement was mirrored almost 35 years later when researchers inserted a cell with transparent liquid, nitrobenzene, in the cavity of a pulsed ruby laser. To their surprise, they observed that the laser output consisted of frequency-shifted components in addition to the radiation at the lasing frequency.[2,3] Similar to Raman scattered light, this new radiation was seen at frequency shifts that correspond to the frequencies of molecular vibrations. Unlike the incoherently scattered light seen by Raman, however, the frequency-shifted emission from the ruby laser exhibited a well-defined propagation direction and coherence. In analogy with the principle of stimulated fluorescence on which the ruby laser was based, the name stimulated Raman scattering was coined to describe this newly discovered phenomenon. The field of coherent Raman scattering was born.

The frequency-shifted components on the ruby laser output appeared at lower energies, and are thus Stokes shifted. On closer inspection, however, coherent components at higher energies were also found. Minck et al. theorized that these anti-Stokes contributions resulted from a cascaded process in the Raman active medium.[4] In this process, the ruby laser wavelength at frequency ω_0 was first shifted to lower frequencies $\omega_0 - \omega_r$ through the stimulated Raman process, where ω_r are the frequencies corresponding to the Raman active vibrations of the molecule. Once a $\omega_0 - \omega_r$, beam has built up in the cavity, it is able to combine with the fundamental frequency and generate new frequency components at $\omega_0 + \omega_r$, i.e., coherent anti-Stokes radiation.[4]

Nonetheless, in Minck et al.'s experiment several stimulated Raman processes took place in the laser cavity simultaneously, which made it difficult to isolate the generation of the anti-Stokes components from the generation of the red-shifted Stokes contributions. Two years later, Maker and Terhune from the Ford Motor Company were able to generate coherent anti-Stokes radiation outside of a laser cavity.[5] In their experiment, the required red-shifted $\omega_0 - \omega_r$ component was generated from the fundamental laser beam through a stimulated Raman process in a cell of a Raman active medium. Both beams were then collinearly focused into the sample. Maker and Terhune observed a clear signal at $\omega_0 + \omega_r$. Moreover, they showed that the strength of the signal depended on the presence of a vibrational resonance in the sample at ω_r, which demonstrated the potential of anti-Stokes generation as a molecular spectroscopic tool.

The efficiency of the coherent anti-Stokes generation process depends on the ability of the sample material to respond to three optical frequencies by producing an oscillatory electronic motion at a fourth frequency, which is a combination frequency of the three incoming optical fields. This material property is called the third-order susceptibility and is written as $\chi^{(3)}$.[6,7] The third-order susceptibility is composed of parts that depend on the presence of a vibrational mode and parts that are purely electronic in nature, which are known as the resonant and nonresonant contributions, respectively. For spectroscopic measurements, the resonant part $\chi_r^{(3)}$ is of interest, which was the subject of extensive study in early experiments on the nonlinear properties of solids and liquids.[8–10] In 1974, Begley et al. summarized the most important advantages of vibrational spectroscopy based on nonlinear anti-Stokes generation.[11] First, the coherent anti-Stokes mechanism offers signals that are over five orders of magnitude stronger relative to spontaneous Raman scattering. Second, this nonlinear technique avoids interference with a one-photon excited fluorescence background that often plagues conventional Raman measurements. By baptizing the technique with the name coherent anti-Stokes Raman spectroscopy (CARS), Begley advertised the method as an attractive tool for rapid vibrational spectroscopy.

The much stronger signals compared to spontaneous Raman scattering has made CARS the method of choice for the rapid identification of chemicals present in flames and combustion processes.[12,13] An added advantage of CARS is that the signal strength is also temperature dependent, which enables an accurate temperature analysis of hot gases and flames.[14,15] Thermometry and chemical analysis of hot gases continues to be one of the major applications of the CARS technique.

The technique received a next boost when the advent of ultrafast picosecond lasers opened up the possibility of directly time-resolving the vibrational relaxation of selective molecular modes.[16] The broad bandwidth pulses that became available when femtosecond lasers entered the laboratories of spectroscopists enabled furthermore the coherent excitation of multiple Raman modes.[17–19] In femtosecond CARS, the time-resolved signal typically displays oscillatory features, which is a direct manifestation of mutual destructive and constructive interferences, often called quantum beats, between the different modes. Using Fourier transform methods, the time-resolved CARS trace can be related to the Raman spectrum. Femtosecond CARS on nonabsorbing substances can thus be seen as a form of Fourier transform Raman spectroscopy.

Although CARS is a third order nonlinear process, the technique is unable to resolve information beyond what is contained in the Raman spectrum, such as the mutual coupling between vibrational modes. To observe such couplings, higher order coherent Raman experiments are required.[20] When applied to systems with electronic resonances, however, femtosecond CARS may reveal information on time-dependent vibronic relaxation, which cannot be probed with spontaneous Raman scattering techniques.[21] These advantages have kept the popularity of time-resolved CARS spectroscopy as a probing tool for the ultrafast molecular dynamics in the condensed phase at a high level.

11.2   The Birth of CARS Microscopy

11.2.1   First Generation CARS Microscopes

In microscopy, signals are collected from many spatially resolved locations in the sample, yielding images that typically consist of several thousands to millions of pixels. Microscopic imaging is thus based on the collection of many individual measurements, either sequentially or in parallel. With such a large number of measurements, optical microscopy relies on a contrast mechanism that is associated with a high-photon flux. To build a microscope based on vibrational contrast, the CARS mechanism is a natural candidate, as the signal yields are much higher than what the spontaneous Raman scattering process can offer. The first Raman microscope was conceived in 1975,[22] but long image acquisition times had hampered the application of this approach for imaging of dynamic samples such as live biological specimens. In the early 1980s, Duncan et al. recognized the potential advantage of CARS microscopy over the existing Raman microscope in terms of imaging speed. In 1982, they constructed the first CARS microscope.[23]

The system built by Duncan et al. was fuelled by two visible picosecond dye lasers that provided the pump $\omega_p \equiv \omega_0$ and Stokes $\omega_S \equiv \omega_0 - \omega_r$ beams for the CARS process (see Fig. 11.1). Before the beams were focused to a 10-μm spot, a scanning mirror applied an adjustable angle to the incident radiation, which enabled lateral motion of the focal spot over a 300-μm range. Unlike the early CARS work of Maker and Terhune, the pump and Stokes beams were not

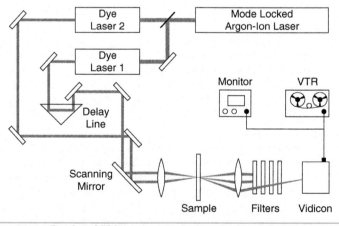

FIGURE 11.1   The first CARS microscope built at the Naval Research Laboratory in 1982 by Duncan et al. Note that a noncollinear beam geometry was used and that the high resolution was attained by the high numerical detection lens. (*Reproduced from Ref. 23, with permission of the Optical Society of America.*)

arranged in a collinear geometry in this early nonlinear microscope. Instead, Duncan et al. chose to adopt a noncollinear arrangement of the beams. Such a geometry had become commonplace for CARS measurements in condensed phase materials, which was motivated by extending the interaction length over which the pump, Stokes and anti-Stokes waves stay in phase during the signal generation process.[24] Indeed, noncollinear excitation geometries extended the range of phase matching between the waves from less than a hundred micrometers to up to several centimeters.

Although the anti-Stokes signal was generated from a relatively large diameter focal spot, the spatial resolution of the first CARS microscope was determined by the collection optics rather than the focusing lens. The CARS light was captured by a microscope objective, filtered by a stack of spectral filters and projected onto a camera placed in the image plane, yielding images with sub-micrometer resolution. The imaging speed of the CARS microscope lived up to the expectations: within only 2 seconds, vibrationally sensitive images of a 200×200 mm area at microscopic resolution were shot, clearly claiming superiority over the much slower Raman microscope. However, the spectral contrast was rather disappointing. Only after use of image subtraction techniques could deuterated lipids in dense liposomes clusters be discriminated from their nondeuterated counterparts.[25,26] The gain in speed relative to the Raman microscope was compromised by a significant loss in vibrational contrast, casting some serious doubts on the practical benefits of CARS microscopy.

11.2.2 Second Generation CARS Microscopes

The low contrast in the first CARS microscopes was caused by the presence of a strong nonresonant background. Scholten et al., who built the first widefield CARS microscope,[27] proposed several possible background rejection mechanisms, among which resonant enhanced CARS[28] and background cancellation by phase mismatching.[29] But was not until 1999, 18 years after the inception of CARS microscopy, that the technique was fortuitously resuscitated by Zumbusch et al. and was saved from becoming a dust collecting curiosity.[30] Key to the success of the second generation of CARS microscopes was the reintroduction of the collinear excitation geometry. Zumbusch et al. realized that when the incident beams are focused by a high numerical aperture microscope objective, the interaction length between the pump, Stokes and anti-Stokes fields, is so short that the waves are unable to run out of phase. Hence, there is no need for a noncollinear phase-matching arrangement of the beams per se.[31] With the collinear arrangement, smaller and cleaner focal volumes are produced, which condenses the location of CARS signal generation to a sub-micrometer spot in the sample. CARS generation in small, phase-matched volumes has two important advantages: (1) the microscope has an intrinsic three-dimensional resolution because of the confinement of the interaction volume in both the lateral and axial dimensions, similar to the two-photon excited fluorescence microscope,[32] and (2) reduction of the focal volume assigns more importance to small vibrationally resonant structures relative to the nonresonant signal generated from the bulk. This latter notion is the major reason why the nonresonant background in Duncan's imaging system was overwhelming whereas it is manageable in Zumbusch's collinear excitation microscope, where the focal volume is almost three orders of magnitude smaller.

Although image acquisition times in the first incarnation of the collinear CARS microscope were too long for practical biological imaging, within 2 years subsequent improvements of the light source, the scanning mechanism and the detectors brought the imaging rate down to a couple of seconds per image.[33,34] Despite the fact that the nonresonant background still imposed contrast challenges, for several imaging applications, most notably the imaging of lipids, the vibrational signal turned out to be strong enough to compose high-contrast images. The first applications to visualizing intracellular lipid bodies[35,36] and membranes[37–40] proved decidedly successful, which spurred the continuously growing popularity of CARS microscopy as a useful visualization tool in cellular biology.

Since 1999, the field of CARS microscopy has grown beyond all expectations. Active CARS research is conducted both on the technological front, which seeks to improve the chemical imaging capabilities, and the applications front. Although CARS imaging has found several applications in material science,[41–43] the majority of CARS applications are within the realm of biological and biomedical research. In this chapter, we will provide examples and prospects of CARS in these latter areas. For a more extensive overview of the developments in the CARS microscopy field since 1999, the reader is referred to several excellent review papers in the literature.[44–47]

11.3 CARS Basics

In order to better appreciate the imaging properties of the CARS microscope, we will briefly explain the basic physical principles that underlie this technique. Our discussion will be largely qualitative, with an emphasis on the physical picture rather than their mathematical descriptions. The different principles on which the CARS technique is based are not all conveniently explained within a single framework. Some properties are easily explained within a classical mechanical picture, whereas the clarification of other properties requires a quantum mechanical framework. This dual picture is the source of some confusion about some of the aspects of CARS. In the following, we will exclusively adopt a classical picture to explain the generation of waves at the anti-Stokes frequency. A similar discussion with more mathematical representation can be found elsewhere.[48–50]

11.3.1 Nonlinear Electron Motions

In the CARS process, light beams are used with optical frequencies ($\sim 10^{13}$-10^{17} Hz), corresponding to wavelengths in the visible and near-infrared range. Nuclei in molecules are unable to respond to an electric field that oscillates at such high frequencies. The electrons surrounding the nuclei, however, will respond to the electric field by oscillating at the frequency of the incoming electromagnetic field. For relatively weak electric fields, the electrons respond linearly to the driving field. Under these conditions, the spatial extent of the electronic oscillation is small and the motion in the potential well is harmonic. For stronger fields, however, the electrons are pulled farther from their equilibrium positions and the cloud picks up anharmonic motions. As shown in Fig. 11.2, the response of the electrons to the incoming field is no longer linear. CARS is based on these anharmonic motions of the electron cloud.

If the electrons are driven at two strong optical frequencies simultaneously, the anharmonically oscillating cloud will contain oscillatory motion at combination frequencies. Of relevance to the CARS process is the electron cloud's ability to shake at the difference frequency between the pump and the Stokes fields, i.e., at the beat frequency $\omega_p - \omega_S$. In practice, such oscillations occur in any molecular sample when the pump and Stokes beams are applied, irrespective of the presence of nuclear resonances at $\omega_p - \omega_S$. Whenever the electrons shake at the beat frequency, the electronic properties of the material will be slightly altered relative to the situation when the light beams are absent. More specifically, the refractive index of the material is modulated at the difference frequency. A changing refractive index implies that a third light wave of frequency ω_{pr} that travels through

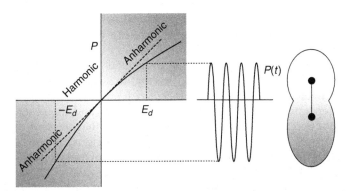

FIGURE 11.2 Polarization of the material as a function of the driving field E_d. For stronger fields, the polarization is no longer linearly proportional to the driving field as a consequence of the anharmonicity of the potential in which the electrons reside. Under these conditions, the oscillation amplitude of the polarization is distorted, which is the source of optical nonlinear signals, including CARS.

the sample will be scattered. The scattered light will naturally be modulated at the difference frequency, which corresponds to scattered light components at the frequencies $\omega_{pr} \pm (\omega_p - \omega_S)$. In the case that the third light wave is of similar frequency as the pump light, we will thus find anti-Stokes radiation at $2\omega_p - \omega_S$.

This type of anti-Stokes scattering is very different from spontaneous Raman scattering in that the scattered light components are all modulated with the same phase, i.e., the anti-Stokes radiation is coherent. This coherence also implies that the waves add up constructively in a specific direction, and destructively interfere in all other directions, producing radiation with a well-defined propagation direction. In the CARS microscope, the direction in which the waves add up, i.e., the phase-matching direction, is the forward propagation direction that is collinear with the incident beams. For comparison, because of the lack of coherence, the signal resulting from the spontaneous Raman process is scattered isotropically.

11.3.2   Resonant and Nonresonant Contributions

The process we have discussed so far is responsible for the generation of the nonresonant background in CARS. Clearly, the occurrence of this contribution is purely electronic and bears no dependence on the presence of vibrational modes. Why, then, is the CARS process sensitive to vibrational resonances that are nuclear in nature? This sensitivity stems from the notion that the electrons are bound by a potential that is defined by the location of the nuclei. At some frequencies, namely, the nuclear resonance frequencies, this potential becomes more malleable. Consequently, the electron cloud becomes more polarizable at these frequencies. In other words, the presence of nuclear modes perturbs the polarizability of the molecule's electron density at well-defined frequencies, which can now be shaken into resonance whenever the difference frequency $\omega_p - \omega_S$ matches the nuclear resonance frequency. The oscillation amplitude of the electron cloud will be larger at such frequencies and produce more scattered light at $2\omega_p - \omega_S$.

From this discussion we also see that the CARS process brings about two types of coherent anti-Stokes fields: one field contribution generated from purely electronic motions and the other field contribution resulting from electron motions that depend on the presence of nuclear modes. The total CARS signal is the square modulus of the coherent sum of this nonresonant and resonant field component. The coherent mixing of the nonresonant and resonant signals makes it particularly difficult to detect each contribution separately. Many research efforts are devoted to suppressing the nonresonant portion of the CARS signal. Polarization sensitive detection,[51,52] interferometric mixing[41,53] and frequency modulation[54] are examples of techniques that reduce the nonresonant background in practical and rapid CARS imaging.

11.4   CARS by the Numbers

11.4.1   Signal Generation in Focus with Pulsed Excitation

Compared to the linear response of materials to optical fields, the third order susceptibility of samples is extremely small. Expressed in electrostatic units, measured values of $\chi^{(3)}$ for condensed phase materials are on the order of $\sim 10^{-14}$ esu.[55] The $\chi^{(3)}$ related changes to the refractive index are of magnitude 10^{-16} cm^2/W, which implies that an optical field with an intensity of 1 W/cm^2 will modify the material's refractive index by only 10^{-16}. Clearly, much stronger optical fields are required to induce a detectable third-order nonlinear signal. The high-peak powers offered by pulsed excitation provide such electric field strengths.

Keeping the average power constant, increasingly higher peak powers are obtained for increasingly shorter pulses. Consequently, the highest third-order signals are obtained for the shortest pulses. The purely electronic (nonresonant) CARS response is indeed maximized for the shortest possible pulse, as it scales as $\sim 1/\tau^2$, where τ is the pulse width.[56] Nonetheless, the ratio of the vibrationally resonant response over the nonresonant signal is decreasing for shorter pulse widths. This is because not all the frequency components of broader bandwidth pulses combine to drive the vibrational modes at $\omega_p - \omega_S$, whereas all combinations of the spectral components contribute to generate the nonresonant response. A balance between maximum signal generation and an optimized resonant-to-nonresonant signal ratio is found for pulse widths of 2 to 5 ps.[33] For such temporal widths, the spectral width of the pulses matches the width of the Raman resonances in condensed phase materials, which constitutes the most favorable excitation condition for CARS imaging based on a single Raman band.

When a 5 ps, 800 nm pulse of 0.1 nJ is focused by a water immersion microscope objective with a numerical aperture of 1.2, the peak intensity in focus amounts to 2×10^{11} W/cm^2. Optical fields that correspond to these intensities bring about detectable changes to the refractive index, especially in the phase-matched direction. Under such conditions, the amount of detected CARS signal generated from pure water at the OH-stretching vibration (~ 3300 cm^{-1}) can be as high as 500 photons per shot.[34] For sub-micrometer sized objects, such as a lipid bilayer visualized at the CH_2-stretching vibration (2845 cm^{-1}), the CARS signals are generally substantially smaller, but still detectable. With similar pulse energies, a bilayer with more than 10^6 CH_2 modes in the focal volume can be visualized in the phase-matched forward direction at ~ 0.1 photons detected per shot. Such photon detection rates are sufficient to produce good quality images recorded with sub-ms pixel dwell times and ~ 80 MHz pulse repetition rates. For planar lipid membranes detection in the epidirection is similarly sensitive, with detection sensitivities of less than 10^6 CH_2 modes in focus at sub-ms pixel dwell times.[57]

11.4.2   Photodamaging

Naturally, higher CARS signals are attained by increasing the pulse power. However, photodamaging concerns put a practical limit on how much power can be applied to the sample. Two types of photodamage are relevant to this discussion. First, light absorption by components in biological materials, which scales linearly with the average illumination power, produces heating of the sample. Using near-infrared radiation, sample heating is generally negligible for average powers less than 10 mW in most biological materials.[58,59] Second, nonlinear excitation of compounds in cells and tissues may induce photochemical changes with possible toxic photoproducts, among which the formation of radicals.[58,60] In CARS studies, nonlinear photodamage is oftentimes the prime source of damage to the sample.[61] By keeping the pulse energies below 1 nJ, nonlinear photodamage in CARS microscopy can be generally avoided for most samples.[56] In practice, imaging with focal intensities of 10 mW from a ~80 MHz pulse train produces excellent CARS signal levels while photodamaging effects are kept to a minimum.

11.4.3   CARS Chemical Selectivity

The CARS imaging microscope has proven to be a very sensitive tool to visualize the distribution of lipids in biological samples. For instance, CARS has been used to follow the growth and trafficking of lipid droplets in a variety of cell types[36,56] and microorganisms,[62] to visualize the agent-induced morphological changes to myelin sheets in the spinal cord[63,64] and to map out lipid deposits in atherosclerotic lesions.[65] All these studies are facilitated by the high density of CH_2 modes in lipids, which produces a CARS strong signal at its symmetric stretch vibration at 2845 cm^{-1}. As illustrated by Fig. 11.3, the lipid CARS response also benefits from having its major signatures in a relatively quiet region of the vibrational spectrum, which prevents spectral interferences with neighboring bands. Other dense CH_2-containing compounds and a concentrated substance like water can also be relatively easy visualized in the high-frequency range (2500 to 3500 cm^{-1}) of the vibrational spectrum. Specificity among lipids and other CH_2-containing compounds can furthermore by obtained through the use of deuterium labels.[25,57,66,67]

The situation in the fingerprint region (~800 to 1800 cm^{-1}) is, however, quite different. Unlike the limited number of molecular modes in the high-frequency region, many molecular groups have their frequencies in the fingerprint region. Indeed, a typical Raman spectrum from a biological sample is characterized multiple partially

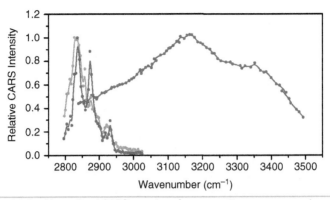

FIGURE 11.3   Normalized CARS spectra of common tissue compounds. Cholesterol (red), the lipid tristearin (green) and water (blue) are shown in the region of the vibrational spectrum that includes the CH and OH stretching vibrations.

overlapping vibrational bands. Such congested spectra may complicate a clear identification of the molecular compounds, and advanced algorithms such as hierarchical cluster analysis are often required to extract the molecular composition from measured spectra.[68] In CARS, matters are even more complicated. Because each vibrational band carries its own frequency-dependent spectral phase, the coherent anti-Stokes Raman spectrum is affected by interferences among the different spectral signatures, in addition to interference with the nonresonant background. As a consequence, the spectral information in CARS spectra from the fingerprint region typically appears featureless and washed out. Much of the interferences can be undone by means of phase retrieval algorithms like the maximum entropy method (MEM), which extracts Raman-like spectra out of congested CARS spectra (see Fig. 11.4).[69,70] With the aid of signal processing tools, CARS spectroscopy in the fingerprint region has several advantages relative to Raman spectroscopy, especially in terms of speed.

For high-speed CARS imaging studies, which rely on the availability of clear signatures to generate image contrast, postacquisition spectral processing is not always an attractive option. Instead, methods have been developed that aim at direct contrast enhancement of a particular signature through optimized excitation and detection conditions. Heterodyne CARS microscopy, which avoids spectral interferences by detecting the CARS field instead of the intensity, is an example of a technique that can recover spectral signatures that are otherwise unsuitable for imaging.[53,71] This approach has been used to image proteins through the CH_3 stretching vibrations, which are usually affected by the spectral interferences with the nearby CH_2 symmetric stretch mode.

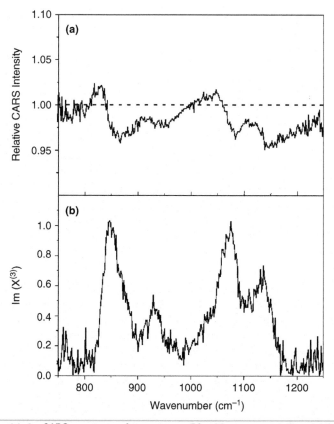

FɪɢᴜʀE 11.4   CARS spectrum of an aqueous 50 mM sucrose solution. (*a*) CARS signal in the 800 to 1200 cm^{-1} region relative to the nonresonant background from the buffer solution. Note the washed out spectral contrast due to intereference with the nonresonant electronic signal. (*b*) Retrieved vibrational Raman spectrum [Im($\chi^{(3)}$)] from the CARS data using the maximum entropy phase retrieval method. Clear Raman signatures are discerned in the retrieved spectra. (*Courtesy of Mischa Bonn, AMOLF, the Netherlands.*)

11.4.4   CARS Sensitivity

With regular CARS microscopy, image contrast can be generated based on concentration variations that are less than 10^6 CH_2 oscillators in focus.[57] These numbers translate to about less than 10^5 lipid molecules in the focal volume, which amounts to (sub-)mM concentrations. In many situations, particularly in drug delivery studies, the target molecule is present in the sample at much lower concentrations in a surrounding, where variations in nonresonant background levels may be substantial. For these studies, more sensitive detection methods are required. Frequency modulation (FM-)CARS microscopy is an example of a detection technique that suppresses the nonresonant background and sensitively extracts the vibrationally resonant CARS signal, producing sensitivities of less than 10^6 Raman oscillators in the presence of a considerable nonresonant background.[54] Techniques like FM-CARS have the potential to detect sub-mM concentrations of Raman active agents in actual tissues.

Is CARS a suitable technique for single molecule vibrational spectroscopy? In case of collective Raman modes, such that can be found in carbon nanotubes, CARS signals can certainly be generated from single structures. In the limit of a single local mode, many of the benefits of CARS disappear. In particular, the coherent addition of emitting Raman oscillators, producing strong signals in phase-matched direction, no longer applies to the single oscillator limit. A theoretical analysis shows that the CARS response is not necessarily stronger than the spontaneous Raman response in this limit, as the emission is essentially incoherent.[72,73]

These considerations illustrate that CARS microscopy in the single molecule limit requires additional enhancement mechanisms to generate detectable signals. Motivated by the success of surface-enhanced Raman scattering (SERS),[74,75] researchers have started studies in which the enhanced local fields associated with surface plasmons of metallic substrates are used to boost the CARS response. Because surface plasmon resonances produce an intrinsic electronic anti-Stokes response (see Fig. 11.5),[76–78] surface-enhanced (SE-)CARS may be more challenging than SERS. Nonetheless, past and recent work has indicated that SE-CARS is experimentally feasible and may offer additional ways of probing the vibrational response of molecules, potentially at the single molecule limit.[79–82]

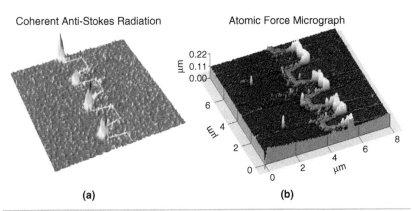

FIGURE 11.5 Coherent anti-Stokes image (a) and atomic force microscopy topograph (b) of a gold nanowire sample. This zig-zag nanowire exhibits alternating plateaus with heights of 20 and 80 nm, respectively. Note that the higher third-order coherent signals from the wire are obtained from the lower plateaus, indicating a stronger electronic plasmon resonance in those regions.

11.5   CARS and the Multimodal Microscope

The CARS imaging system is composed of a fast-scanning micro-scope and an ultrafast light source. For generating images quickly, picosecond lasers have been shown to optimize the CARS contrast in the microscope relative to femtosecond excitation.[33] This is particularly true for generating CARS contrast from a relatively narrow (>10 cm^{-1}) line in the Raman spectrum. Optimum contrast is obtained if the spectral width of the laser pulses complies with the width of the target Raman band. However, for broader vibrational bands—most notably the OH-stretching vibration of water, and, to a lesser extent, the lipid CH_2-stretching spectral range—femtosecond laser beams can be used as well.

The pump and Stokes fields necessary for the CARS process are usually derived from separate light sources. For instance, two syn-chronized ultrafast Ti:sapphire lasers can be used to deliver the pump and Stokes beams.[33,83] Alternatively, a synchronously pumped optical parametric oscillator offers a convenient system for producing stable pump and Stokes pulse trains (see Fig. 11.6). Besides generating CARS, such ps/fs pulse trains are also conducive for inducing accom-panying nonlinear signals in the focal volume of the objective lens. In particular, two-photon-excited fluorescence (TPEF) and light result-ing from second harmonic generation (SHG) are readily observed in a CARS microscope. Hence, by using additional detectors and appropri-ate spectral filtering, a multitude of nonlinear signals can be monitored

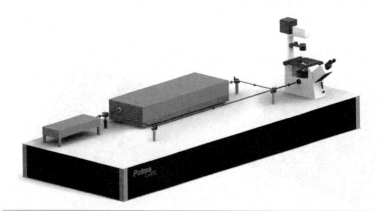

FIGURE 11.6   Rendering of a typical multimodal microscope based on a picosecond Nd:Vanadate laser, a synchronously pumped optical parametric oscillator (OPO) and an optical microscope. In such a scheme, the pump beam for the CARS process is delivered by the OPO and the Stokes beam by the Nd:Vanadate laser. The pump and Stokes beam are overlapped in space and time and collinearly directed to the microscope. Scanning of the focal spot is accomplished either by scanning the sample stage or by angle scanning the incident beams.

simultaneously. Indeed, in one of the first rapid CARS imaging studies of living cells, the CARS technique was combined with simultaneous two-photon-excited fluorescence lifetime microscopy (FLIM).[34] Other examples include multimodal imaging based on CARS, TPEF, SHG, and sum frequency generation (SFG) microscopy.[65,84]

The CARS imaging microscope enables a complete multimodal investigation of tissues based on endogenous contrast. In addition to the chemical selectivity offered by CARS, simultaneously detected SHG signals reveal tissue collagen patterns and TPEF signatures report on endogenous fluorophores such as elastin fibers and nicotinamide adenine dinucleotide (NAD) metabolic agents.[85,86] The CARS microscope thus constitutes the ultimate nonlinear imaging platform.

11.6   CARS in Tissues

The major difference between thin samples (μm sized) and tissues is the presence of significant light scattering in the latter. Light scattering is a consequence of variations in the linear refractive index in the sample. In tissues, refractive index variations result from large structures such as extracellular fibers and smaller structures such as intracellular organelles in an otherwise aqueous environment. In addition, light absorption is also relevant for thicker tissues. Both light scattering and absorption give rise to signal loss in tissues relative to thin samples. The consequences of these linear optical effects on the CARS signal will be discussed below.

11.6.1   Focusing in Tissues

Optimum CARS signals are obtained when the incoming light is condensed into a tight focal spot. Naturally, light scattering in tissues compromises the formation of a clean focal spot. The reason for this is twofold. First, scattering of excitation light in the tissue leads to loss of amplitude in the vicinity of the focal volume.[87–89] Second, the phase of the incident waves will be compromised upon arrival in the focal region. The focal volume, which exists by virtue of constructive interference of light waves in one point in space,[90] is very sensitive to the coherence of the incoming light. Loss of phase coherence implies that the waves will be unable to completely interfere to produce a tight focal spot. Hence, in the presence of scattering, the focal volume will be smeared out and contain a lower excitation density. CARS signals are directly affected and generally much lower if tissue scattering is significant.

Using NIR radiation, appreciable CARS signals up to 0.25 mm can generally be produced in most tissue types. Special long working distance objectives can often be used to increase this penetration depth by a factor of two. Beyond 0.5 mm, light scattering severely complicates the formation of a sufficiently tight focal spot for CARS generation. To increase the number of in-phase photons that arrive in the focal volume at greater depths, higher excitation powers can be used. Such an approach has been used to accomplish deep-tissue imaging with two-photon-excited fluorescence,[91] and is, in principle, also possible for CARS microscopy. Absorption and linear heating of the sample are the limiting factors for applying more power to achieve strong signals at greater depths. The photodamage threshold in pigmented skin, for instance, is found to be 500 W/cm^2.[92] Generally, when keeping sample illumination dosages below 50 mW per beam, imaging well below the damage threshold can be achieved.

Besides lower signals as a consequence of random scattering throughout the tissue, scattering also affects the CARS imaging properties when scattering objects in focus affect the signal generation process. Figure 11.7 shows an example of how linear refractive index differences in focus compromise the image quality. The image of a paraffin oil droplet in water is severely affected by the refractive index difference between the droplet and its aqueous surrounding $\Delta n \sim 0.15$. Phase distortions of the incident light along with phase mismatching of the CARS radiation in focus both contribute to the distorted image. When the surrounding is replaced with dimethyl sulfoxide, a fluid with a refractive index closer to that of paraffin oil $\Delta n = 0.03$, the image appears relatively undistorted. This simple example shows that linear refractive index differences will always affect image appearance in turbid media like tissues. Figure 11.8 shows that scattering may affect CARS in a more significant way than two-photon-excited fluorescence microscopy. In this regard, a proper

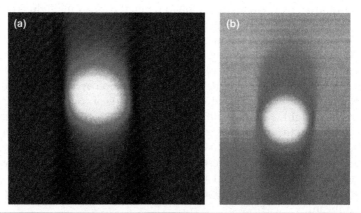

FIGURE 11.7   Effect of $\chi^{(1)}$ scattering on coherent $\chi^{(3)}$ signals. (*a*) CARS image of a dodecane droplet in water. Clear vertical shadow edges can be seen in nonresonant signal from the water, which is a direct consequence of linear scattering of light at the dodecane/water interfaces. (*b*) Paraffin droplet in d-DMSO. Because the refractive index differences between paraffin and d-DMSO are minimal, the shadowing effect is much reduced.

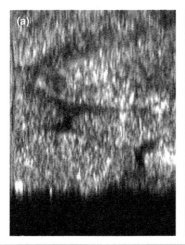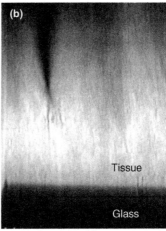

FIGURE 11.8 Effect of linear scattering on quality of CARS tissue images. (a) Two-photon-excited autofluorescence xz-image of chicken breast tissue, excited at 800 nm using a 1.2-NA water immersion lens. Total scan depth is 80 μm. (b) Simultaneously detected CARS signal from water. Note that the penetration depth of CARS is less than for two-photon-excited fluorescence, and that the coherent third-order signal is more sensitive to linear scattering at dense objects. Linear scattering is evident from the apparent shadowing streaks in the image.

understanding of these effects will help the interpretation of the images.

11.6.2   Backscattering in Tissues

In addition affecting the incident light and the CARS signal generation process, scattering also acts on the generated CARS light. Although postgeneration does not decrease the number of CARS photons produced, it changes the propagation direction of the CARS emission by redistributing it over a large cone angle. Since the unaffected CARS signal is predominantly propagating in the forward direction,[93,94] postgeneration scattering will generally lower the amount of photons that can be captured in the forward direction and increase the number of photons that can be intercepted in the backward direction. The latter notion is particularly relevant for CARS imaging in thick tissues when forward detection is limited due to the opacity of the tissue and the signal can only be detected in the backward direction.

In Fig. 11.9 the amount of CARS light that is scattered back from a tissue phantom is plotted as a function of phantom thickness. It is clear that for tissues thicker than a couple of hundred micrometers, a significant fraction of the forward propagating light is redirected into the epidirection.[95] Backward scattering of forward propagating CARS

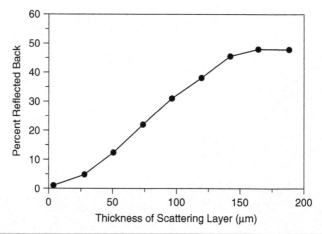

Figure 11.9   Percent of CARS signal detected in the backward direction as a function of the thickness of the scattering layer. The model scattering layer was a 10 percent intralipid emulsion. Backscattering of forward propagating coherent radiation is thought to be the major mechanism that contributes to the contrast in CARS tissue images.

light is the major mechanism that enables the collection of appreciable signals in the epidetection channel. The same principle has also been shown to be important in SHG imaging.[96,97]

11.6.3   Typical Endogenous Tissue Components

Which important tissue components can be straightforwardly visualized with CARS? Several CH_2-rich structures can be imaged with good contrast in the fast imaging CARS microscope. Figure 11.10 shows an

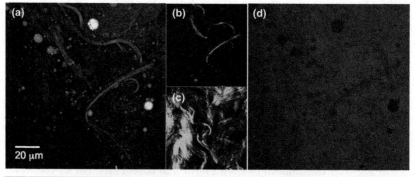

Figure 11.10   Typical tissue components seen with CARS microscopy. (a) CARS image of human dermis at 2868 cm^{-1}, showing strong signals from lipid and elastin and a faint resonant signal from dermal collagen. (b) TPEF image of elastin. (c) SHG image of collagen. (d) CARS image taken at 2993 cm^{-1}, off-resonance with the CH-stretching vibrations of lipid and structural protein, showing a clear contrast change with image (a).

image taken in the dermis of human skin ex vivo. The dermis is rich in structural fibers such as collagen and elastin, which can both be visualized with CARS, as is evident from the figure. Similar observations have been made in arterial tissue.[84] Alternatively, elastin can be visualized by two-photon-excited fluorescence and collagen by second harmonic generation. The CARS contrast of these structural fibers may be useful if molecular spectroscopic information from the fibers is desired. In addition, tissue fat generates a very clear contrast, because of the high density of CH_2 modes. CARS microscopy is the method of choice for studies that require visualization of fat in tissues, which has been put to a good use in biomedical imaging studies concerned with obesity-related fat accumulation in mammary tissues[98] and atherosclerotic lipid deposits in arterial tissue.[65] More examples of lipid images will be given in the next section.

When addressed at the OH-stretching frequency, water also produces strong signals in the CARS imaging microscope. Visualizing tissue water at rapid image acquisition times is useful for following water diffusion and real-time hydration dynamics. The ability to monitor water diffusion is not only relevant to tissue biology, but has also found applications in food science. In Fig. 11.11, for instance, the hydration process of water in cheese is mapped as a function of time, which reveals important information on how hydration depends on fat content.

11.7   CARS Biomedical Imaging

11.7.1   Ex Vivo Nonlinear Imaging

CARS is an excellent tool for examining tissues ex vivo without the need for labeling tissue components. The label-free approach enables investigation of tissue structures that are intact and not compromised by labeling protocols. Examining intact tissue is particularly important for disease-related research, where the biochemical and morphological characteristics of the diseased tissue need to be preserved for a proper analysis. Standard staining protocols are known to severely alter the morphology and integrity of the tissue, as well as to affect the presence of tissue fat. CARS is particularly suitable to image lipids in intact tissues, as illustrated by the biomedical imaging examples below.

Lipid Quantification in Breast Cancer Tissue

Recent nuclear magnetic resonance (NMR) studies have shown that the concentration of NMR-visible lipids in breast cancer tissue is significantly lower compared to healthy tissue.[99,100] The origin of this signature of cancer is unknown, although it has been suggested that a depletion of intracellular lipid droplets in cancer cells may play a major role. Lipid droplets, (sub)-micrometer-sized bodies of neutral lipids, are a natural component of mammary epithelia.[14] In cancer cells,

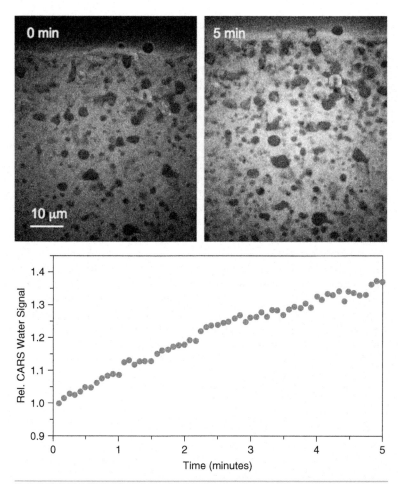

FIGURE 11.11   Hydration dynamics in cheese. Snapshots of cheese taken at OH-stretching vibration of water (3150 cm^{-1}) during a hydration experiment. Dark holes correspond to lipid clusters, which are off-resonance at this Raman shift. Brighter signals indicate higher water concentration over time. A slight swelling of the cheese is also seen. Bottom graph shows the relative increase of the CARS water signal over a time frame of 5 minutes. (*Courtesy of Friesland Foods Corporate Research, the Netherlands.*)

arrest of cell differentiation suppresses the formation of lipid droplets and decreases the corresponding lipid pool.[101–103] CARS is ideally suited to quantify the intracellular lipid droplet correlation and to correlate that with cell malignancy. Figure 11.12 shows that the number of lipid droplets in breast cancer cells in cell culture as visualized with CARS decreases with increasing malignancy of the cell. The challenge is to observe a similar trend in intact tissues. Figure 11.13 shows CARS imaging results from rat mammary tissue, a model system for human breast tissue. In combination with SHG imaging, the

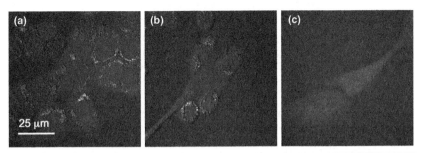

FIGURE 11.12   CARS images of (*a*) live nonmalignant (MCF-12A); (*b*) mildly malignant (MCF-7), and (*c*), malignant (MBA-MB-231) breast cancer cells. The bright spots are the lipid droplets. Note that the droplet concentration in malignant cells is lower than in normal cells.

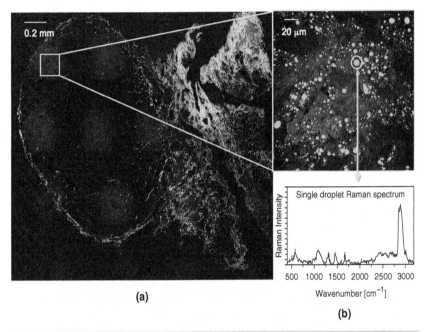

FIGURE 11.13   Label-free image of lipids in mammary tumor tissue. (*a*) large area (1.2 × 1.2 mm) image of lipid (CARS, red) and collagen (SHG, green). (*b*) top: high-resolution image of tumor cells. The CH_2 contrast from cells is indicated in red and the CH_2 contrast from lipids has been color-coded green. (*b*) bottom: Raman spectrum of single lipid droplet. The CH_2 stretching vibrations in the 2850 to 3000 cm^{-11} range are easily identified.

tumor region can be clearly identified. The inset shows furthermore that intracellular lipids can be visualized clearly in the three-dimensional tissue. Such imaging studies are further complemented by confocal Raman spectroscopic measurements for detailed chemical analysis of the lipid droplet content.

Lipids in Atherosclerotic Lesions

Atherosclerosis is a disease that affects the walls of arterial blood vessels, forming pools of lipid-rich macrophages, smooth muscle cells, lipids, and components of the extracellular matrix. These lesions develop a fibrous encapsulation, which can become increasingly thin and may rupture as the lesion matures. Rupture of the fibrous cap causes the release of the inflammatory elements into the lumen, which, in turn, may obstruct blood flow.[104,105] Multimodal CARS microscopy is ideally suited to characterize the different stages of the lesion. A CARS-based methodology may eventually grow into a fiber-based diagnostic tool for early diagnosis of this arterial disease. Several studies on carotid arteries of Yorkshire pigs have underlined the potential of CARS microscopy for atherosclerosis research.[65,84] Here we present the results of an imaging study on a mouse model system. The advantage of the ApoE-deficient mouse model is that it enables the study of atherosclerotic plaques as a function of multiple controllable parameters.

In Fig. 11.14 a millimeter-sized piece of the aortic arch is shown for a mouse with disabled kidney function on a normal (Chow) diet. Several lesions of different degree of severity are recognized by the elevated levels of lipids. The lesions display different concentrations of macrophages, which are clearly identified in the zoomed-in CARS

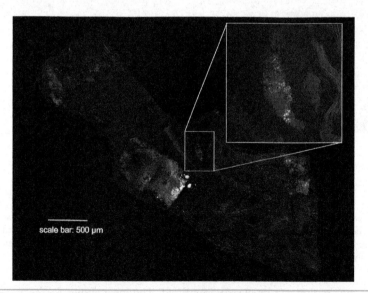

scale bar: 500 µm

Figure 11.14 Large area composite CARS/SHG image of aorta of an ApoE deficient mouse. En face images were obtained from the luminal side at the lipid 2845 cm⁻¹ signature band. CARS contrast is indicated in red/orange and SHG contrast in blue. Inset shows small atherosclerotic lesion at an early stage. Strong CARS signal from elastin of the blood vessel wall are also observed.

images. In combination with two-photon-excited fluorescence contrast of elastin and SHG contrast of collagen, a comprehensive mapping of the major arterial tissue components can be accomplished.[65] Such three-dimensional chemical maps of intact arterial tissue are invaluable for a better understanding of atherosclerotic lesion development as a function of kidney function and diet.

11.7.2   In Vivo Nonlinear Imaging

For in vivo imaging applications, the issue of photodamage is particularly relevant. To limit tissue illumination in a particular location, fast scanning is imperative. When imaging fat in superficial tissue layers, the CARS signals are generally high enough to allow for video-rate imaging. At these rapid image acquisition times, localized tissue heating is minimized and photodamage can be reduced. Figure 11.15 shows the feasibility of in vivo imaging in a study concerned with visualizing lipid components in the mouse skin.[95] Important tissue structures such as lipid lamellae of the stratum corneum, sebaceous glands, dermal adipocytes and the fat-containing cells of the subcutaneous layer are readily visualized with video-rate CARS. With imaging depths of up to several hundred micrometers, CARS microscopy constitutes a powerful method for investigating endogenous tissue structures in superficial layers without any form of labeling. In vivo biomedical imaging applications are just starting to emerge, but it is clear that CARS microscopy has the potential to significantly contribute to basic scientific research and diagnostics of superficial tissues.

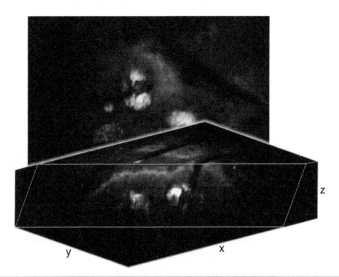

FIGURE 11.15   Three-dimensional CARS image with lipid contrast from mouse skin in vivo. Lipid-rich sebaceous glands near the hair follicle are clearly recognized. Data stack measures $800 \times 640 \times 125 \ \mu m^3$.

11.8   What Lies at the Horizon?

Since the inception of the second generation of CARS microscopes, the development of CARS microscopy has been characterized by steady technological advances. With robust laser technology and commercial CARS microscopes that are beginning to find the market, a new era of CARS microscopy has arrived in which the applications will be the driving force. The CARS technique has already proven to be a foremost imaging tool for lipids in cells and tissue. The lipid imaging capability will find many relevant applications in tissue biology, some of which are discussed in this chapter. Biomedical applications also push the technology toward the integration of optical fibers and fiber-based probes. There is no doubt that such imaging extensions will become available for CARS microscopy in the near future, solidifying the technique's membership to the family of reliable biomedical imaging tools.

Acknowledgments

The author likes to thank Dr. Vishnu Krishnamachari, Alex Nikolaenko, Maxwell Zimmerley, Mercedes Lin, Hyunmin Kim, and Ryan Lim for their help in preparing the figures.

References

1. K. S. Krishnan and C. V. Raman, "A New Type of Secondary Radiation," *Nature*, **121**:501–502, 1928.
2. G. Eckhardt, R. W. Hellwarth, F. J. McClung, S. E. Schwarz, and D. Weiner, "Stimulated Raman Scattering from Organic Liquids," *Physical Review Letters*, **9**:455–457, 1962.
3. R. W. Hellwarth, "Analysis of Stimulated Raman Scattering of a Giant Laser Pulse," *Applied Optics*, **2**:847–853, 1963.
4. R. W. Minck, R. W. Terhune, and W. G. Gado, "Laser-Stimulated Raman Effect and Resonant Four-Photon Interactions in Gases H_2, D_2 and CH_4," *Applied Physics Letters*, **3**:181–184, 1963.
5. P. D. Maker and R. W. Terhune, "Study of Optical Effects due to an Induced Polarization Third Order in the Electric Field Strength," *Physical Review*, **137**: A801–818, 1965.
6. J. A. Armstrong, N. Bloembergen, J. Ducuing, and P. S. Pershan, "Interactions Between Light Waves in a Nonlinear Dielectric," *Physical Review*, **127**:1918–1939, 1962.
7. D. A. Kleinman, "Nonlinear Dielectric Polarization in Optical Media," *Physical Review*, **126**:1977–1979, 1962.
8. M. D. Levenson, "Interference of Resonant and Nonresonant Three-Wave Mixing in Diamond," *Physical Review B*, **6**:3962–3965, 1972.
9. M. D. Levenson, "Feasibility of Measuring the Nonlinear Index of Refraction by Third-Order Frequency Mixing," *IEEE Journal of Quantum Electronics*, **QE-10**:110–115, 1974.
10. J. J. Wynne, "Nonlinear Optical Spectroscopy of $\chi^{(3)}$ in $LiNBO_3$," *Physical Review Letters*, **29**:650–653, 1972.
11. R. F. Begley, A. B. Harvey, and R. L. Byer, "Coherent Anti-Stokes Raman Spectroscopy," *Applied Physics Letters*, **25**:387–390, 1974.

12. I. E. Harris and M. E. McIlwain, "Coherent Anti-Stokes Raman Spectroscopy in Propellant Flames," in: *Fast Reactions in Energy Systems*, C. Capellos and R. F. Walker (eds.), Reidel, Boston, 1981.
13. R. J. Hall and A. C. Eckbreth, "Coherent Anti-Stokes Raman Spectroscopy: Applications to Combustion Diagnostics," in: *Laser Applications*, R. K. Erf (ed.), Academic, New York, 1982.
14. A. C. Eckbreth, "CARS Thermometry in Practical Combustors," *Combustion and Flame*, **39**:133–147, 1980.
15. A. C. Eckbreth, P. A. Bonczyk, and J. F. Verdieck, "Combustion Diagnostics by Laser and Fluorescence Techniques," Progress in Energy and Combustion Science, **5**:253–322, 1979.
16. A. Laubereau and W. Kaiser, "Vibrational Dynamics of Liquids and Solids Investigated by Picosecond Light Pulses," *Reviews of Modern Physics*, **50**:607–665, 1978.
17. R. Leonhardt, W. Holzapfel, W. Zinth, and W. Kaiser, "Terahertz Quantum Beats in Molecular Liquids," *Chemical Physical Letters*, **133**:373–377, 1987.
18. S. Mukamel, "Femtosecond Optical Spectroscopy: A Direct Look at Elementary Chemical Events," *Annual Review of Physical Chemistry*, **41**:647–681, 1990.
19. K. A. Nelson and E. P. Ippen, "Femtosecond Coherent Spectroscopy," *Advances in Chemical Physics*, **75**:1–35, 1989.
20. Y. Tanimura and S. Mukamel, "Femtosecond Spectroscopy of Liquids," *Journal of Chemical Physics*, **99**:9496–9511, 1993.
21. M. Schmitt, G. Knopp, A. Materny, and W. Kiefer, "Femtosecond Time-Resolved Coherent Anti-Stokes Raman Scattering for the Simultaneous Study of Ultrafast Ground and Excited State Dynamics: Iodide Vapour," *Chemical Physics Letters*, **270**:9–15, 1997.
22. M. Delhaye and P. Dhamelincourt, "Raman Microprobe and Microscope with Laser Excitation," *Journal of Raman Spectroscopy*, **3**:33–43, 1975.
23. M. Duncan, J. Reintjes, and T. J. Manuccia, "Scanning Coherent Anti-Stokes Raman Microscope," *Optics Letters*, **7**:350–352, 1982.
24. A. C. Eckbreth, "BOXCARS: Crossed-Beam Phase-Matched CARS Generation in Gases," *Applied Physics Letters*, **32**:421–423, 1978.
25. M. D. Duncan, "Molecular Discrimination and Contrast Enhancement Using a Scanning CARS Microscope," *Optics Communications*, **50**:307–312, 1984.
26. M. D. Duncan, J. Reintjes, and T. J. Manuccia, "Imaging Biological Compounds Using the CARS Microscope," *Optical Engineering*, **24**:352–355, 1985.
27. T. Scholten, "Coherent Anti-Stokes Raman scattering (CARS): Technique and Applications to Biophysical Studies; the Potentials of CARS Microscopy," University of Twente, the Netherlands, Enschede, 1989.
28. I. Chabay, G. K. Klauminzer, and B. S. Hudson, "Coherent Anti-Stokes Raman Spectroscopy (CARS): Improved Experimental Design and Observation of New Higher-Order Processes," *Applied Physics Letters*, **28**:27–29, 1976.
29. T. A. H. M. Scholten, G. W. Lucassen, F. F. M. d. Mul, and J. Greve, "Nonresonant Background Suppression in CARS Spectra of Dispersive Media Using Phase Mismatching," *Applied Optics*, **28**:1387–1400, 1989.
30. A. Zumbusch, G. Holtom, and X. S. Xie, "Vibrational Microscopy Using Coherent Anti-Stokes Raman Scattering," *Physical Review Letters*, 82:4142–4145, 1999.
31. M. Müller, J. Squier, C. A. D. Lange, and G. J. Brakenhoff, "CARS Microscopy with Folded BoxCARS Phasematching," *Journal of Microscopy*, **197**:150–158, 2000.
32. W. Denk, J. H. Strickler, and W. W. Webb, "Two-Photon Laser Scanning Fluorescence Microscopy," *Science*, **248**(4951):73–76, 1990.
33. J. X. Cheng, A. Volkmer, L. D. Book, and X. S. Xie, "An Epi-Detected Coherent Anti-Stokes Raman Scattering (E-CARS) Microscope with High Resolution and High Sensitivity," *Journal of Physical Chemistry B*, **105**:1277–1280, 2001.
34. E. O. Potma, W. P. de Boeij, P. J. M. van Haastert, and D. A. Wiersma, "Real-Time Visualization of Intracellular Hydrodynamics," *Proceedings of the National Academy of Science U.S.A.*, **98**:1577–1582, 2001.
35. J. X. Cheng, Y. K. Jia, G. Zheng, and X. S. Xie, "Laser-Scanning Coherent Anti-Stokes Raman Scattering Microscopy and Applications to Cell Biology," *Biophysical Journal*, **83**:502–509, 2002.

36. X. Nan, J. X. Cheng, and X. S. Xie, "Vibrational Imaging of Lipid Droplets in Live Fibroblast Cells with Coherent Anti-Stokes Raman Scattering Microscopy," *Journal of Lipid Research*, **44**:2202–2208, 2003.

37. J. X. Cheng, A. Volkmer, L. D. Book, and X. S. Xie, "Multiplex Coherent Anti-Stokes Raman Scattering Microspectroscopy and Study of Lipid Vesicles," *Journal of Physical Chemistry B*, **106**:8493–8498, 2002.

38. M. Muller and J. M. Schins, "Imaging the Thermodynamic State of Lipid Membranes with Multiplex CARS Microscopy," *Journal of Physical Chemistry B*, **106**:3715–3723, 2002.

39. G. W. Wurpel, J. M. Schins, and M. Muller, "Direct Measurement of Chain Order in Single Phospholipid Mono- and Bilayers with Multiplex-CARS," *Journal of Physical Chemistry B*, **108**:3400–3403, 2004.

40. E. O. Potma and X. S. Xie, "Detection of Single Lipid Bilayers with Coherent Anti-Stokes Raman Scattering (CARS) Microscopy," *Journal of Raman Spectroscopy*, **34**:642–650, 2003.

41. S. H. Lim, A. G. Caster, O. Nicolet, and S. R. Leone, "Chemical Imaging by Single Pulse Interferometric Coherent Anti-Stokes Raman Scattering Microscopy," *Journal of Physical Chemistry B*, **110**:5196–5204, 2006.

42. E. O. Potma, X. S. Xie, L. Muntean, J. Preusser, D. Jones, J. Ye, S. R. Leone, W. D. Hinsberg, and W. Schade, "Chemical Imaging of Photoresists with Coherent Anti-Stokes Raman Scattering (CARS) Microscopy," *Journal of Chemical Physics B*, **108**:1296–1301, 2004.

43. B. G. Saar, H. S. Park, X. S. Xie, and O. D. Lavrentovich, "Three-Dimensional Imaging of Chemical Bond Orientation in Liquid Crystals by Coherent Anti-Stokes Raman Scattering Microscopy," *Optics Express*, **15**:13585–13596, 2007.

44. J. X. Cheng, "Coherent Anti-Stokes Raman Scattering Microscopy," *Applied Spectroscopy*. **91**:197–208, 2007.

45. J. X. Cheng and X. S. Xie, "Coherent Anti-Stokes Raman Scattering Microscopy: Instrumentation, Theory and Applications," *Journal of Physical Chemistry B*, **108**:827–840, 2004.

46. C. L. Evans and X. S. Xie, "Coherent Anti-Stokes Raman Scattering Microscopy: Chemical Imaging for Biology and Medicine," *Annual Review of Analytical Chemistry*, **1**:883–909, 2008.

47. A. Volkmer, "Vibrational Imaging and Microspectroscopies Based in Coherent Anti-Stokes Raman Scattering Microscopy," *Journal of Physics D*, **38**:R59–R81, 2005.

48. S. Maeda, T. Kamisuki, and Y. Adachi, "Condensed Phase CARS," in: *Advances in Nonlinear Spectroscopy*, R. J. H. Clark and R. E. Hester (eds.), John Wiley and Sons Ltd., New York, 1988, pp. 253–297.

49. R. W. Boyd, *Nonlinear Optics*, Academic Press, San Diego, 2003.

50. E. O. Potma and X. S. Xie, "Theory of Spontaneous and Coherent Raman Scattering," in: *Handbook of Biological Nonlinear Optical Microscopy*, B. R. Masters and P. T. C. So (eds.), Oxford University Press, New York, 2008, pp. 164–185.

51. J. X. Cheng, L. D. Book, and X. S. Xie, "Polarization Coherent Anti-Stokes Raman Scattering Microscopy," *Optics Letters*, **26**:1341–1343, 2001.

52. F. Lu, W. Zheng, and Z. Huang, "Heterodyne Polarization Coherent Anti-Stokes Raman Scattering Microscopy," *Applied Physics Letters*, **92**:123901, 2008.

53. E. O. Potma, C. L. Evans, and X. S. Xie, "Heterodyne Coherent Anti-Stokes Raman Scattering (CARS) Imaging," *Optics Letters*, **31**:241–243, 2006.

54. F. Ganikhanov, C. L. Evans, B. G. Saar, and X. S. Xie, "High-Sensitivity Vibrational Imaging with Frequency Modulation Coherent Anti-Stokes Raman Scattering (FM-CARS) Microscopy," *Optics Letters*, **31**:1872–1874, 2006.

55. F. Kajzar and J. Messier, "Third-Harmonic Generation in Liquids," *Physical Review, A*, **32**:2352–2363, 1985.

56. X. Nan, E. O. Potma, and X. S. Xie, "Nonperturbative Chemical Imaging of Organelle Transport in Living Cells with Coherent Anti-Stokes Raman Scattering Microscopy," *Biophysical Journal*, **91**:728–735, 2006.

57. L. Li, H. Wang, and J. X. Cheng, "Quantitative Coherent Anti-Stokes Raman Scattering Imaging of Lipid Distribution in Coexisting Domains," *Biophysial Journal*, **89**:3480–3490, 2005.

58. A. Hopt and E. Neher, "Highly Nonlinear Photodamage in Two-Photon Fluorescence Microscopy," *Biophysical Journal*, **80**:2029–2036, 2001.

59. A. Schönle and S. W. Hell, "Heating by Absorption in the Focus of an Objective Lens," *Optics Letters*, **23**:325–327, 1998.

60. K. Konig, P. T. C. So, W. W. Mantulin, and E. Gratton, "Cellular Response to Near-Infrared Femtosecond Laser Pulses in Two-Photon Fluorescence Microscopy," *Optics Letters*, **22**:135–136, 1997.

61. Y. Fu, H. Wang, and J. X. Cheng, "Characterization of Photodamage in Coherent Anti-Stokes Raman Scattering Microscopy," *Optics Express*, **14**:3942–3951, 2006.

62. T. Helllerer, C. Axäng, C. Brackmann, P. Hillertz, M. Pilon, and A. Enejder, "Monitoring of Lipid Storage in Caenorhabditis Elegans Using Coherent Anti-Stokes Raman Scattering (CARS) Microscopy," *Proceedings of National Academy of Science U.S.A.*, **104**:14658–14663, 2007.

63. T. B. Huff and J. X. Cheng, "In Vivo Coherent Anti-Stokes Raman Scattering Imaging of Sciatic Nerve Tissues," *Journal of Microscopy*, **225**:175–182, 2007.

64. T. B. Huff, Y. Shi, Y. Yan, H. Wang, and J. X. Cheng, "Multimodel Nonlinear Optical Microscopy and Applications to Central Nervous System," *IEEE Journal of Selected Topics in Quantum Electronics*, **14**:4–9, 2008.

65. T. T. Le, I. M. Langohr, M. J. Locker, M. Sturek, and J. X. Cheng, "Label-Free Molecular Imaging of Atherosclerotic Lesions Using Multimodal Nonlinear Optical Microscopy," *Journal of Biomedical Optics*, **12**:054007, 2007.

66. G. R. Holtom, B. D. Thrall, B. Y. Chin, H. S. Wiley, and S. D. Colson, "Achieving Molecular Selectivity in Imaging Using Multiphoton Raman Spectroscopy Techniques," *Traffic*, **2**:781–788, 2001.

67. E. O. Potma and X. S. Xie, "Direct Visualization of Lipid Phase Segregation in Single Lipid Bilayers with Coherent Anti-Stokes Raman Scattering (CARS) Microscopy," *ChemPhysChem*, **6**:77–79, 2005.

68. M. Diem, M. Romero, S. Boydston-White, M. Miljkovi, and C. Matthäus, "A Decade of Vibrational Micro-Spectroscopy of Human Cells and Tissue, (1994–2004)," *Analyst*, **129**:880–885, 2004.

69. E. M. Vartiainen, H. A. Rinia, M. Muller, and M. Bonn, "Direct Extraction of Raman Line-Shapes from Congested CARS Spectra," *Optics Express*, **14**:3622–3630, 2006.

70. H. A. Rinia, M. Bonn, M. Müller, and E. M. Vartiainen, "Quantitative CARS Spectroscopy Using the Maximum Entropy Method: The Main Lipid Phase Transition," *ChemPhysChem*, **8**:279–287, 2007.

71. M. Jurna, J. P. Korterik, C. Otto, and H. L. Offerhaus, "Shot Noise Limited Heterodyne Detection of CARS Signals," *Optics Express*, **15**:15207–15213, 2007.

72. G. I. Petrov, R. Arora, V. V. Yakovlev, X. Wang, A. V. Sokolov, and M. O. Scully, "Comparison of Coherent and Spontaneous Raman Microspectroscopies for Invasive Detection of Single Bacterial Endospores," *Proceedings of National Academy of Science U.S.A.*, **104**:7776–7779, 2007.

73. C. A. Marx, U. Harbola, and S. Mukamel, "Nonlinear Optical Spectroscopy of Single, Few and Many Molecules: Nonequilibrium Green's Function QED Approach," *Physical Review, A*, **77**:022110, 2008.

74. M. Moskovits, "Surface-Enhanced Spectroscopy," *Reviews of Modern Physics*, **57**:783–826, 1985.

75. A. Otto, I. Mrozek, H. Grabborn, and A. Akermann, "Surface-Enhanced Raman Scattering," *Journal of Physics Condensed Matter*, **4**:1143–1212, 1992.

76. D. S. Chemla, J. P. Heritage, P. F. Liao, and E. D. Isaacs, "Enhanced Four-Wave Mixing from Silver Particles," *Physical Review B*, **27**:4553–4558, 1983.

77. M. Danckwerts and L. Novotny, "Optical Frequency Mixing at Coupled Gold Nanoparticles," *Physical Review Letters*, **98**:026101–026104, 2007.

78. H. Kim, D. K. Taggart, C. Xiang, R. M. Penner, and E. O. Potma, "Spatial Control of Coherent Anti-Stokes Emission with Height-Modulated Gold Zig-Zag Nanowires," *Nano Letters*, in press, 2008.

79. C. K. Shen, A. R. B. d. Castro, and Y. R. Shen, "Surface Coherent Anti-Stokes Raman Spectroscopy," *Physical Review Letters*, **43**:946–949, 1979.

80. T. Ichimura, N. Hayazawa, M. Hashimoto, Y. Inouye, and S. Kawata, "Local Enhancement of Coherent Anti-Stokes Raman Scattering by Isolated Gold Nanoparticles," *Journal of Raman Spectroscopy*, **34**:651–654, 2003.
81. T. Ichimura, N. Hayazawa, M. Hashimoto, Y. Inouye, and S. Kawata, "Tip-Enhanced Coherent Anti-Stokes Raman Scattering for Vibrational Nanoimaging," *Physical Review Letters*, **92**:220801, 2004.
82. T. W. Koo, S. Chan, and A. A. Berlin, "Single-Molecule Detection of Biomolecules by Surface-Enhanced Coherent Raman Scattering," *Optics Letters*, **30**:1024, 2005.
83. E. O. Potma, D. J. Jones, J.-X. Cheng, X. S. Xie, and J. Ye, "High-Sensitivity Coherent Anti-Stokes Raman Scattering Microscopy with Two Tightly Synchronized Picosecond Lasers," *Optics Letters*, **27**(13):1168–1170, 2002.
84. H. W. Wang, T. T. Le, and J. X. Cheng, "Label-Free Imaging of Arterial Cells and Extracellular Matrix Using a Multimodal CARS Microscope," *Optics Communications*, **281**:1813–1822, 2008.
85. W. R. Zipfel, R. M. Williams, R. Christie, A. Y. Nikitin, B. T. Hyman, and W. W. Webb, "Live Tissue Intrinsic Emission Microscopy using Multiphoton-Excited Native Fluorescence and Second Harmonic Generation," *Proceedings of National Academy of Science U.S.A.*, **100**:7075–7080, 2003.
86. W. R. Zipfel, R. M. Williams, and W. W. Webb, "Nonlinear Magic: Multiphoton Microscopy in the Biosciences," *Nature Biotechnology*, **21**:1369–1377, 2003.
87. X. Deng and M. Gu, "Penetration Depth of Single-, Two-, and Three-Photon Fluorescence Microscopic Imaging through Human Cortex Structures: Monte Carlo Study," *Applied Optics*, **42**:3321–3329, 2003.
88. R. Drezek, A. Dunn, and R. Richards-Kortum, "Light Scattering from Cells: Finite-Difference Time Domain Simulations and Goniometric Measurements," *Applied Optics*, **38**:3651–3661, 1999.
89. A. K. Dunn, V. P. Wallace, M. Coleno, M. W. Berns, and B. J. Tromberg, "Influence of Optical Properties on Two-Photon Fluorescence Imaging in Turbid Samples," *Applied Optics*, **39**:1194–1201, 2000.
90. B. Richards and E. Wolf, "Electromagnetic Diffraction in Optical Systems II: Structure of the Image Field in an Aplanatic System," *Proceeding of the Royal Society A*, **253**:358–379, 1959.
91. F. Helmchen and W. Denk, "Deep Tissue Two-Photon Microscopy," *Nature Methods*, **2**:932–940, 2005.
92. M. Rajadhyaksha, R. R. Anderson, and R. H. Webb, "Video-Rate Confocal Scanning Laser Microscope for Imaging Human Tissues In Vivo," *Applied Optics*, **38**:2105–2115, 1999.
93. J.-X. Cheng, A. Volkmer, and X. S. Xie, "Theoretical and Experimental Characterization of Coherent Anti-Stokes Raman Scattering Microscopy," *Journal of Optical Society of America B*, **19**:1363–1375, 2002.
94. E. O. Potma, W. P. d. Boeij, and D. A. Wiersma, "Nonlinear Coherent Four-Wave Mixing in Optical Microscopy," *Journal of Optical Society of America B*, **17**:1678–1684, 2000.
95. C. L. Evans, E. O. Potma, M. Puoris'haag, D. Cote, C. Lin, and X. S. Xie, "Chemical Imaging of Tissue In Vivo with Video-Rate Coherent Anti-Stokes Raman Scattering (CARS) Microscopy," *Proceeding of National. Academy of Science U.S.A.*, **102**:16807–16812, 2005.
96. F. Légaré, C. Pfeffer, and B. R. Olsen, "The Role of Backscattering in SHG Tissue Imaging," *Biophysical Journal*, **93**:1312–1320, 2007.
97. O. Nadiarnykh, R. B. LaComb, and P. J. Campagnola, "Coherent and Incoherent SHG in Fibrillar Cellulose Matrices," *Optics Express*, **15**:3348–3360, 2007.
98. T. T. Le, C. W. Rehrer, T. B. Huff, M. B. Nichols, I. G. Camarillo, and J. X. Cheng, "Nonlinear Optical Imaging to Evaluate the Impact of Obesity on Mammary Gland and Tumor Stroma," *Molecular Imaging*, **6**:205–211, 2007.
99. I. Barba, M. E. Cabanas, and C. Arus, "The Relationship between Nuclear Magnetic Resonance-Visible Lipids, Lipid Droplets, and Cell Proliferation in Cultured c6 cells," *Cancer Research*, **59**:1861–1868, 1999.

100. L. L. Moyec, R. Tatoud, M. Eugene, C. Gauville, I. Primot, D. Charlemagne, and F. Calvo, "Cell and Membrane Lipid Analysis by Proton Magnetic Resonance Spectroscopy in Five Breast Cancer Cell Lines," *British Journal of Cancer*, **66**:623–628, 1992.

101. E. J. Delikatny, W. A. Cooper, S. Brammah, N. Sathasivam, and D. C. Rideout, "Nuclear Magnetic Resonance-Visible Lipids Induced by Cationic Lipophilic Chemotherapeutic Agents are Accompanied by Increased Lipid Droplet Formation and Damaged Mitochondria," *Cancer Research*, **62**:1394–1400, 2002.

102. K. Glunde, V. Raman, N. Mori, and Z. M. Bhujwalla, "RNA Interference-Mediated Choline Kinase Suppression in Breast Cancer Cells Induces Differentiation and Reduces Proliferation," *Cancer Research*, **65**:11034–11043, 2005.

103. P. N. Munster, M. Srethapakdi, M. M. Moasser, and N. Rosen, "Inhibition of Heat Shock Protein 90 Function by Ansamycin Causes the Morphological and Functional Differentiation of Breast Cancer Cells," *Cancer Research*, **61**:2945–2952, 2001.

104. J. Narula and H. W. Straus, "Imaging of unstable Atherosclerotic Lesions," *European Journal of Nuclear Medicine and Molecular Imaging*, **32**:1–5, 2005.

105. R. Virmani, F. D. Kolodgie, A. P. Burke, A. V. Finn, H. K. Gold, T. N. Tulenko, S. P. Wrenn, and J. Narula, "Atherosclerotic Plaque Progression and Vulnerability to Rupture: Angiogenesis as a Source of Intraplaque Hemorrhage," *Artheriosclerosis Thrombosis Vascular Biology*, **25**:2054–2061, 2005.

Index

CPSIA information can be obtained
at www.ICGtesting.com
Printed in the USA
LVOW02*2329040516
486734LV00001B/1/P